Lecture Notes on Data Engineering and Communications Technologies

Volume 21

Series editor

Fatos Xhafa, Technical University of Catalonia, Barcelona, Spain
e-mail: fatos@cs.upc.edu

The aim of the book series is to present cutting edge engineering approaches to data technologies and communications. It publishes latest advances on the engineering task of building and deploying distributed, scalable and reliable data infrastructures and communication systems.

The series has a prominent applied focus on data technologies and communications with aim to promote the bridging from fundamental research on data science and networking to data engineering and communications that lead to industry products, business knowledge and standardisation.

More information about this series at http://www.springer.com/series/15362

Srikanta Patnaik · Xin-She Yang
Madjid Tavana · Florin Popentiu-Vlădicescu
Feng Qiao
Editors

Digital Business

Business Algorithms, Cloud Computing and Data Engineering

 Springer

Editors
Srikanta Patnaik
Department of Computer Science
  and Engineering, Faculty
  of Engineering and Technology
SOA University
Bhubaneswar, Odisha
India

Xin-She Yang
School of Science and Technology
Middlesex University
London
UK

Madjid Tavana
Business Systems and
  Analytics Department
La Salle University
Philadelphia, PA
USA

Florin Popentiu-Vlădicescu
University Politehnica of Bucharest
Bucharest
Romania

Feng Qiao
Shenyang Jianzhu University
Shenyang
China

ISSN 2367-4512   ISSN 2367-4520 (electronic)
Lecture Notes on Data Engineering and Communications Technologies
ISBN 978-3-319-93939-1   ISBN 978-3-319-93940-7 (eBook)
https://doi.org/10.1007/978-3-319-93940-7

Library of Congress Control Number: 2018944343

Printed on acid-free paper

This Springer imprint is published by the registered company Springer Nature Switzerland AG
The registered company address is: Gewerbestrasse 11, 6330 Cham, Switzerland

Preface

Professionals and Business Consultants have predicted that "Digital Business" shall dominate the Computing and Business arena for the next 10 years rather than simply Data Mining or Business Intelligence. It is one of the niche areas of Computer Science and Business Management. The term "Digital Business" is used in a context of digital transformation, disruptive technologies and holistic business integration/optimization/convergence. However, its dimension is much more than that. A key part of this domain is information employed at work, which requires a holistic approach of information management and connecting value chain to derive more throughput out of the entire business ecosystem.

As digital technologies offer new ways to connect, collaborate, conduct business and build bridges between people, it touches the core of all business functions. Digital technologies have also challenged existing business models and their present form of operation. One of the key driving forces is the capacity of innovation and the commercialization of information and communication technology. Digital business encompasses the entire business model, such as marketing, ICT, social as well as core business activities and its convergence. From a sheer technology perspective, it has gone beyond, such as Cloud Computing, Internet of Things, Mobile platform, Big Data, etc. This edited volume covers the three distinct areas such as Cloud Computing, Digital Mess, i.e. Internet of Things and Business Algorithms which integrate them all.

All most every industrial sector and business enterprises are now focusing on digitization of core systems and processes, along with development of new frameworks and models. Digital Mess or, Internet of Things (IoT) has become the core of many applications by combining affordable embedded sensors and actuators. This digital transformation process has given rise to processing of massive amount of heterogeneous data, known as Big Data through cloud computing.

This volume covers various business algorithms to solve various issues and challenges, faced by the business organizations, including Big Data, Digital Business Transformation, Cloud Computing and IoT along with some scientific and web-based application areas. This book also provides future research directions for these areas for researchers.

We are thankful to the Editor-in-Chief of the Springer Book Series on *Lecture Notes on Data Engineering and Communications Technologies* Prof. Fatos Xhafa and also Springer Executive Editor Dr. Thomas Ditzinger, for his kind acceptance of this volume in his series. We are also thankful to all the reviewers, for their support in reviewing the chapters to bring out the volume in time. I am also thankful to all my co-editors Prof. Xin-She Yang, Dr. Madjid Tavana, Dr. Florin Popentiu-Vlădicescu and Prof. Feng Qiao, for their support and encouragement.

I am sure that the readers shall get immense ideas and knowledge from this volume.

Bhubaneswar, India   Prof. (Dr.) Srikanta Patnaik

Editorial

The evolution of new technologies such as IoT, cloud computing and widespread use of smart phones has initiated the faster adoption of digitization wave by organizations and enterprises in business world. This current technological advancement has given rise to a revolutionary era, which has lead to the transformation of traditional business world into digital form. Moreover, digital transformation of businesses has created many new opportunities with the integration of digitization process across the horizontal and vertical value chains of the organization ranging from digitization of planning and other processes to digitization of the product itself and respective solutions. All these transformations and disruptions together form a digital ecosystem where both products and processes are digitized and are adopted by traditional firms and industries and thus present new obstacles and challenges.

This book has presented a comprehensive collection of works from potential researchers and scholars who have shared their works and findings related to this area. It addresses a wide range of IoT-based problems, frameworks, solutions and applications related to smart environments and applications, where vast amount of data is being generated. Again, as this research area is highly multidisciplinary in nature, the chapters considered in this book are the ones that have addressed recent trends and challenges while maintaining innovativeness. Further, this book is technically organized to contain 18 chapters that are broadly categorized into four major sections namely (i) Digital Business Transformation, (ii) Cloud Computing (iii) IOT & Mobility (iv) Information Management & Social Media. A brief overview of each chapter is given below:

Chapter "Towards Algorithmic Business: A Paradigm Shift in Digital Business" gives a brief idea about what is algorithmic business along with an overview about how it is related to IoT and digital transformation. It also discusses how algorithms are impacting current and future business applications.

Chapter "A Decentralized Business Ecosystem Model for Complex Products" first studies decentralized business ecosystem for complex products by considering different scenarios and identifies the objectives, significant actors and their relevant interaction patterns for developing architecture to support the model. The new ecosystem they introduce not only supports transactions between complex products but also supports both value creation and sharing between the components.

Chapter "Compliance Management in Business Processes" addresses the compliance management issue in business processes by first giving an overview of the concept and possible pros and cons in the absence of compliance management system. They further discuss various compliance management activities and finally the compliance management functionalities (CMFs).

Chapter "Sustainable Cloud Computing Realization for Different Applications: A Manifesto" investigates the challenges in sustainable cloud computing along with the current developments for different applications. The authors propose a taxonomy for application management in sustainable cloud computing and map existing works to the taxonomy for identifying research gaps.

Chapter "Auction Based Scheme for Resource Allotment in Cloud Computing" proposes a dynamic model for resource allocation in cloud computing using winner determination scheme for computing cost and achieving optimal resource allocation. The authors consider the requirement of both the users and the cloud service providers for calculating the final cost.

Chapter "M-Cloud Computing Based Agriculture Management System" presents a mobile-based cloud computing framework to solve problems related to agriculture for increasing productivity. The framework attempts to solve problems concerning agriculture faster by gathering and sharing environmental as well as location-based information among farmers thus modernizing agriculture.

Chapter "Detection and Analysis of Drowsiness in Human Beings Using Multimodal Signals" proposes a framework for analysing and detecting the drowsiness state in a human being that monitors alertness of mind and body while driving. The authors consider here multi-modal signals such as visual information and biosignals (EEG) for implementing the framework for the analysis purpose.

Chapter "Enhancing Security and Privacy in Enterprises Network by Using Biometrics Technologies" proposes a novel security mechanism using biometric trait of employees in the encryption process of data exchanged by enterprises. The proposed approach ensures security and confidentiality in enterprise network to provide stronger security measures.

Chapter "Lightweight Context-Based Web-Service Composition Model for Mobile Devices" discusses a new lightweight context-based model for web service composition of the mobile device itself using POS tagger. The authors use POS tagger to identify service requests from users in NLQ form and the responses are further composed in the mobile device itself by web service providers.

Chapter "On Weighted Extended Cumulative Residual Entropy of k-th Upper Record" considers a shift-dependent version of generalized cumulative residual entropy and discusses its advantages and applications for k-th upper record and establishes a non-parametric estimator with its asymptotic normality along with characterization of the results.

Chapter "Impact of Mobility in IoT Devices for Healthcare" proposes an IoT-based platform that helps in providing personalized tips, tracing, reminders and educational contents about medical conditions proactively. The authors also discuss about mobility in IoT and its convergence with respect to mobile cloud, devices and social media.

Chapter "Multiple Mobile Elements Based Energy Efficient Data Gathering Technique in Wireless Sensor Networks" studies the limitations of the existing data gathering techniques in WSN and proposes a related framework that introduces Mobile Elements. Further, the authors use data fusion at cache point to ensure data gathering efficiency by reducing instances of transmissions. Also, the proposed technique is claimed to be efficient by the authors in many aspects such as packet delivery ratio, lesser delay, reduced overhead optimum energy consumption and decreased packet drop.

Chapter "Online Social Communities" presents a comprehensive list of online social communities and classifies them according to their usage. Various ONSC are further discussed from the user participation lifecycle perspective along with disjoint and overlapping communities over both static and dynamic networks.

Chapter "The 'Verticals', 'Horizontals', and 'Diagonals' in Organisational Communication: Developing Models to Mitigate Communication Barriers Through Social Media Applications" proposes a three-tier communication paradigm for addressing the communication barrier problem among organizational workforce. This paradigm builds interpersonal as well as intrapersonal skills and mitigates performance anxieties to enhance overall productivity. The authors examine these aspects and mitigate the communication barrier through social media applications.

Chapter "Subjective Interestingness in Association Rule Mining: A Theoretical Analysis" investigates the nature of subjective interestingness among associations of several items of a supermarket considering the manager's expectation and customer's purchase patterns. The authors observe the limitations and propose a two-dimensional framework that presents matching methodology horizontally and granularity of user knowledge across vertical axis. They also make an attempt to identify relevant research gaps in this context and pose significant research questions.

Chapter "Identifying Sentiment of Malayalam Tweets Using Deep Learning" presents a comparative study to identify sentiments in Malayalam tweets by using various deep learning methods. The findings are then compared with several baseline methods such as SVM and RKS-RLSC, etc. for the effectiveness of methods. F1-score, precision, recall and accuracy are considered by the authors as evaluation measures.

Chapter "Twitter Based Sentiment Analysis of GST Implementation by Indian Government" provides a mathematical analysis of sentiments of Indian public over GST implementation and the author's observation over the impact made by the tax structure is carried out on Twitter data that has been collected over the period of GST implementation phase as well as the Pre-GST and Post-GST phases of the Indian economic reformation scenario.

Chapter "Event Detection Using Twitter Platform" discusses several event detection techniques that are used on Twitter for prediction, detection and managing disaster-based events. Further, the chapter summarizes the performance of various data collection, analysis and event detection tools over Twitter data.

Prof. Srikanta Patnaik
Prof. Xin-She Yang
Dr. Madjid Tavana
Dr. Florin Popentiu-Vlădicescu
Prof. Feng Qiao

Contents

About the Editors

Dr. Srikanta Patnaik is a Professor in the Department of Computer Science and Engineering, Faculty of Engineering and Technology, SOA University, Bhubaneswar, India. He has received his Ph.D. (Engineering) on Computational Intelligence from Jadavpur University, India in 1999 and supervised 12 Ph.D. theses and more than 30 M.Tech. theses in the area of Computational Intelligence, Soft Computing Applications and Re-Engineering. He has published around 60 research papers in international journals and conference proceedings. He is author of two textbooks and edited 12 books and few invited book chapters, published by leading international publisher like Springer-Verlag, Kluwer Academic, etc. He was the Principal Investigator of AICTE-sponsored TAPTEC project "Building Cognition for Intelligent Robot" and UGC-sponsored Major Research Project "Machine Learning and Perception using Cognition Methods". He is the Editors-in-Chief of *International Journal of Information and Communication Technology* and *International Journal of Computational Vision and Robotics* published from Inderscience Publishing House, England and also Editors-in-chief of Book Series on "Modeling and Optimization in Science and Technology" published from Springer, Germany.

Xin-She Yang is Reader/Professor in Computational Modelling and Simulation at School of Science and Technology, Middlesex University. He is the winner of Shaanxi Province Distinguished Talented Professorship award at Xi'an Polytechnic University in 2011. He is an Adjunct Professor of Reykjavik University, Iceland, and a Guest Professor of both Harbin Engineering University and Shandong University, China. Before he joined Middlesex University, he was a Senior Research Scientist at UK's National Physical Laboratory and Cambridge University after obtaining his DPhil in Applied Mathematics from University of Oxford.

He has authored/edited 19 books with Wiley, Elsevier, World Scientific, Dunedin Academic and Springer, and published more than 300 papers. His *h-index* is 66 with more than 32000 citations. He is the Editor-in-Chief of *Int.*

J. Mathematical Modelling and Numerical Optimisation (IJMMNO, Inderscience, Scopus), serves as an editorial board member of several international journals, including Elsevier's *Journal of Computational Science* (JoCS, SCI-indexed), ISRN-Applied Mathematics, IJAI, SJI and *Int. J. Bio-Inspired Computation* (IJBIC, SCI-Indexed), and the editor of OCP Science book series. He is also the Chair of the IEEE CIS task force on Business Intelligence and Knowledge Management, and Director of *International Consortium for Optimization and Modelling in Science and Industry*. He is the Editor-in-Chief of the journal: *Int. Journal of Mathematical Modelling and Numerical Optimisation* (IJMMNO).

His main research interests include Design Optimization, Applied and Industrial Mathematics, Computational Modelling and Simulation. He is the recipient of 1996 Garside Scholar Award of Oxford University. He has been on program committees of over 40 international conferences, a workshop organizer of 10 workshops and acted as a program co-chair and/or advisory chair of more than 20 international conferences. Also, he has been acting as an examiner for Ph.D. candidates in China, India, Malaysia, Italy, Spain, New Zealand and UK, and an external evaluator for funding councils such as Austria Science Fund, Kazahkstan Science Foundation and Icelandic Science Foundation. He has developed new metaheuristic algorithms such as firefly algorithms, bat algorithm and cuckoo search. His research was highlighted in *Nature Science* Update (Aug 2000), *New Scientist* (Nov 2004), Science Magazine online news (Aug 2010) and *Science Daily* (May 2010). He has given more than 56 invited lectures in 22 countries, including 15 invited plenary/keynote talks at the international conferences such as SEA2011, Mendel'12, BIOMA2012, ICISCA2014, EU/ME2014 and ICCS2015.

Dr. Madjid Tavana is Professor and Distinguished Chair of Business Analytics at La Salle University, where he serves as Chairman of the Business Systems and Analytics Department. He also holds an Honorary Professorship in Business Information Systems at the University of Paderborn in Germany. Dr. Tavana is Distinguished Research Fellow at the Kennedy Space Center, the Johnson Space Center, the Naval Research Laboratory at Stennis Space Center, and the Air Force Research Laboratory. He was recently honored with the prestigious Space Act Award by NASA. He holds an MBA, PMIS, and Ph.D. in Management Information Systems and received his Post-Doctoral Diploma in Strategic Information Systems from the Wharton School at the University of Pennsylvania. He has published 13 books and over 250 research papers in international scholarly academic journals. He is the Editor-in-Chief of *International Journal of Applied Decision Sciences, International Journal of Management and Decision Making, International Journal of Communication Networks and Distributed Systems, International Journal of Knowledge Engineering and Data Mining, International Journal of Strategic Decision Sciences, and International Journal of Enterprise Information Systems*.
Email: tavana@lasalle.edu
Web: http://tavana.us.

Prof. Florin Popentiu-Vlădicescu graduated in Electronics and Telecommunications from University Politechnica of Bucharest in 1974, holds a Ph.D. in Reliability since 1981. He has been appointed Director of the "UNESCO Chair in Information Technologies Department" at University of Oradea. Also, he is Associated Professor with University "Politechnica" of Bucharest, Faculty of Automatic Control and Computer Science. He is the founder of the first "UNESCO Chair of Information Engineering", in UK, established at City University London, in 1998. He published over 100 papers in international journals and conference proceedings. Also, he is author of one book and co-author of four books.

He has worked for many years on problems associated with software reliability and has been Co-Director of two NATO Research Projects. Also, he is on the advisory board of several international journals, including "Reliability and Risk Analysis: Theory & Applications" and is a reviewer to "ACM Computing Reviews". He is currently Associated Editor to the *International Journal of Information and Communication Technology* (IJICT)—Inderscience Publishers. He is an independent expert to the European Commission—H2020 programme, for Net Services—Software and Services, Cloud. He is currently Visiting Professor at "ParisTech". He also lectures at the Technical University of Denmark. He was elected Fellow of the Academy of Romanian Scientists in 2008 and Director of the Doctoral School "Engineering sciences" for the period 2011–2015. Details about his CV are presented on the following sites:

http://www.staff.city.ac.uk/~pop/

http://www.staff.city.ac.uk/~pop/list.html.

Prof. Feng Qiao received his B.Eng. in Electrical Engineering and M.S.E. in Systems Engineering from the Northeastern University, Shenyang, China, in 1982 and 1987, respectively; and his Ph.D. in Intelligent Modelling and Control from the University of the West of England, Bristol, UK in 2005. During the period between 1987 and 2001, he worked at the Automation Research Institute of Metallurgical Industry, Beijing, China, as a Senior Engineer in Electrical and Computer Engineering. He is currently a Professor at Faculty of Information and Control Engineering, Shenyang JianZhu University (SJZU), Shenyang, China. His research interests include modelling and simulation, fuzzy logic systems, neural networks, nonlinear systems, stochastic systems, sliding mode control, robust control, adaptive control, process control, structural vibration control, fault diagnosis and robotic manipulation. Currently, He is acting as the Editor-in-Chief of the *International Journal of Simulation and Process Modelling*, a member of editorial board of the *International Journal of Modelling, Identification and Control*, and he serves on many international conferences as IPC and OPC members.

Part I
Digital Business Transformation

Towards Algorithmic Business: A Paradigm Shift in Digital Business

Pragyan Nanda, Sritam Patnaik and Srikanta Patnaik

Abstract The emergence of cloud computing, IoT, and increasing number of smart phones laid the foundation for digital business. In the current scenario of digital transformation, digitization of enterprises has become a survival strategy for organizations to cope with the fast changing and uncertain business environment. Even if digital transformation is essential for the growth of traditional business in today's transitional phase and enables countless opportunities, but still the major challenges commonly faced here are what is need to be done and how it should be done to address business related issues. Moreover, since these days volume of data is increasing exponentially, but raw data has no value until analyzed or utilized. Algorithmic business is one of the solutions to afore mentioned said challenges. It involves the use of smart algorithms in providing important business insights, defining company processes handling customer services, analyzing business data and making important decisions. This chapter presents a brief overview about algorithmic business, discusses various aspects, deployment strategies, challenges, opportunities such as algorithmic market place etc. Speed and scale are some of the primary advantages of deploying algorithmic business in enterprises.

Keywords Digital business · Business algorithmic · Process automation
Smart algorithms

P. Nanda (✉) · S. Patnaik
Faculty of Engineering and Technology, Department of Computer Science and Engineering,
SOA University, Bhubaneswar 751030, Odisha, India
e-mail: n.pragyan@gmail.com

S. Patnaik
e-mail: patnaik_srikanta@yahoo.co.in

S. Patnaik
School of Computer Engineering, Department of Electrical and Computer Engineering,
National University of Singapore, Singapore 117583, Singapore
e-mail: sritampatnaik@u.nus.edu

© Springer International Publishing AG, part of Springer Nature 2019
S. Patnaik et al. (eds.), *Digital Business*, Lecture Notes on Data Engineering
and Communications Technologies 21, https://doi.org/10.1007/978-3-319-93940-7_1

1 Introduction

In today's digital era, the rapid advancement of technologies such as cloud computing, IoT along with the extensive usage of smart phones has initiated rapid establishment of the digital transformation phase, thus, leading to evolution of digital business. Digital businesses posses many advantages over traditional business models as it opens new business opportunities by creating advanced models that merges both the physical and the digital worlds together. This involves digitization of almost all systems and processes lying across vertical as well as horizontal value chains [14, 17, 26]. The digitization process is usually accomplished in three major phases: (i) the first phase involves digitization of processes and systems at base level such as vertical value chain digitization which includes digitization of several processes ranging from designing and planning of products to development and efficient management of manufactured end products. This digitization process is further extended to integration of various processes across horizontal value chain which again ranges from several suppliers of raw materials to intermediate suppliers to final customers through adoption of various tracking and tracing systems. (ii) Second phase involves digitization of the product itself as well as various services associated with the product such as addition of different types of sensors to products for collecting data related to products and analyzing the retrieved data for refinement of products to improve customer satisfaction. (iii) Finally, the third phase provides several digital business models and solutions integrated with data-driven platforms for digital disruptions, thus generating a complete digital ecosystem. Now this complete digital ecosystem can be integrated entirely into existing traditional business environments to allow faster communication with introduction of interactive frameworks such as on-line purchase and sell channels embedded with intelligent algorithms to makes fast and optimal decisions in crucial situations to generate additional revenues. These beneficiary visions have enforced organizations for adapting digitization phases in their firms and factories. And due to this fast adaption of digital era, organizations are producing huge amount of data every day.

Tackling this massive collection of data, also widely known as Big Data, is one of the many new challenges that business organizations are facing along with many advantages and opportunities. This flooding of organizations with huge amount of incorrect and inconsistent data occurs because enormous amount of data are collected from different sources and in different formats. While storing and accessing this huge amount of collected data, lots of complexity is involved due to the variations in media types such as texts, images, audios and videos. Since these data are generated, collected and processed in real time world during various scenarios, so maintaining pace between data generation, collection and processing is again a tough task and leads to missing of parts of data further leading to retrieval of inaccurate information. Also this huge volume of data is a combination of both structured and unstructured contents from heterogeneous sources, analyzing which is another difficult task. Other challenges include authenticity of the source of the data, freshness of the data collected as updated and latest data will provide more accurate information about status of various processes, manpower and labour, customer requirements, sales and pricing etc. This Big Data plays the foundation role for all types of organizations ranging from small scale to large scale ones that have joined the race of being digitized. Moreover, digitization of enterprises has hardly left any aspect of modern business environment untouched, thus making it a survival challenge for organizations to cope with continuously changing and unfavourable circumstances.

As discussed previously, the digital transformation process is changing traditional business world dramatically by globalizing the competition among organizations and raising the bar of customer expectations. Also, advanced technologies ranging from smart mobile devices, Internet of Things (IoT) and cloud-based services has become an inevitable part of every individual's life as well as every business including large-scale business firms to the smallest shops. The vast adoption of smart phones and devices has allowed all these devices to connect and communicate with each other through internet and hence the concept is coined as the Internet of Things (IoT). Again, almost all industrial sectors are in-process of deploying digital transformation to explore new horizons of growth leading to adoption of data analytic solutions and IoT [5, 9, 12, 16]. And this extensive adoption generates a huge amount of data on daily basis and analysing which provides relevant information that impacts the underlying business strategies and processes significantly. Further, the extracted information can be utilized to generate significant insights for making crucial decisions in several scenarios through various data analytic techniques and algorithms. In a way, it can be summarized that decision support systems embedded with data analytics for IoT will next drive several business models for different organizations.

Thus, this wide adaptability of advanced technologies such as IoT, cloud-based services and smart mobiles has put immense pressure on all sorts of organizations, as a result of which the organizations can either adapt to the digital change and evolve or face severe operational as well as financial risks leading to the downfall of the organization. "Digital business" explicitly means the wide and unavoidable adaption of digital business ecosystem discussed above by almost all organizations. Further, this explosive adaption of digital ecosystem has opened a universe of applications where organizations can directly interact with real-time partners and consumers in both business-to-business and business-to-consumer scenarios respectively. Moreover, this inescapable adoption has lead to a paradigm shift from traditional business to digital business by re-defining how organizations will grow, sustain, compete and prosper in the digital era. And those organisations that fail to adapt to the rapid change will hardly survive. However, digital business revolves around three basic characteristics namely (i) agility: the organization must be capable of adapting regulatory and market changes quickly by including digitization of back-end servers as well as applications (ii) speed: the organization must be able to anticipate real-time customer demands evolving with changing markets and respond rapidly to avoid loss of opportunities (iii) and finally being innovative: organizations have to be innovative by adding value to products and services being delivered for boosting sales and gain loyalty for competitive advantages. But for transforming traditional business into digital business, an organization has to make huge changes from top-to-bottom including digitization of the smallest data to change of technological architectures to infrastructures. While the primary characteristics of traditional business are stability and security, the above mentioned characteristics form the underlying principles of the digital transformation of organizations. Thus adaption of digital transformation by the organization involves digitization of several internal as well as external operations to gain maximum flexibility and benefits in competitive environments.

Some of the major advantages of the paradigm shift is recurring of profit from revenue, from subscriptions in e-commerce based services, services providing buy and sell market without infrastructures like Uber and AirBNB, providing algorithm-driven strategies for adding value to customer experiences like navigation based apps providing shortest routes to destination with traffic details, enhancing customer experiences through multiple channels, addition of sensors to capture real-time data etc. These advantages justify the need of integrating digital transformation into traditional business for staying competitive in a continuously changing environment.

This chapter presents a brief overview about algorithmic business, covering various aspects, deployment strategies, challenges, opportunities such as algorithmic market place etc. Speed and scale are some of the primary advantages of deploying algorithmic business in enterprises. Also huge amount of data can be utilized to provide significant insights in a cost-effective manner. Some of the important challenges explored in this chapter are the possibility of erroneous decisions made by the algorithms due to existence of bugs in algorithms. Also, since algorithms are not context-sensitive, they can be biased on the basis of data used.

2 Emergence of Algorithms in Business

Some of the important outcomes of this digital transformation are increase of profits with increase of digital channels for business, increase in the generation of revenue due to improved performance of internal as well as external operations, increase in allies and partnerships between organizations, re-defining competition as survival strategy, reduction of overall costs involved in procurement to logistics to promotions, creation of new markets and business by crossing pre-defined industrial boundaries. Thus, digital business creates a digital ecosystem by merging physical world with a digital network of several physical processes and applications that are embedded with smart technologies for sensing and interacting with each other as well as external environment. This wave of digital transformation forms the base of fourth industrial revolution where completely automated and smart products and industries will be brought into existence by merging different spheres. These smart systems will be capable of making intelligent and decentralized decisions without human interventions by creating virtual copies of various processes along with their updated status for communicating with other automated systems. Further, these intelligent systems collect data, analyze and generate relevant information and insights to be used for making decisions and taking actions in several sectors like fabrication of products, manufacturing, logistics, supply chain management and risk analysis etc. through inter-connected machines communicating with intelligent interactive services over global networks.

Fig. 1 Action as a function of data

Although in current scenario, digital transformation plays a major role the growth of business but achieving complete digital transformation and complete automation of systems for realization of smart products and services is still an ongoing process and how to do is still considered to be difficult. And the answer to this difficulty lies in adaption of intelligent algorithms for development of smart applications [17, 21, 25]. Algorithms, as we already know, are a set of rules encapsulated together to act on a given data for generating the expected outcome. These algorithms utilizing knowledge and intelligence to provide various solutions and insights for the given scenarios and data can be packaged into functions, applets or even products and services. This can be shown in the figure below (Fig. 1):

The above figure is a generalized representation of algorithm where algorithms are defined as functions performed over data provided to generate a set of actions as outcomes. While developing smart and automated applications, products, and services; complex algorithms with strong mathematical foundations will form the underlying principle to drive decision making and automation process of business. This adaption of smart algorithms in industries and business for using the encapsulated knowledge to generate insights from data and make decisions about consequent actions is coined as 'Algorithmic Business' by Peter Sondergaard, senior vice president at the analyst firm Gartner. The analyst firm Gartner highlights the importance of algorithms in business in a symposium conducted in 2015 by mentioning that the in future the value of business will be in algorithms and not in data as data itself has no meaning. Intelligent algorithms are required to generate patterns and insights from the data to make decisions and perform actions. This digital transformation of organizations combined with intelligent algorithms brings a revolution in the way business was being done before. The entire transformation revolves around technological changes, customers' response behaviour to the change and new business opportunities opened as a result of this revolution. It involves how to utilize technological enhancements to generate the best possible outcomes from analysis of customers' response to these enhancements. Realization of this revolution is possible by the application of intelligent algorithms which takes outcomes generated from analytic algorithms as inputs for making optimal decisions. Moreover, these outputs generated mathematically from algorithms can be used repeatedly to improve results and insights. This improvement in business due to adaption of intelligent algorithms, also adds speed and scale to business by processing transactions quickly and increasing number of simultaneous and interrelated connections.

3 Digital to Algorithmic Transformation

In this internet era, organizations are steadily moving into algorithmic transformation from digital transformation. Also, IOT has opened the way for innovativeness through algorithmic transformation to generate information and insights from big data since data itself has no meaning unless it is analyzed in some way or the other due to its complexity [1, 3, 5, 16, 22]. This particular context of doing something meaningful with big data needs intelligent algorithms for extracting information from both structured and unstructured data i.e., the focus of organizations has shifted from enormous data to the algorithms using them. It is not like we are currently not using algorithms, in fact we are already surrounded by algorithms in different forms and scenarios, but in future algorithms will form the base of most of the solutions generated for solving specific problems and converting it into decisions and actions. Although "algorithmic business" has not been employed universally as a term, still it forms the underlying principle of many significant applications. For example, Google's driver-less car uses a proprietary algorithm for connecting the physical objects with sensors to collect data and combine everything into software for transportation application. Similarly, high frequency trading uses a unique algorithm to drive higher return generating decisions. Also another secret algorithm of Google, page-rank algorithm used in the search engine is responsible for making it a valued brand name. Moreover, some of the other organizations that has already adopted intelligent algorithms that has added value to the organization include Amazon, Netflix, airline industries and even global retailers. Therefore, it can be inferred that further development of cognition-based software will lead to autonomous interactions between machines and intelligent algorithms will play the role of competitive and powerful weapons by taking digital transformation to the next level which is considered as algorithmic transformation.

As we know, algorithms provide a mechanism for capturing important information and insights from given data which can further be packaged into a reusable form. And usage of these intelligent algorithms in reusable form for adding value to business. As discussed above, algorithmic business revolves around these algorithms and monetizes these algorithms for generating revenue in different ways such as providing licences for using unique algorithms [18]. For example, in a food industry, to implement automated replenishment system, some just-in-time algorithm used for logistics by some organization can be licensed instead of developing it from scratch at the expense of huge cost. These smart algorithms are responsible in differentiating organizations from their competitors in common market places. Also this license providing or buying and selling of intelligent algorithms will create new markets and opportunities with exponential growth in significant increment in revenue generation. Moreover, sophisticated algorithms will define new products and services along with business model and strategies for maximizing organization value and revenue.

4 Deployment Strategies of IoT

The rise in internet wave has paved way for implementation of digital business and new models and patterns are being generated for smooth adoption by organizations. Physical industries are also being digitized rapidly for moving into the next level of revolution i.e., automation. But the key principle underlying this transformation is adoption of Internet of Things (IoT) for merging the physical and digital worlds. Researchers are currently working on many hybrid solutions for implementation of respective real-time environments. Moreover, organizations will be more dependent on algorithms for all sort of significant operations. Algorithmic business involves this use of smart algorithms in providing important business insights, defining company processes handling customer services, analyzing business data and making important decisions. Therefore, smart and innovative algorithms are required to analyse big data generated by various organizations, provide insights into it in order to automate various core processes of digital business [9, 11, 13]. In other words, we can say that the future of digital business can be determined by algorithms. And these algorithms playing the driver role in the digital world can create a whole set of new business opportunities with proper support.

Although algorithms usually described as a set of instructions for doing certain jobs and have always been there but in the fast changing business scenario they are being centrally utilized to make competitive business decisions without human interventions. Innovative algorithms along with advanced technologies and novel business models can help in the development of new customized products and services to support various industries in deploying digital transformation effectively. Even if digital transformation is essential for the growth of traditional business in today's transitional phase and enables countless opportunities, but still the major challenges commonly faced here are what is need to be done and how it should be done to address business related issues. Moreover, since these days volume of data is increasing exponentially, but raw data has no value until analyzed or utilized.

As discussed previously, smart things consists of elements combining working principles with algorithms from both physical and digital worlds. In case of manufacturing industries, keeping account of inventories available in warehouses is a erroneous and expensive task but with the deployment of smart containers and shelves equipped with sensors and transmitters, inventory details can be measured and communicated with almost zero marginal cost. Thus development of intelligent algorithms to identify and measure things can be considered as a deployment challenge. Moreover this collected data can either be made freely available or can be leveraged to maximize profits. Another current algorithmic business models widely being used is implementation of demand and supply algorithm over real-time data by Uber and other taxi-based services. These organizations use heat mapping for providing real-time analytics for demand and supply of taxis. A heat map represents data graphically by showing individual values using matrix through colors. This detects high demand with low supply situations automatically and increases.

Fleisch et al. have identified three basic trends of new digital business models in their research [10], which involves (i) increase in integration of users and customers across value chains to open source some tasks for utilizing users in customization of contents and re-distributing them further, (ii) service-oriented models are also followed to provide run-time as well as after-sales services to customers for maintaining customer loyalty relationships, (iii) and finally, analytical models consisting of intelligent algorithms for core competence is also applied for collection of past data about customer experiences and utilizing it for future refinement of product design and structuring of price. Although previously, industrial sectors have been classified into digital and non-digital industries, the emergence of internet has been responsible in bringing major breakthroughs in both sectors. While adoption of IoT is not only reviving old business models but also generating new business patterns in existing digital sectors like Facebook, Google, YouTube, eBay, Amazon etc., in case of non-digital sectors, internet has helped in simplifying processes with cost reduction while not compromising the quality and variety of products. But this adoption of entire organizations involves a lot of challenges.

Some of the deployment challenges for digital business include management of high resolution data in the digital world as huge amount of high resolution data with multiple dimensions is being generated from several areas of physical industries such as images, texts, audios, videos etc. Intelligent algorithms are being used for processing these high resolution data along with hardware. When queries are being processed in the digital world, queries need to be processed at lightning speed and should be capable of generating anything ranging from texts, documents to images and videos as result in real-time environment as per requirement. Google, for an example has utilized the high resolution data through smart algorithms to provide advertisements to users on the basis of their search behaviour for optimizing profit. This dynamic processing of high resolution data involves use of higher resolution providing control circuits with sensors and actuators which is much higher and finely grained than that of physical world. IoT makes this possible in physical world by merging both digital and physical worlds together with a vision of digitization of each and every object and its position in the physical world. This can be achieved with the deployment of mini-computers for making smart objects that can collect information about its surrounding and capable of communicating the same with other smart objects. Since these mini-computers are barely visible, so people communicate with the smart object.

Fleisch et al. have also [10] further discussed a digital business model that intermingles the patterns of both physical and digital business patterns into a hybrid layered construct. This hybrid construct known as value-creation model shown below in Fig. 2, consists of five major layers which are usually involved in any abstract IoT based applications.

The details of the layers are given as follows: (i) the first layer of the value-creation model consists of the physical thing itself, responsible for fulfilling its purpose for the user in the immediate surroundings; (ii) the second layer consists of sensors and actuators loaded as elements in mini-computers that measures local data and generate local services as benefits to the user; (iii) the third layer is

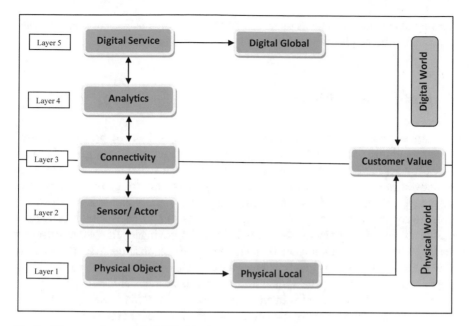

Fig. 2 Value-creation model of digital business

responsible for connectivity, as the elements of second layer are connected to internet providing global accessibility; (iv) since the previous layers cannot deliver any benefit themselves, the next layer is responsible for collecting, storing, classifying and processing the sensor data. It further searches and integrates consequent web services to find the corresponding actuator elements; (v) In the final year, the entire data obtained id structured digitally and packaged into a suitable form, which can be a web-based service or application and globally accessible. These patterns generated by the digital business model, cannot be separated from smart things and thus deliver benefits to the users accessing them at negligible costs.

All the above mentioned layers are dependent on each other and cannot be created independently as deployment of IoT is not merely addition of isolated layers but also addition of processes containing intelligent algorithms for providing maximum benefits to the user. The three major steps that lead to algorithmic business include (i) identifying the unique and smart algorithms or a set of rules that smartly makes the business work (starting from the trade secret of the products to data parsers to intelligent scheduling of processes for achieving goals) towards faster response and revenue maximization. (ii) Assignment of ownership to these algorithms, i.e. who should take responsibility for the algorithms. (iii) Finally third is classification of the algorithms into public and private assets since algorithms can be considered for monetizing information assets if an organization is having skill sets to generate information from the collected data. In other words, we can say that this changes the way of considering data as the centre of transforming business of organizations; instead algorithms play the central role in the transformation.

5   Business Algorithms

The organizations doing business online receive huge benefits by practising Search Engine Optimization while those that are not practising are at the losing end. SEO not only provides better search results to users using it but also helps e-commerce based businesses in considering many aspects to stay on the top for gaining competitive advantages. It helps businesses in maintaining user friendly and faster websites to form a strong customer base. As customer satisfaction contributes a lot to in increasing the number of customers as well as assisting in improving loyalty of the customers leading to building reputation and brand name of the organization. With the increase in website traffic driven by loyal customer base online businesses improve their ranking in search engines and are visible in top result pages thus enhancing the growth of the business. It further helps in exploring new markets. Since web world is currently the fastest ever growing market, it also provides better conversion rates thus converting visitors into subscribers and customers. SEO also helps local customers to reach local businesses and bypass competition among local businesses through SEO and online promotional strategies [19, 31]. These online search engines work on the basis of underlying algorithms that retrieve the Search Engine Results Page (SERP) using ranking of pages relevant to the search query matching strategy. Also sometimes, on the basis of matching keywords of sponsors and advertisers, results are generated. Apart from this RSS feeds, newsletters, and social media newsfeed also play a major role in building customer base by driving traffic to sites.

While most of the businesses that are based on algorithmic approaches are already reaping benefits up to certain extent from employing the above mentioned strategies for return on investment, some of the benchmarks such as page ranking algorithms for SEO of Google, Google pay per click strategy, recommender algorithms of Amazon, Edge News feed algorithms of Face book, and pricing algorithm of UBER are a few to name. Some of these algorithms are discussed in this section.

5.1   Google's Algorithms

Google is the most widely adopted search engine that can provides immense benefits by understanding the page ranking strategy for web pages of businesses. Mostly the users of Google search don't leave the first page of the resulting pages. Therefore, Google makes enhancements to its ranking algorithms on regular basis. And these small changes usually impact the ranking of web pages highly, making businesses keep track of these enhancements and improve their pages accordingly to stay on top. The most successful businesses enhance their online presence to be highly ranked by

Google. Some of the important Google algorithms namely Google Panda, Penguin, Humming bird, Pigeon, Mobile, Rank-Brain, Possum and Fred [7, 23, 28, 32] are briefly presented below.

5.1.1   Google Panda

Google's algorithms usually rely on 200 or more unique clues such as key terms present on websites, recency of contents, regional information of the user, ranking of the potential pages and many more, that help to predict what the user is searching for. Therefore, online businesses hire SEO marketers or agencies for monitoring these updates and identifying the changes required to be made in their websites to meet Google's search criteria and thus tailoring the same to their online presence in order to drive more traffic to their websites. The earliest algorithm developed by the company was Google Panda in 2011 [23, 28] designed for penalizing low quality websites with poor content by preventing them from appearing in top results of the search instance. This update not only ensured the originality of contents and articles but also the value they bear for the viewers by adopting some metrics. One of the metrics the underlying algorithm used for judging the quality of the content of the website is the duration a user stays on the website before returning to the main page. Google Panda uses quality scores obtained above as a factor for further ranking the pages.

5.1.2   Google Penguin

Next, another algorithmic update launched by Google is Penguin in 2012, which aimed at fighting against spam, vague links and penalizing deceitful websites [7, 23, 28, 32]. Usually, these websites use unfair practices for SEO to gain high page rankings. Penguin penalizes these websites by down-ranking these sites once the anchor texts are seem to be over-optimized and links appear to be manipulative. While Panda inspects the contents of the entire site, Penguin works in more real-time mode thus inspecting page-specific contents and penalizes the web page, once something suspicious is identified. The website can recover its ranking only after the penalty is removed. In 2016, both Panda and Penguin became Google's core algorithms. Further, two more updates Google Pirate and Google Top Heavy were launched by Google in the year 2012, where the former plays the role of a filter and filters out sites with copyright issues and the later penalizes sites with heavy advertisements respectively.

5.1.3   Google Humming Bird

Again, in 2013, Google launched two more algorithmic updates named as Google Payday and Google Humming Bird. While Google Payday cleans up commonly spammed terms for search such as insurance, various loans, pornography etc., the Google Humming Bird update is responsible for preventing keyword stuffing along with low-quality contents. The role of the Humming Bird was different from that of Panda and Penguin [23, 28]. It was designed as an interpreter that looks for the meaning of the search query instead of the individual words and provides more accurate result of pages containing the intended meaning of the searched query. The underlying principle applied here is natural language processing (NLP) based upon synonyms, semantic-based indexing and co-occurring terms.

5.1.4 Google Pigeon

Also, Google again launched the next algorithmic update Pigeon in 2014 which considered location as a vital part of the search query. This algorithm forms tie-ups between the core algorithms as well as local algorithms and ranks local results by SEO factors. This helped a lot to review-oriented sites thus improving their visibility on the basis of SERP [23, 28]. Some more algorithmic updates launched by Google include Mobilegeddon in 2015 that ranks mobile-friendly pages highly while not affecting the usual usage of desktops and tablets.

5.1.5 Google RANKBRAIN and FRED

Again in 2015, one more update RANKBRAIN was launched by Google that uses artificial intelligence, mathematical equations and semantics of language in order to learn the search habits of the user [28]. Similarly, Possum was launched in 2016, which utilizes user's location to provide local results thus benefitting the businesses located closer to the user with higher visibility. Along with location parameter, this update removes duplicate entries from search results. And last but not the least, the recent most algorithmic update is FRED, launched in 2017, which targets sites breaking Google's guidelines by overloading pages with ads and low-quality content [28]. It also targets sites with sole intention of ad revenue generation and bearing little value to users. Understanding the guidelines provided by these algorithms will definitely help various businesses to grow.

5.2 Amazon's Algorithm

Amazon the e-commerce giant uses intelligent search and recommendation algorithms to reach potential customers and grow its business. Most of the customers search the product they are interested in directly on Amazon instead of Google. Now, businesses are focusing on modifying their web presence not only on Google but also on Amazon by adopting new marketing strategies to increase visibility. The increased visibility of businesses will in turn increase the number of buyers of several product ranges. In order to increase their web exposure on Amazon and maximize their opportunities, businesses need to understand the working principle of Amazon's search algorithm. However, Google algorithms and that of Amazon's differ subtly in their base purpose of conceptualization. The Google algorithms were developed with the intent of returning the searched contents to users, so as to keep people returning to the pages again and again, thus generating profit from clicks and ads. Whereas, Amazon algorithms mainly targets customers, for selling products although it behaves as a search engine to certain extent. Further, it matches best prices with products considering user preferences to attract customers and grow business.

5.2.1   Amazon A9

In comparison to Google algorithms, the Amazon algorithm A9 is quite straightforward and considers those factors for ranking that immediately exist on the page. The algorithm predicts customer relevancy, studies their buying behaviour and provides them competitive prices of products along with other value added buying options. Some other factors considered by Amazon's algorithm that helps in increasing visibility of brands include sales ranking of similar products (i.e. increased sales in a particular category will lead to higher ranks), sellers past record, customer reviews on the purchased products, clarity in image and information of the products, completeness of listings etc. Also, Amazon's algorithm recommends products to users on the basis of products usually bought together using collaborative filtering which may create an interest in the customer to purchase the same and thus adding profits [24, 34, 30, 33].

5.3   Face Book's News Feed Algorithm

Face book uses News Feed algorithm for emphasizing public posts that people mostly find interesting and interact through likes, shares, and comment depending on user's preferences and ongoing trends over the network. A careful observation of the algorithm can provide brands and other businesses to create potential contents to attract customer conversations from the social media [2, 6, 8, 27]. Businesses are required to create contents having higher value that can generate comments and shares for the posts so as to increase the ranking of the content in News Feed of users. Brands need to encourage users to interact with each other to keep posting comments and shares, thus making brands top-stories. This will result in forming a strong customer base and popularizing the brand within a broader reach of opportunities to grow their business.

The News Feed algorithm earlier know as EdgeRank algorithm that determines the top-most stories scoring higher ranks to appear in the user's News Feed. Edges are the individual actions performed by the users ranging from posting a status update, tagging photos to posting comments on other user's status update and everything.

These edges are then ranked by the algorithm to filter out each user's top-most stories. The EdgeRank algorithm assigns rank score on the basis of three major criteria: (i) Affinity Score: that considers "how connected a user is" by observing the number of mutual friends with other users, and calculating affinity score for each explicit action such as clicking, liking, commenting and sharing etc. (ii) Edge Weight: Each edge has been assigned a default weight such as comments have higher weight as compared to likes. These weights of edges add to the rank score of the top stories of the users and finally (iii) Time Decay: Since EdgeRank is evaluated dynamically, older edges lose points and newer ones get higher scores. This earlier version of EdgeRank algorithm has been gone through lots of enhancements and modifications according the requirement of the users.

Some other businesses that are based on algorithmic business framework include Uber, the ride sharing Transport Company. Although it claims that its surge pricing algorithm is based upon the demand and supply concept [4] but the underlying operational strategy still follows black-box algorithms about which not much has been disclosed publicly. However, the other businesses such as eBay and Airbnb that adopt on dynamic pricing algorithms are transparent to public.

6   Applications

In today's digital era, many of the organizations are currently developing certain applications that are being built using intelligent algorithms for processes and services to be delivered to customers. Moreover, Instead of leaving the decisions to be made by human beings, on the basis of the information generated by analyzing the massive amount of data generated from heterogeneous sources, these algorithms take the responsibility of making dynamic decisions for actions to be made using real-time values and studying different aspects and consequences of the decision. Hence, these intelligent algorithms are considered as the centre for the deployment of digital business [9, 11, 13, 15, 20, 29]. Although, the term "Algorithmic Business" is gaining attention recently after being coined by Peter Sondergaard, senior vice president, Gartner; but still the algorithmic model for business has been practically in use since a long time, enabling disruptive business. Some of the major organizations that are currently using intelligent algorithms in various applications for enhancing their business include:

6.1   Manufacturing Industries

Manufacturing industries are installing smart factories equipped with autonomous systems, intelligent tools and machineries containing sensors and actuators that collects real-time data and optimizes several automated processes. And, intelligent algorithms remain behind everything as the underlying principle.

6.2 Health Care Industries

Intelligent algorithms developed in clinical sciences have helped to build new insights about several diseases thus making them more curable. These insights are generated by analyzing huge amount of graphical as well as textual data acquired from individual patients. Further, development of advanced algorithms in clinical science has enabled healthcare industry to convert the information they have extracted into assets. Also, these advanced algorithms can further be monetized to generate revenue.

6.3 Smart Products

Various new smart products being developed and ranging from clothes, fitness products, to kitchen products and accessories being manufactured are embedded with micro-chips containing sensors and actuators that communicate with human beings through intelligent interactive systems and make decisions about improving experience and performance.

6.4 Retail and E-Commerce Industries

Retail sector has also not been left untouched from the algorithmic wave of digital era. In fact, retail sector is the one that has widely embraced algorithmic business by solving several problems using intelligent algorithms. Retail industries are applying intelligent algorithms in logistics and supply chain to find best routes for supply of raw materials as well as delivery of final products to save time. While real-time data of currents status are collected and analyzed by these algorithms, load limits, traffic congestions, vehicle specifications and limits are some of the constraints being considered while applying the intelligent algorithms in retail sectors. Some other retail based applications of smart algorithms involve merchandising of products in both offline and online displays to stimulate customer interests, optimal pricing of products, designing of promotion strategies and discount offers etc. by analyzing not only real-time interactive data about customer preferences but also about selling horizon and storage capacity as constraints. However, maximizing revenue, quick response and customer satisfaction are considered as the major goals of the retail sector while designing these intelligent algorithms. Further these designed intelligent algorithms can be treated as assets and protected from other retailers to gain competitive advantage; also it can also be monetized for profit gains. Also e-commerce based websites such as Alibaba and Amazon are top users of algorithmic model for business for automating various processes involved such as managing inventory, recommending products to customers on the basis of their preferences, providing a comparative view of various sellers using matching algorithms and lastly for implementing pricing strategies. Further, the adopted intelligent algorithms dynamically adjust prices of products depending on past market trends, demands and experiences.

6.5   Smart Cities

With the development of smart products and smart transports leading to smarter ways of living, smart environment, smart economy and smart governance emerges the concept of smart cities. These cities are expected to be heavily equipped with internet, telecommunication as well as sensor networks to facilitate IoT services. Smart cities commit to consider improvisation of life quality along with efficient consumption of energy and safety of citizens as major focus and work towards adaption of intelligent algorithms for bringing it into reality.

6.6   Education Industries

Also education industry has adopted smart algorithms to some extent by designing smart assessment tests for students and taking initiatives in delivering quality education after weaknesses of students are identified.

6.7   Airline Industry

Intelligent algorithms are also being used in airline sector since a long period to set passenger fares. While designing these algorithms a set of variables are considered on the basis of certain factors such as choice of seat class by the passenger from the set—economy and business, location of the sit in the aircraft, time of the year that is being travelled, day of the week that is being travelled, duration left before the travelling date, etc. Optimal pricing of tickets are dynamically set by these smart algorithms considering the above factors as well as certain constraints by constantly monitoring the demand for tickets as well as availability of tickets while keeping revenue maximization and customer satisfaction as objectives.

6.8   Human Resources

Further, intelligent algorithms are designed evaluating the suitability of candidates for specific jobs in Talent Acquisition sector. Smart algorithms for various psychometric tests can be designed for assessing IQ levels to leadership, flexibility in working with a team to leading a team etc. to assigning the candidate a deserving position.

6.9 Transportation

Bus and on-demand taxi services are also applying intelligent algorithms are making communication easier along with enhancing their business. Taxi service companies like Uber do not even own any vehicle or infrastructure. Instead, they widely use smart algorithms for connecting passengers with drivers available in the passengers' locality. Here, the smart algorithms are considering requirements of passengers and time of the day (peak hour or regular hour) as price deciding factors, while availability of vehicles is being considered as constraints for optimizing the price.

6.10 Stock Trading

In this digital era, trading systems are also adopting intelligent algorithms for deciding successful trading strategies. These algorithms design new investment and trading strategies by automating the process of continuous monitoring of market trends and applying modern modeling tools and techniques for predicting investment options.

7 Challenges

Although algorithmic business offers many benefits to an organization but some inevitable challenges also exist there which need to be addressed for successful implementation of algorithmic business. Some of these challenges can be enumerated as (i) the first major challenge consists of modernization of infrastructure and resources to cope with the fast transformation, (ii) a revolutionary change in the traditional architectures in all aspects (to enhance, in-memory computation, processing speed etc.) to support higher scalability with quicker performance for generating real-time insights can be considered as the next challenge (iii) Moreover, making such enormous changes in almost all business sectors, will lead to the rising demand in building of new skill sets pools for handling these new changes. (iv) further, in case of a supply chain scenario, algorithms may not be capable of negotiating with suppliers to the extreme extent that human can attain to get lowest rates for procuring supplies of materials or intermediate products (v) again, not being context sensitive as the response of a customer in a feedback form, collected about a certain product or service depends on the current mental state of the customer. Therefore, the response may not be considered as a standard as improvisation based on that feedback may or may not help the organization, (vi) next analysis of security risks associated with emerging architectures and business models along with managing them as soon as they are detected is a must since it might lead to higher and critical risks as everything revolves around confidential information, (vii) another major downside of algorithmic business can be certain unavoidable circumstances that can be created as a result of algorithms centrally controlling business and machines. Suppose some real-time data is incorrect or inconsistent then the algorithms may become biased and behave unexpectedly making the entire system unstable and can cause severe damages both professionally as well as financially. Therefore, preventive practices must be designed to take potential decisions and consequent actions to nullify such effects. (viii) Also there is always the risk of hackers conning the algorithms to attain their objectives. Identifying such risks and applying measures to mitigate these risks for protecting the organization is also an issue. (ix) Finally, since algorithmic business is based upon usage of algorithms or a set of rules and the algorithms are yet to learn about ethics, we still cannot rely on algorithms solely leaving entire control to smart algorithms.

8 Conclusion

As almost all the enterprises and organizations are open-handedly embracing the technical advancements developments emerged for the analysis of big data, algorithms are gradually strengthening their position in economical enhancement of any organization. Algorithmic business is slowly gaining attention of academic as well as industry oriented researchers with growth of smart live styles and smart business. Algorithms are providing major breakthroughs in the way business is being done by shifting the centre from data to algorithms. They not only help in understanding the occurrences of certain events on the basis of historical data but also predict future uncertain events and design rules and actions to be taken to avoid risks. This can be considered as a significant research area covering a wide range of application areas and the challenges they are facing in deploying the same. In this chapter, we have tried to provide a base framework by exploring several aspects of algorithmic business ranging from its arrival to its rise and deployment strategies to potential applications and challenges.

References

1. Applin SA, Fischer MD (2015) New technologies and mixed-use convergence: how humans and algorithms are adapting to each other. In: 2015 IEEE international symposium on technology and society (ISTAS), pp 1–6. IEEE
2. Bucher T (2017) The algorithmic imaginary: exploring the ordinary affects of Facebook algorithms. Inf Commun Soc 20(1):30–44
3. Chaffey D (2015) Digital business and E-commerce management. Pearson Education Limited
4. Chen L, Mislove A, Wilson C (2015) Peeking beneath the hood of uber. In: Proceedings of the 2015 Internet Measurement Conference, pp 495–508. ACM
5. Cocchia A (2014) Smart and digital city: a systematic literature review. Smart city. Springer International Publishing, pp 13–43
6. Cotter K, Cho J, Rader E (2017) Explaining the news feed algorithm: an analysis of the News Feed FYI blog. In: Proceedings of the 2017 CHI conference extended abstracts on human factors in computing systems, pp 1553–1560. ACM

7. Datta P, Vaidhehi V (2017) Influencing the PageRank using link analysis in SEO. Int J Appl Eng Res 12(24):15122–15128
8. DeVito MA (2017) From editors to algorithms: a values-based approach to understanding story selection in the Facebook news feed. Digit Journal. 5(6):753–773
9. Earley S (2015) Analytics, machine learning, and the internet of things. IT Prof 17(1):10–13
10. Fleisch E, Weinberger M, Wortmann F (2015) Business models and the internet of things. Interoperability and open-source solutions for the internet of things. Springer, Cham, pp 6–10
11. Gartner (2013) Gartner says the internet of things installed base will grow to 26 billion units by 2020, Dec 2013. www.gartner.com/newsroom/id/2636073
12. Gubbi J, Buyya R, Marusic S, Palaniswami M (2013) Internet of things (IoT): a vision, architectural elements, and future directions. Future Gen Comput Syst 29(7):1645–1660
13. Gartner (2014) Gartner's 2014 hype cycle for emerging technologies maps the journey to digital business, Aug 2014. www.gartner.com/newsroom/id/2819918
14. Hermann M, Pentek T, Otto B (2016) Design principles for industrie 4.0 scenarios. In: 2016 49th Hawaii international conference on system sciences (HICSS), pp 3928–3937. IEEE
15. Jia X, Feng Q, Fan T, Lei Q (2012) RFID technology and its applications in internet of things (IoT). In: 2012 2nd international conference on consumer electronics, communications and networks (CECNet), pp 1282–1285. IEEE
16. Komninos N (2016) Smart environments and smart growth: connecting innovation strategies and digital growth strategies. Int J Knowl Based Dev 7(3):240–263
17. Kellner T (2013) Analyze this: the industrial internet by the numbers & outcomes. Ge report. www.gereports.com/post/74545267912/analyze-this-the-industrial-internet-by-the. Accessed 30 Nov 2014
18. Laney DB (2017) Infonomics: how to monetize, manage, and measure information as an asset for competitive advantage. Routledge
19. Latzer M, Hollnbuchner K, Just N, Saurwein F (2016) The economics of algorithmic selection on the Internet. Handbook on the economics of the internet, chapter 19, p 395
20. Lee I, Lee K (2015) The internet of things (IoT): applications, investments, and challenges for enterprises. Bus Horiz 58(4):431–440
21. Levy H (2017) The arrival of algorithmic business. smarter with Gartner. https://www.gartner.com/smarterwithgartner/the-arrival-of-algorithmic-business/
22. Leonhardt D, Haffke I, Kranz J, Benlian A (2017) Reinventing the IT function: the role of IT agility and IT ambidexterity in supporting digital business transformation
23. Mashrafi M (2017) Investigate the effect of semantic search on search engine opti-mization. Doctoral dissertation, Cardiff Metropolitan University
24. Mishra R, Kumar P, Bhasker B (2015) A web recommendation system considering sequential information. Decis Support Syst 75:1–10
25. Newell S, Marabelli M (2015) Strategic opportunities (and challenges) of algorithmic decision-making: a call for action on the long-term societal effects of 'datification'. J Strateg Inf Syst 24(1):3–14
26. Perera C, Liu CH, Jayawardena S, Chen M (2014) A survey on internet of things from industrial market perspective. IEEE Access 2:1660–1679
27. Rader E, Gray R (2015) Understanding user beliefs about algorithmic curation in the Facebook news feed. In: Proceedings of the 33rd annual ACM conference on human factors in computing systems, pp 173–182. ACM
28. Rogers R (2018) Aestheticizing Google critique: A 20-year retrospective. Big Data Soc 5(1):2053951718768626
29. Tao F, Zuo Y, Da Xu L, Zhang L (2014) IoT-based intelligent perception and access of manufacturing resource toward cloud manufacturing. IEEE Trans Ind Inf 10(2):1547–1557
30. Xu C, Wang C, Gong L, Li X, Wang A, Zhou X (2017) Acceleration for recommendation algorithms in data mining. High performance computing for big data: methodologies and applications, p 121
31. Xu Z, Sugumaran V, Yen NY (2018) Special issue: algorithmic and knowledge-based approaches to assessing consumer sentiment in electronic commerce. Electron. Commer Res 18(1):1–1

32. Zhang S, Cabage N (2017) Search engine optimization: comparison of link building and social sharing. J Comput Inf Syst 57(2):148–159
33. Zhou M, Ding Z, Tang J, Yin D (2018) Micro behaviors: a new perspective in e-commerce recommender systems. In: Proceedings of the eleventh ACM international conference on web search and data mining, pp 727–735. ACM
34. Zhuhadar L, Kruk SR, Daday J (2015) Semantically enriched massive open online courses (moocs) platform. Comput Hum Behav 51:578–593

A Decentralized Business Ecosystem Model for Complex Products

Mirjana Radonjic-Simic and Dennis Pfisterer

Abstract Consumers looking for complex products, demand highly personalized combinations of individual products and services to satisfy a particular need. While online marketplaces work well for individual products and services (or a predefined combination of them), they fall short in supporting complex products. The complexity of finding the optimal product/service combinations overstrains consumers and increases the transaction and coordination costs for such products. Another issue related to contemporary online markets is the increasing concentration around platform-based ecosystems such as, e.g., Amazon, Alibaba or eBay. That increases the "positional power" of these platforms, putting them in the position, where they can dictate the rules and control access directing towards de facto centralization of previously decentralized online markets. To address these issues, we propose a novel business ecosystem model for complex products—a model of a strictly decentralized exchange environment purposefully designed to support complex products in a way to lower transaction and coordination costs and alleviate the adverse effects of growing platform power. This chapter introduces our ecosystem model by describing its primary artifacts: (1) the ecosystem structure, and (2) the ecosystem architecture. The ecosystem structure maps the activities, actors, their roles, and the essential value creation pattern required to support different complex product scenarios. It integrates various actors (i.e., individuals, companies, communities, autonomous agents, and machines), and enables them to constitute and enrich their ecosystem without any central instance of control or governance (i.e., underlying platform). The ecosystem architecture describes the building blocks of our ecosystem model, and the relationships among them considered essential to support the ecosystem structure on the operational level. It is represented by a highly scalable and strictly decentralized software-system architecture that supports arbitrarily complex products given existing domain-knowledge, relevant for commercial transactions in a particular domain.

M. Radonjic-Simic (✉)
Baden-Wuerttemberg Cooperative State University Mannheim, Mannheim, Germany
e-mail: mirjana.radonjic-simic@dhbw-mannheim.de

D. Pfisterer
Institute of Telematics, University of Lübeck, Lübeck, Germany
e-mail: pfisterer@itm.uni-luebeck.de

S. Patnaik et al. (eds.), *Digital Business*, Lecture Notes on Data Engineering
and Communications Technologies 21, https://doi.org/10.1007/978-3-319-93940-7_2

The feasibility of the proposed software-system design is demonstrated based on a prototypical implementation and an evaluation use case scenario.

1 Introduction

Recent developments towards a highly interconnected world of people, processes, data, and things [1] unlock new possibilities for commercial exchange and give rise to novel types of products and services. These products and services, as well as complex combinations of them, can be personalized in a way to satisfy individual needs and contextual requirements of consumers.

Consider the use case of a couple that plans an evening with friends, for example visiting a concert at the city's theater and having dinner at a nice Italian restaurant. Besides buying tickets and reservation of a table, this also includes finding a parking space close to both locations and booking a babysitter that looks after the children. As a consumer, the couple is looking for a *complex product* by demanding a highly personalized combination of products and services to satisfy a particular need.

We define *complex products* as arbitrary combinations of individual products or services that meet particular consumer needs based on a consumer-defined context. Thereby, consumer-defined context encompasses a broader range of information providing a comprehensive description of consumers' needs. Examples are personal preferences and constraints such as budget, quality, schedule, and payment modalities. For our couple, in addition to just specifying concert details (e.g., classic, musical or other), this could be engaging a babysitter from a well-rated babysitting agency, making the table reservation with certain restrictions (e.g., availability of a seaside terrace) and parking slot with less than 200 m walking distance.

To make an informed decision, our couple (i.e., a consumer of this complex product) needs to know where and how to find viable offerings, to compare them, and aggregate all relevant information manually and put them in the context of their preferences and requirements. That complexity and high consumer involvement lead to the decision making according to the principle of adverse selection [2], where consumers substitute an "optimal" buying decision with a "good enough". That increases the transaction and coordination costs for complex products. To approach this issue, specialized online marketplaces (i.e., platforms) exist that support consumers in finding complex products for certain domains. Examples are travel platforms such as *Opodo*[1] or *Expedia*[2] that provide pre-defined combinations of flights, hotels, rental cars, and insurances. However, such platforms are limited to supporting consumers in creating and requesting arbitrary complex products, which need to fulfill particular consumer-defined criteria. The aforementioned use case already spans different service domains and includes contextual requirements affecting the complex product as a whole.

---

[1]http://www.opodo.com.
[2]http://www.expedia.com.

Another issue related to contemporary online markets is the increasing concentration of supply around mega platforms such as, e.g., Amazon,[3] Alibaba.[4] This underpins the positional power of these platforms, putting them in a monopoly position [3] where they can dictate the rules, control access, and offerings, and thus leads to a de facto centralization of the previously decentralized offerings on the Internet.

To address these two issues, we propose a novel business ecosystem model for complex products—a model of a strictly decentralized exchange environment purposefully designed to support complex products in a way to lower transaction and coordination costs and alleviate the adverse effects of growing platform power. The contributions of this chapter, therefore, are the following:

- *We analyze complex product scenarios and elicit the business, organization, architecture, and technology-related objectives relevant to the modeling of a business ecosystem for complex products.*
- *We outline its structure by identifying activities, actors, their roles, and the essential value creation pattern required to support complex product scenarios.*
- *We design and develop a highly scalable and strictly decentralized architecture to support the operational level of the proposed ecosystem structure on the operational level.*
- *We prototypically implement the proposed architecture and demonstrate its feasibility in the context of an evaluation use case scenario.*

This chapter is organized as follows. First, Sect. 2 presents the background of business ecosystems and briefly describes our approach and employed methodology. Next, Sect. 3 analyzes complex product scenarios and discusses the overall objectives and requirements relevant to the modeling of a business ecosystem for complex products. After that, Sect. 4 reviews the related work and compares it to the elicited requirements. Section 5 outlines the proposed business ecosystem model by describing its concept and ecosystem structure. Thereafter, Sect. 6 introduces the ecosystem architecture by specifying its entities and functional components and evaluates its feasibility in the context of a use case scenario. Finally, Sect. 7 summarizes this chapter and gives an outlook for the future research.

2 Background and Methodology

In this section, we will first look at business ecosystems and different modeling approaches. Afterwards, in Sect. 2.2, we briefly describe the theoretical framework of this work and employed methodology.

---

[3]http://www.amazon.com.
[4]http://www.alibaba.com.

2.1   Business Ecosystems

The term business ecosystem is first introduced by Moore [4] defining it as "an economic community supported by a foundation of interacting organizations and individuals—the organism of the business world. This economic community produces goods and services of value to customers, who are themselves members of the ecosystem". The member organism also includes supplier's competitors, institutions, and other stakeholders, which together define business networks (i.e., ecosystems) that create value through a process of cooperation and competition [5, 6]. This explicit dependence of involved actors who depend on each other for their mutual effectiveness and survival is considered the essential feature of business ecosystems that distinguishes them from other organizations of economic actors like, e.g., supply chains or value networks [7].

The literature on business ecosystems suggests that two determinants are essential for successful business ecosystems [5, 8, 9]. First, how value is created to attract and retain ecosystem's participants and motivate them to engage and provide growth potential. And second, how created value is distributed and shared among involved participants. Consequently, a business ecosystem construct is considered as a network-centric structure [10] composed of entities and components that describe how value is created and captured [11].

There are several approaches on how to conceptualize business ecosystems, such as BEAM [9], MOBENA [12], 6c [8], VISOR [11], and concepts of Value Design [13]. BEAM and MOBENA focus on the strategic aspects suggesting a network-centric approach to value creation and capture across the ecosystem. In contrast, Value Design emphasizes a more actor-centric view of value creation in a dynamic co-evolving network of different actors. However, as with the previous two, Value Design is in a conceptual stage and falls short to provide insights for practical implementation. VISOR and 6c are more explicit on the operational level but predominately concentrated on the technical aspects of an ecosystem as a platform-based construct that utilizes technology for value creation and distribution.

A common pattern of the aforementioned approaches is that they prioritize the "focal actor" and the network of actors tied to it (i.e., platform). Ecosystem-as-structure [14], in contrast, prioritizes the value proposition of the ecosystem (i.e., the focal value proposition). It suggests that the modeling process should start with a definition of the focal value proposition and continue with designing an alignment structure of activities and actors, which need to interact and exchange value for the focal value proposition to materialize. Accordingly, and as illustrated in Fig. 1, the core elements that underlie the ecosystem construct are activities, actors, positions, and links. Activities define the discrete actions to be undertaken, actors represent the entities that undertake these activities, positions, and links specify where in the flow of activities actors are located, and how they need to interact and exchange value. These elements mutually depend on each other, and collectively they characterize the configuration of activities and actors required for the focal value proposition to be realized.

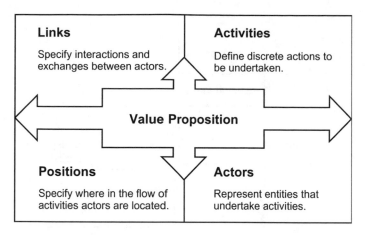

Fig. 1 Elements of an ecosystem construct [14]

2.2 Methodology

The methodology employed to develop this work follows the Design Science Research (DSR) method [15]. We apply the problem-centered approach since the primary goal of this work is to address the problem related to increasing transaction costs for complex products and the adverse effects of the growing platform power. As illustrated in Fig. 2, the used methodology aligns with five steps.

In Step 1, we analyze complex product scenarios using the BOAT framework [16] as a method for analysis of networked business scenarios. Thereby, we distinguish between business (B), organization (O), architecture (A), and technology (T)—related objectives, which a business ecosystem model must meet to support complex product scenarios.

In Step 2, we operationalize these objectives and formulate business goals and requirements relevant to the design of the ecosystem model that is represented by two artifacts: the ecosystem structure and underlying architecture.

In Step 3, we outline an ecosystem structure employing the ecosystem-as-structure modeling approach [14] (cf. Sect. 2). Based on the stated business goals, we first formulate the value proposition of the envisioned ecosystem. Then, we outline the activities, actors, and interactions necessary for the value proposition to realize.

Based on that, in Step 4, we design a blueprint of the underlying ecosystem architecture to support such an ecosystem structure. The blueprint describes the necessary components of the architecture, the externally visible properties of these components and the relationship among them. Finally, in Step 5, the proposed ecosystem architecture is prototypically implemented and demonstrated in the context of an evaluation scenario.

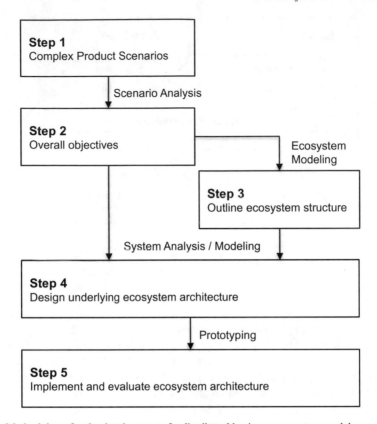

Fig. 2 Methodology for the development of a distributed business ecosystem model

3 Overall Objectives

The aforementioned use case of a couple planning an evening with friends represents a class of different scenarios in which consumers are looking for complex products to satisfy personalized needs and demands. In this section, we will discuss and analyze such complex product scenarios and point out their main characteristics. These characteristics will also serve as the base from which to elicit the business goals (cf. Sect. 3.2) and requirements (cf. Sect. 3.3) on a business ecosystem model that supports such scenarios.

3.1 Complex Product Scenarios

Complex product scenarios denote settings in which consumers and providers engage together to achieve specific goals represented through the transactions of complex

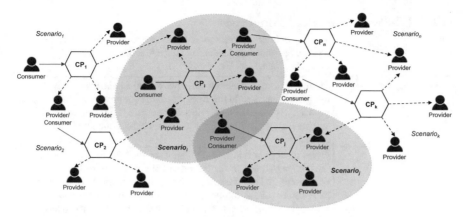

Fig. 3 Visualization of complex product scenarios

products. Complex product scenarios are considered *context-centric, distributed and transaction-oriented business scenarios*:

- *context-centric*—as initiated by consumer's demand for a personalized combination of products and services to satisfy a particular need. In the case of our couple, that is a specific set of services (e.g., a babysitter, tickets, table reservation and a parking slot), that need to meet constraints such as, e.g., a babysitter from a certified and well-rated agency and a parking slot with a short walking distance to both locations.
- *distributed*—as consumer's demand might span different business domains (e.g., events, gastronomy, babysitting, parking), complex product scenarios might require many diverse cross-domain providers to deliver distinct parts of the required product/service combination.
- *transaction-oriented*—as its central goal is the commercial transaction of a particular complex product. In contrast to a commercial transaction of an individual product or service, the transaction of a complex product is considered a bundle of single transactions embedded in one enclosing transaction.

From the market-level perspective, complex product scenarios represent two-sided interactions between market participants (consumers and providers). As illustrated in Fig. 3, the involved market participants are linked via complex products (CPs) they intend to consume (represented by the solid lines) and products/services they intend to provide (dotted lines). As shown, market participants might initiate and participate in different scenarios being a consumer or a provider or both at the same time. For example, one of the providers involved in *Scenario$_i$* might also initiate *Scenario$_j$* taking the role of a consumer asking for CP_j.

In addition to the identified characteristics of complex product scenarios, Tables 1 and 2 summarize further characteristics derived from the scenario analysis, which we conducted using the BOAT method [16].

Table 1 Classification of complex product scenarios

Purpose	To support interactions between consumers and providers to achieve a particular goal represented by the transaction of a particular complex product
Involved parties types	Involved parties (consumers and providers) can be everybody or everyone connected to the Internet asking (or providing) for complex products. These can be individuals, companies, institutions, and software agents and machines acting on behalf of a provider or consumer
Time scope	Dynamic and short lasting. Dynamic, as the selection of potential interaction partners (potential providers), is determined by consumers, and short lasting, because the interactions of the involved partners are transaction-oriented. That is, as soon as the transaction is settled the relationship between involved partners comes to an end

Table 2 Further characteristics of complex product scenarios

Aspect	Characteristics
Business	**Drivers** For a consumer to increase efficiency in making transactions of complex products (lower coordination and transaction costs) For a provider to improve the accessibility of own offerings and the level of personalization by providing products/services in consumer defined context
Organization	**Networked structure** (actor-to-actor) to integrate a vast number of involved consumers and providers (i.e., actors) since complex product scenarios might span different product/service domains and require many heterogeneous providers to deliver distinct parts of the requested product/service combination
Architecture	**Business functions** Facilitating the formation of an actor-to-actor network and support the interactions among actors within a networked organizational structure Enabling the composition of complex products and integration of consumer-defined context Supporting the transaction of complex products by enabling activities related to the transaction process [20], that divides into phases: information (formulating, publishing offerings demand), negotiation (matching and ordering), settlement (realizing the transaction regarding payment and delivery), and after-sales (reviews and possible dispute resolution)
Technology	**Supportive technology** Internet and Web Technology as the basis for communication and interaction between involved parties within an actor-to-actor network Transaction process management technology to support the identified business functions and related processes Data management technology to support specific aspects of organizing, storing and retrieving information within the actor-to-actor networks User interface technology to enable composing complex products and formulating complex products requests

We first classify complex product scenarios regarding their scope, involved parties, and applicable time scope, as presented in Table 1. Then, we look at further characteristics taking the perspective of a business ecosystem as an organizational structure to support such scenarios. The analysis starts with formulating the business drivers for involved participants, followed by considerations how complex scenarios need to be organized to meet identified business drivers, and which business functions are required to support the organizational structure. The analysis closes with technology-related details deemed necessary for the realization of such a structure. Table 2 presents the identified characteristics and their relation to each analyzed aspect.

3.2 Business Goals

The identified characteristics of complex product scenarios impose a set of specific objectives for the organization and underlying architecture of a business ecosystem aiming to support such scenarios. Based on that, and our previous work in [17, 18] we formulate the primary business goals of a decentralized business ecosystem model for complex products. The business goals ($BG1, …, BG5$) are summarized and briefly discussed in Table 3.

3.3 Requirements

Having formulated the business goals of a decentralized business ecosystem for complex products, in this section, we use them as the rationale to derive the most relevant requirements to be met by the underlying ecosystem architecture. Thereby, we follow the definition of business ecosystem architecture as a blueprint or a description of the building blocks of an ecosystem and how they relate to each other, what they do, and how they interact [19].

Table 4 summarizes the derived requirements ($R1, …, R8$) and describes them briefly.

4 Related Work

This section reviews a set of approaches and concepts for commercial exchange environments and compares them considering the stated requirements ($R1, …, R8$). And thus, identifies the missing gap to be closed by our business ecosystem model for complex products.

Electronic marketplaces (e-marketplaces) refer to a market-oriented organization of commercial exchange, aiming to increase market transparency and lower

Table 3 Business goals of a decentralized business ecosystem model for complex products

#	Business goal	Description
BG_1	Enabling transactions of complex products in a trustful and reliable manner	For consumers, it means increasing efficiency, less time and effort in making informed buying decisions, and for providers, gaining higher accessibility of their offers and the level of personalization related to the consumer-defined context
BG_2	Facilitating decentralization and scalability	To alleviate the adverse effects of the growing platform power, and increase the flexibility of integration for a broader range of different product/service domains and a vast number of involved consumers and providers
BG_3	Allowing trade in any business domain	Since complex products can be arbitrary complex combinations of individual products and service spanning different domains, the ecosystem has to be open for any business domain
BG_4	Enabling cross-domain transactions	Since complex products can encompass any composition of products or services, the ecosystem as a business environment has to allow cross-domain transactions
BG_5	Promoting actor-to-actor economy principles	The ecosystem has to support activities related to the formation of an actor-to-actor network, activities among the actors and ensure an adequate level of trust and security, seen as very important for actor-to-actor transactions [41]

transaction and coordination costs [20]. Established e-marketplaces also called "platforms" operate as centralized solutions with a focus on the supply and the availability of individual products and services. Regarding the complex product composition and integration of contextual information ($R1$, $R2$), most platforms only support compositions of products and services within their domain boundaries (i.e., industry, domain, or type of products/services) or they offer a pre-defined combination of them. In general, e-marketplaces work well for the transaction of individual products and services providing reliable support in matching, ranking, and settlement activities for both transaction partners (consumers and providers). They are centralized solutions and fall short to support transactions of complex products ($R1$, $R3$) that need to fulfill a specific consumer-defined context ($R2$). Furthermore, they do not support cross-domain transactions among different platforms ($R4$).

Decentralized peer-to-peer (P2P) marketplaces refer to the concept of a dis-intermediated exchange environment that brings potential buyers and sellers together to engage with each other directly, without any intermediary [21]. Concepts introduced by Eymann [21], Schmees [22], Serban et al. [23], Xia et al. [24], Hausheer

Table 4 Requirements on the architecture of a decentralized ecosystem for complex products

Rationale	#	Requirement	Description
BG_1, BG_3	R_1	Complex product composition	Capability to enable consumers to compose complex products as arbitrary combinations of individual products and services
BG_1, BG_3	R_2	Integration of contextual information	Capability to integrate consumer-defined context, which can encompass much more information than those related to the decision criteria and include the broader range of constraints and requirements
BG_2, BG_3	R_3	Distributed transactions	Capability to support transactions of products/services from different providers in a single enclosing transaction (from the consumer's point of view)
BG_4	R_4	Cross-domain transactions	Capability to support transactions of products/services from different business-domains in a single enclosing transaction (from the consumer's point of view)
BG_1, BG_3, BG_5	R_5	Advanced matching	Capability to support matching a large number of fragmented, heterogeneous consumers and providers efficiently, by keeping transaction and coordination costs low
BG_1, BG_3, BG_5	R_6	Advanced ranking	Capability to support consumers in making informed decisions by applying ranking mechanisms that consider context-related constraints to create the "best-fit" list of offers
BG_5	R_7	Sophisticated reputation and feedback mechanism	Capability to support consumers in making informed decisions by considering sophisticated information about potential trading partners
BG_2, BG_5	R_8	Actor-to-Actor network	Capability to enable the forming of an actor-to-actor network and supporting the interactions among actors in a direct and decentralized manner

and Stiller [25] propose different technological solutions, i.e., frameworks and architectures, on how to organize "market without makers". Yet, they do not support the full transaction process in a cross-domain and distributed manner. Some of them are related to a particular product/service domain, e.g., [23], others are limited on the exchange of services [26] or even only on digital goods [22], and [25] supports only certain trading forms such as, e.g., auction-based or supply-oriented trade. In contrast, the projects BitMarkets [27], OpenMarket [28], and OpenBaazar [29] aim to shift the whole trade onto decentralized P2P environments. These decentralized P2P marketplaces support providers to create and run online stores to sell products and services and connect these stores directly to each other on a global network, which is scalable and without a central point of control (or point of failure). Potential consumers can search, connect, and buy products or services directly from providers. While all of these solutions support the whole transaction process in a cross-domain and distributed manner (preferably using criptocurrency), they are predominately supply-oriented and thus, limited in supporting consumers making informed decisions regarding complex products ($R1$, $R2$, $R3$).

The Intention Economy (IE), by Searls [30], represented through the Project Vendor Relationship Management (VRM), refers to a trading environment that focuses on a buyer's intention to conduct a transaction with potential sellers (i.e., vendors). The main idea of this approach is to make buyers independent in their relationships with the supply-side of markets [31], by providing VRM tools, which support buyers to describe their needs, creating a personal request for proposal (pRFP) and making it visible for the interested vendors. The matching of demand and supply is made by many specialized, domain-specific platforms, which support identifying the best and final offers (BAFO) for the corresponding pRFP, as well as, settlement of the transactions. Even though the VRM tools in collaboration with related platforms enable highly personalized requests, they do not support the integration of the contextual information ($R2$) neither they support cross-domain and distributed transactions ($R3$, $R4$) in a direct manner ($R8$).

Web of Needs (WoN) proposed by Kleedorfer et al. [32], refers to a framework for a distributed and decentralized e-marketplace on top of the Web. The vision of WoN is to standardize the creation of owner-proxies, which describe the demand or supply and represent the intention of a user to enter into a commercial transaction. Owner-proxies contain all information relevant to begin a transaction (e.g., information about the owner, description of the demand or supply). Ones created, owner-proxies are published to the Web by sending them to WoN nodes, which are distributed and interconnected. Published, owner proxies are made aware of each other by independent matching services, which collect information similar to Web search services by crawling through all WoN nodes. In case, potential transaction partners are identified, the matching service sends hint messages to the matched owner-proxies. The owner applications than initiate the transaction process by sending the contact message to the potential transaction partners. WoN enables a rich description of user's demand, advanced matching services and identification of potential trading partners, but it does not support the transaction process as a whole. Also, there is no evidence that WoN integrates any ranking or reputation mechanisms (*R6, R7*), or supports cross-domain and distributed transactions (*R3, R4*).

Concluding remarks on related work: Contemporary solutions for commercial exchange are mature in supporting individual products and services (thus, fulfilling *R3, R5, R6, R7*). However, they are mostly centralized solutions and fail short to support complex products (*R1, R2, R4*) in a strictly decentralized and direct manner (*R8*). Other approaches, address some of these shortcomings, but they do not represent a comprehensive solution on its own. Either they provide tools that need to be integrated such as IE, or they address some of the defined requirements, as the case with the P2P marketplaces and WoN. To our best knowledge, we therefore, conclude that there is no integrated organizational model for the commercial exchange, which meets all of the requirements (*R1, ..., R8*).

5 Outlining a Decentralized Business Ecosystem for Complex Products

This section outlines the proposed business ecosystem model that represents a strictly decentralized exchange environment, purposefully designed to support complex products. And thus, fulfilling the business goals and requirements stated in Sect. 3.

5.1 Model and Concept

The proposed business ecosystem model supports complex products in a way to lower transaction and coordination costs, and it follows the principle of the disintermediation to alleviate the adverse effects of growing platform power.

As visualized in Fig. 4, the proposed model defines a business ecosystem for complex products on three different levels:

- on the level of the *Market Space*,
- on the level of the *Ecosystem Structure*, and
- on the underlying *Ecosystem Architecture*.

Market Space visualizes the ecosystem as a strictly decentralized business environment and illustrates the supported market interactions among involved actors—consumers and providers linked via complex products they demand and products/services they provide.

Ecosystem Structure maps the overall value creation pattern within the ecosystem defined by the value proposition formulated as an *end-user enabled ecosystem for trading complex products directly and reliably*. Based on the business goals and requirements (cf. Sects. 3.2 and 3.3), the ecosystem structure is considered as an alignment structure of essential activity flows and actors (Activities and Actors). It

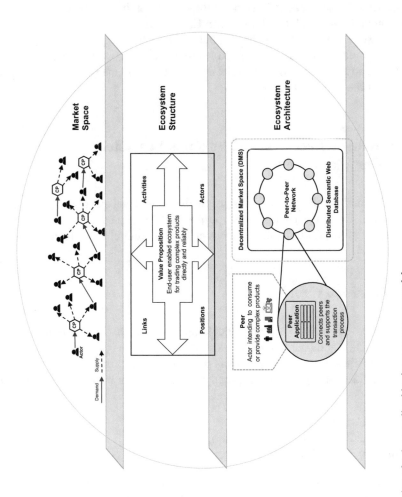

Fig. 4 Conceptual visualization of a decentralized business ecosystem model

illustrates how actors and activities need to link and align with the overall value creation to satisfy the individual and shared motivation across the ecosystem (Positions and Links). Next section (Sect. 5.2) provides a detailed description of the proposed ecosystem structure and its elements.

Ecosystem Architecture describes the building blocks of the proposed ecosystem and the relationship among them considered essential to support the ecosystem structure on the operational level. The proposed ecosystem architecture is represented by a software-system architecture that enables the overall value creation pattern of the ecosystem as a whole. Section 6 presents it in more detail, describing its core entities and inner workings, and evaluates its feasibility using a prototypical implementation in the context of a use case scenario.

5.2   Ecosystem Structure

The value proposition of a decentralized business ecosystem for complex products is to enable end-users (consumers and providers) to establish an open ecosystem in which they can trade complex products directly and reliably. This section outlines the ecosystem structure required for the stated value proposition to materialize.

As shown in Fig. 5, the proposed ecosystem structure presents an alignment structure of actors, activities, and interactions among them. It encompasses the four core elements *Activities*, *Actors*, *Positions*, *Links*, which (when linked together), illustrate how value is expected to be created and shared across the ecosystem. Following, we explain and describe each of these elements in more detail.

5.2.1   Activities

There are three primary activity groups that a decentralized business ecosystem for complex products needs to support in order for its value proposition to realize *Demand and Supply, Transaction Process, and Ecosystem Foundation*:

- *Demand and Supply* activity group maps activities related to the description and publishing of offerings on the provider side, as well as, activities related to the composing of complex products on the consumer side.
- *Transaction Process* activity group maps activity flows necessary for the conducting and monitoring of the transaction process as a whole (cf. Sect. 3.1) but, also for the discrete actions related to the different phases (i.e., information gathering, ordering/contracting, settlement, and after-sales).
- *Ecosystem Foundation* activity group maps activity flows to build the foundation, essential for setting up and functioning of the ecosystem. It includes forming and running the actor-to-actor network, providing hardware (e.g., hosting) and software resources (e.g., tools) and domain knowledge inevitable for the commercial transactions in a specific domain (e.g., product/service ontologies, vocabularies).

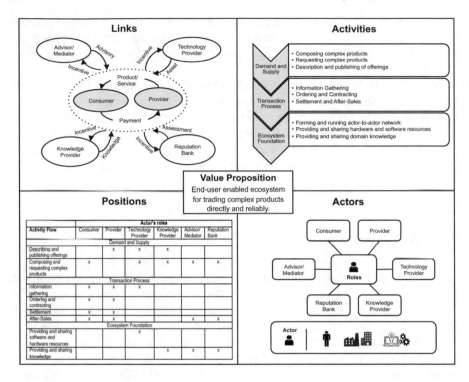

Fig. 5 Structure of the business ecosystem for complex products

5.2.2 Actors

Actors in a business ecosystem for complex products can be individuals, companies, institutions, associations, and autonomous actors such as software agents or machines (cf. Sect. 3.1). As shown in Fig. 5, there are (at least) six roles that actors can take *Consumer*, *Provider*, *Technology Provider*, *Knowledge Provider*, *Reputation Bank*, and *Advisor/Mediator*.

Consumer and *Provider* are considered shaper roles as they shape the value proposition and thus the birth of the ecosystem. Others are enabler roles since they undertake the afore-described activity flows. They are enabling the overall value creation and contributing to the ecosystem's ability to provide comprehensive services to the shaper roles.

Table 5 summarizes identified roles briefly describing their function and motivation, i.e., expected value from the participation across the ecosystem. Roles can overlap and might be taken concurrently, as they do not exclude each other (cf. Sect. 3.1).

Table 5 Roles and expected value

Role	Description	Motivation/Expected value
Consumer	Asking for a complex product	To satisfy a particular need defined through the transaction of a complex product in a reliable and trustful manner
Provider	Offering products or services in one particular domain or many domains	To earn revenue per product/service sold, receive payment; increase the visibility of own offerings and the level of customization by utilizing the contextual information provided by the consumer (contextualization)
Technology provider	Providing technology assets (software or hardware resources) to support the transactions of complex products	To contribute to the ecosystem foundation; earn revenue by guaranteeing availability only to paying users (incentive) and to leverage the user's feedback for improvement and development of technology assets
Knowledge provider	To provide structured knowledge for a particular trading domain	To contribute to the shared knowledge base; earn revenue by providing paid knowledge-based services (incentive)
Advisor/Mediator	Offering knowledge and expertise to support buying decisions and/or resolve disputes	To earn revenue through advisory, consultancy and consumer's feedback (incentive)
Reputation Bank	Assessing ecosystem's participants (i.e., actors) regarding their reliability, solvency, and worthiness	To capture reviews about current transactions needed for qualified assessments and establishing an adequate level of trust among actors

5.2.3 Positions

For the value proposition to be realized, the definition of the required activities and actors who need to undertake these activities is necessary, but not sufficient. To create value, actors need to align with the activities and take a particular position in the overall value creation.

As illustrated in Fig. 5, positions provide an overview of where in the flow of activities identified actors need to be located. As shown, single roles can contribute to several activities, and single activities might require several roles to engage. For example, to support consumers in composing complex products as arbitrary combinations of individual products and services, several roles are necessary to participate. Besides the consumer who initiates the process, the technology and knowledge providers are required to provide tools and knowledge to enable the composition of complex products and the integration of the contextual information. Depending on the complexity and level of personalization, the complex product composition might also involve further actors. For example, advisors can support composing the required product/service combination, and reputation bank might provide information about reputation and worthiness of the possible providers. Hence, support consumers proceed in a more targeted manner, narrowing the selection of the potential providers already at the beginning of the transaction process.

5.2.4 Links

Links between actors, as shown in Fig. 5, illustrate how actors need to interact with each other and specify the transfers between them. The nodes represent actors performing a particular role and the arrows the essential interactions indicating the "value exchanged" between these roles. Collectively they represent the overall pattern of exchanges to archive the stated value proposition as the shared motivation of all actors. But also, they specify the content of transfers required to satisfy the individual motivation of each actor, i.e., the expected value from the participation in the ecosystem (cf. Table 5).

Note that Fig. 5 maps only tangible exchanges such as payment or product/service delivery. For an in-depth discussion, the reader is referred to [17].

6   Ecosystem Architecture

This section presents the ecosystem architecture designed to facilitate the ecosystem structure as outlined in the previous section. The proposed architecture supports the identified alignment structure of the actors, activities, and interactions among them, and thus addresses the requirements ($R1, \ldots, R8$) of a purposefully designed business ecosystem for complex products.

As an explanation of our architecture, we first provide a high-level overview in Sect. 6.1, followed by the description of its core entities and functional components in Sects. 6.2 and 6.3, and inner workings in Sect. 6.4. Finally, we evaluate the feasibility of our ecosystem architecture based on a prototypical implementation and use case evaluation in Sect. 6.5.

6.1   High-Level Overview

The proposed ecosystem architecture represents a blueprint of a software-system required to enable the overall value creation pattern in a strictly decentralized business ecosystem. Figure 6 depicts the software-system design showing its core entities *Peer*, *Peer Application* and *Decentralized Market Space* (*DMS*).

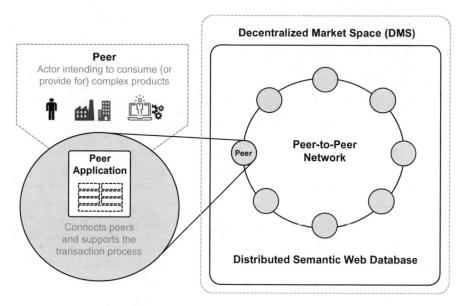

Fig. 6 Core entities of the ecosystem architecture

Peer represents an actor intending to consume (or provide for) a complex product and might take different roles as previously described in Table 5. Peers join the peer-to-peer network and form a strictly decentralized business environment, in which they can trade complex products directly and without any central instance of the control.

Peer Application represents the access point for a peer to participate in the ecosystem. It enables peers to join (and leave) the ecosystem, to connect with other peers of interest by allowing interactions among them and supports all activities related to the composition and transaction of complex products.

Decentralized Market Space (*DMS*) enables addressing peers of interest across a strictly decentralized ecosystem. It incorporates the underlying peer-to-peer network and a Distributed Semantic Web Database. The database holds information about providers and their offerings, and the knowledge necessary for transactions in a particular domain (e.g., product/service ontologies and vocabularies).

6.2   Peer Application

The responsibility of the peer application is to enable end-users (i.e., peers) to constitute and enrich their business ecosystem without any central instance of the control or governance. That includes supporting peers in tasks and activities related to joining and leaving the ecosystem, active participation in the ecosystem as well as engaging in different roles (cf. Table 5).

Fig. 7 Functional
components of the peer
application

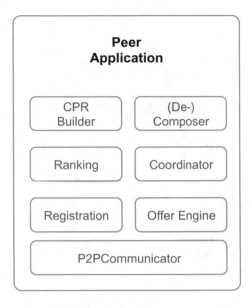

For a peer looking for a complex product (i.e., consumer role), the tasks of the
peer application are:

- to transform the consumer's intention into a complex product request,
- to distribute these requests to potential providers,
- to receive (partial) complex product offers and re-combine them into multiple
  complete complex product proposals,
- to rank them according to the consumer's context and requirements, and
- to coordinate the settlement of single transactions in one enclosing transaction.

On the other side, for a peer looking to provide a product or service (i.e., provider
role), the user application needs:

- to support the description and registration of the offered products/services,
- to receive product/service requests,
- to create offers, and
- to coordinate the transaction settlement.

Figure 7 illustrates the functional structure of the peer application. Based on our
prior work [33, 18] the functional structure of the peer application incorporates following components *Complex Product Request Builder (CPR Builder), (De-) Composer, Ranking, Coordinator, Registration, Offer Engine and P2PCommunicator*.

The *CPR Builder* component enables consumers to compose an individual prod-
uct/service combination, integrate the relevant contextual information and create a
complex product request. As complex product can be composed of products and
services from very different domains, such complex requests need to be described

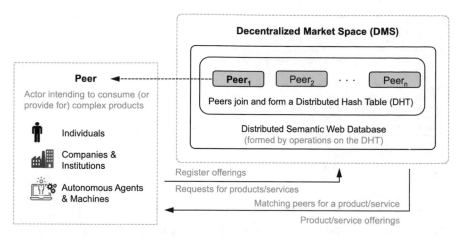

Fig. 8 Architecture of the decentralized market space

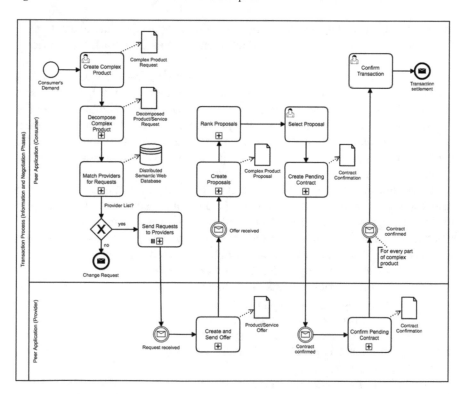

Listing 1.1 An exemplary description of a provider (excerpt)

in a domain-agnostic way so that the peer application can be used for arbitrary product/service domains. Based on our work published in [33], the CPR Builder uses the Resource Description Framework (RDF) [34], to get a rich description of

```
PREFIX dms:<http://dms.com/marketspace/#>
PREFIX gr:<http://purl.org/goodrelations/v1#>
PREFIX rdf:<http://www.w3.org/1999/02/22-rdf-syntax-ns#>
PREFIX rdfs:<http://www.w3.org/2000/01/rdf-schema#>

SELECT ?dname ?sname ?peerid
WHERE {
?domain rdf:type 'domain' .
?domain rdfs:label ?dname .
?domain rdfs:label 'ticketing' .
?domain rdfs:comment ?desc .
?domain bns:hasEntity ?entity .
?entity gr:hasBusinessFunction 'sell'.
?entity rdfs:label ?sname .
?entity gr:BusinessEntity ?be .
?be dms:hasPeerID ?peerid .
}
ORDER BY ?sname
LIMIT 1000
```

Listing 1.2 An ordered list of providers offering in domain ticketing

the requested data and integrates existing domain-related vocabulary and ontologies (i.e., Domain Knowledge). For detailed information about the CPR Builder concept, the reader is referred to [33].

The *(De-)Composer* component splits up the complex product request (described in RDF) into single product/service requests and communicates them to the providers. After receiving single product/service offers, the Composer recombines them into multiple complex product proposals and passes them to the Ranking component. Using the consumer's context (e.g., the current location or other sources of data), the Ranking component ranks all complex product proposals in a way that the best matching is on top of the list.

In case there is a viable complex product proposal, the *Coordinator* component is responsible for supervising the transaction settlement of such proposal. It integrates models for micro-contracting and distributed transaction processing, e.g., [35], Coordinator, and enables monitoring of the transaction process and takes responsibility for transaction completion, as well as, for failure recovery and rollback.

The *Registration and Offer Engine* components support the registration of providers and the offering process. That includes publishing a provider description, providing a price tag for each product/service request and sending offers back to the requesting peer (i.e., consumer). The offering process can also encompass more functionalities depending on the type of the provider (e.g., individuals, companies or institutions) and the kind of products/services they offer (e.g., technology or knowledge provider).

Finally, the *P2PCommunicator* component enables peers to access (join and leave) the ecosystem as a strictly decentralized peer-to-peer network. It integrates discovery and matching mechanisms for finding peers of interest (e.g., matching providers for product/service request as shown in Listing 1.2) and enables the direct, peer-to-peer communication across the ecosystem.

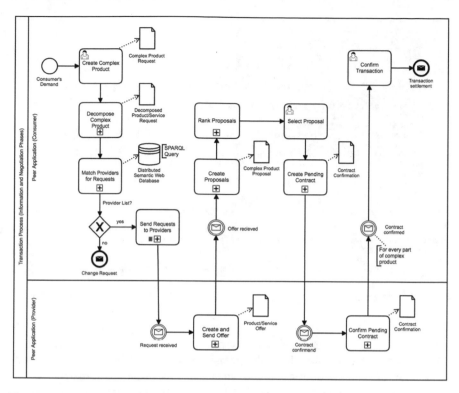

Fig. 9 Transaction process for complex products (excerpt)

6.3 Decentralized Market Space (DMS)

The responsibility of the DMS is to enable addressing peers of interest across a strictly decentralized ecosystem. That includes resolving information necessary for the matching peers for a particular product/service, sending requests for products/services and receiving product/service offerings in a direct, peer-to-peer manner.

As shown in Fig. 8, peers join and form a Distributed Hash Table (DHT). Those intending to provide for complex products register as a provider by publishing the information about themselves and description of domains, in which they offer products or services. This information is encoded as an RDF document and stored in a Distributed Semantic Web Database. Listing 1.1 presents an exemplary RDF description of a provider. The description uses Turtle syntax (Terse RDF Triple Language [34]) and an ontology for electronic commerce (*Goodrelations*[5]). The triples state that *The English Theatre Frankfurt* registers as a *BusinessEntity* who intends to provide offerings in the domain *Ticketing*.

---

[5]http://www.heppnetz.de/projects/goodrelations/.

```
@prefix ns0: <http://dms.com/marketspace/#> .
@prefix gr: <http://purl.org/goodrelations/v1#> .
@prefix ns1: <http://purl.org/tio/ns#> .
@prefix xsd: <http://www.w3.org/2001/XMLSchema#> .
@prefix rdfs: <http://www.w3.org/2000/01/rdf-schema#> .

<http://dms.com/ticketing> ns0:hasOffering <http://dms.com/ticketing/offerings/
    JEKYLLandHYDE> .
<http://dms.com/ticketing/offerings/JEKYLLandHYDE>
    a gr:Offering ;
    gr:name "JEKYLL_and_HYDE"@en ;
    gr:description "A_gothic_musical_thriller_with_a_music_by_Frank_Wildhorn_and_book_
        and_lyrics_by_Leslie_Bricusse"@en ;
    gr:includes <http://www.heppnetz.de/ontologies/tio/examples.rdf#ticket1> ;
    ns1:offeredByEntity <http://dms.com/entity/TheEnglishTheatreFrankfurt> ;
    gr:hasPriceSpecification [
    a gr:UnitPriceSpecification ;
    gr:hasCurrency "EUR"@en ;
    gr:hasCurrencyValue "45.00"^^xsd:float ;
    gr:validThrough "2018-03-04T23:59:59"^^xsd:dateTime
    ] .

<http://www.heppnetz.de/ontologies/tio/examples.rdf#ticket1>
    a ns1:TicketPlaceholder ;
    rdfs:label "JEKYLL_and_HYDE"@en ;
    ns1:accessTo <http://data.linkedevents.org/event/jekylandhydemusical> .
```

Listing 1.3 An exemplary description of a product/service offering

The distributed database is formed by operations on the distributed hash table (DHT). Each DMS instance joins the peer-to-peer network and publishes a self-description of its offered products and services. As a result, each member of the network (i.e., peer) provides resources for the distributed database and hence stores a fragment of the global database. As pointed out in [18], this allows inherent scalability as an increasing number of DMS instances automatically contribute more resources in the underlying peer-to-peer network. One possible implementation of the DMS instance is utilizing existing distributed semantic databases such as DecentSparql [36], which provides full SPARQL 1.1 support and enables scalability, decentralization, fault-tolerance and infrastructure-free operation.

For the peer application to send product/service requests to the peer-to-peer network, this demands that it knows at least one single DMS instance. A request for providers offering a product or service in a domain sent to a single DMS instance returns a list of all peers that provide in that specific domain. Listing 1.2 shows an exemplary SPARQL query (SPARQL 1.1 query language for RDF [37]) that returns an ordered list of registered peers with offerings in the domain "ticketing".

After getting a list of potential providers, the peer application can then send the request for products/services to those providers who are the part of the result set, and thus, address the peers of interest in order to receive offerings for a particular complex product. Listing 1.3 shows an exemplary description of an offering.

6.4 Inner Workings: Transaction Process for Complex Products

Having described the entities and components of the ecosystem architecture, in this section, we provide some details about how these interact to support the transactions of complex products and thus create value for involved peers (i.e., consumers and providers).

Figure 9 shows the transaction process, taking the perspective of a peer performing the consumer role. It illustrates the main activities of the peer application to support the transaction phases: information and negotiation. Additionally, it indicates some message exchanges representing RDF documents, which are produced during or as a result of execution of these activities. The interaction with the Distributed Semantic Web Database is depicted by an RDF store. It represents a SPARQL query mechanism for activities to retrieve or update stored information that persists beyond the scope of the transaction process (cf. Listing 1.2).

Extending the exemplary description of a provider (as shown in Listing 1.1), Listing 1.3 describes an offering for a musical "Jekyll and Hyde" offered by that provider. As with all exchanged messages in the proposed architecture, it is an RDF document encoded in Turtle, and it uses aforementioned ontology *GoodRelations* and the domain related *Ticket (TIO)* ontology.[6]

6.5 Demonstrator

To evaluate the feasibility of the proposed ecosystem architecture, we present a demonstrator that implements the components of the Peer Application and DMS in a testbed environment. It demonstrates the handling of the evaluation scenario and covers the primary activities of consumer and provider roles for the transaction process phases illustrated in Fig. 9.

For the evaluation scenario, we chose the aforementioned use case of the couple (i.e., consumer) asking for a complex product that includes: buying tickets for a concert in the city-theatre, reservation of a table at an Italian restaurant, finding a parking space close to both locations and engaging a well-rated babysitter. The complex product spans four different product/service domains (i.e., babysitting, gastronomy, parking, and ticketing) and has to consider contextual information regarding the schedule, location, and ratings of a particular service.

For the implementation, we used established technologies and existing ontologies and vocabularies, but we also developed custom ontologies (e.g., for babysitting and parking) to describe the products/services in our evaluation scenario.

The peer application is implemented as a web application using HTML, JavaScript, and CSS. It enables the composition of complex products by integrat-

---

[6]http://www.heppnetz.de/ontologies/tio/ns.

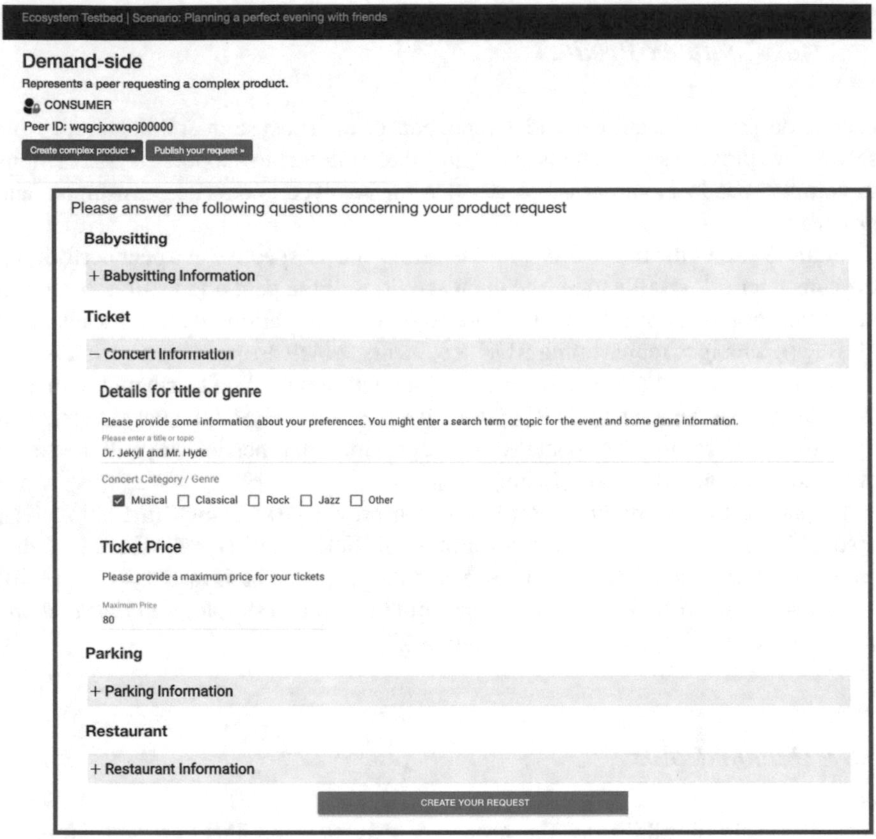

Fig. 10 A composition of a complex product and entering data

ing the CPR Builder component as a web service. The implementation of the CPR Builder and the related concepts of demand and supply ontologies are presented in [33]. Figure 10 shows a screenshot of the user interface for complex product composition and entering data relevant to the publishing of the resulting complex product request.

The peer-to-peer communication is realized using the WebRTC [38]. Figure 11 shows a sample of messages exchanged among peers in the negotiation phase of the transaction process. On the left, it presents the demand-side (i.e., a peer requesting complex product and getting offerings for distinct services from different providers). On the right, it gives an overview of the supply-side (i.e., addressed peers) providing for the scenario-domains babysitting, gastronomy, parking, and ticketing, as well as, shows a registered babysitting agency receiving requests and answering these by sending offerings.

For the instantiation of DMS (i.e., DHT-based peer-to-peer network and Semantic Web database) we use Chord [39] as an implementation of a distributed hash table, and

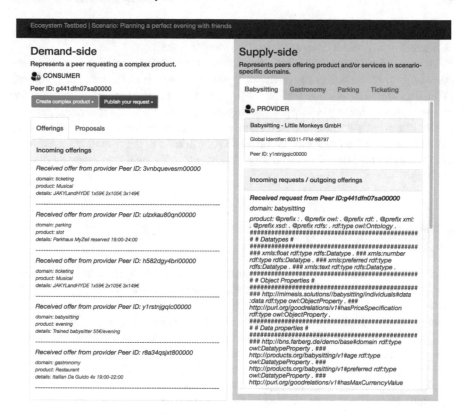

Fig. 11 A sample of exchanged messages among involved peers

Apache Jena components TDB and Fuseki SPARQL Server [40]. Since the focus of the demonstration was to show the phases: information gathering, matching providers for services and negotiation (i.e., communication among peers), we used the TDB as a persistent RDF store on a single machine, and the Fuseki SPARQL Server for the query and update. Further implementations need to replace it with a decentralized Semantic Web database solution (e.g., DecentSparql [36]). The description of the implemented data model and other sample resources are provided at the following URL.[7]

In summary: The use case evaluation shows that the proposed ecosystem architecture is feasible as the demonstrator has been able to handle the evaluation scenario, and thus meets the stated requirements (cf. Table 4). As demonstrated, our architecture supports the composition of arbitrary complex products, which span different domains ($R1, R2, R4$), given existing structured domain-knowledge relevant for commercial transactions in that domain. It facilitates the underlying peer-to-peer network

---

[7]http://www.sim2it.de/digital-business-book-chapter-samples.

of consumers and providers and supports matching among peers to address providers of interest in a direct and decentralized manner (*R5, R8*).

However, the demonstrator does not incorporate all functionality required to support the transaction process as a whole. What is missing is the functionality related to the distributed transaction settlement (*R3*), and the generic ranking algorithms for varying types of consumer-defined ranking constraints (*R6*). Moreover, our demonstrator is currently limited to the consumer and provider roles and does not provide any reputation or feedback mechanisms (*R7*) considered as the central responsibility of the "Reputation Bank" as the designated role to offer qualified assessments (cf. Table 5).

7   Conclusions and Outlook

This chapter introduced a strictly decentralized business ecosystem model for complex products. The main aim of our business ecosystem model is to support complex product scenarios, settings in which consumers and providers engage together to achieve specific goals represented through the transactions of complex products.

Two artifacts define the proposed business ecosystem model. First, the ecosystem structure, that integrates various actors and enables them to constitute and enrich their ecosystem without any central instance of the control. The proposed structure assembles actors, activities, and essential interactions among them, that when linked together illustrate the ecosystem's overall value creation pattern. Second, the ecosystem architecture, that is represented by a highly scalable and strictly de-centralized software-system architecture. It supports arbitrarily complex products, (given existing domain-knowledge relevant for commercial transactions in particular domain), and enables distinct activities related to different phases of the transaction process.

To evaluate the feasibility of our ecosystem architecture, we implemented a demonstrator that instantiates the proposed software-system design on an evaluation use case scenario. The findings suggest, that our demonstrator can handle the evaluation scenario and thus, generally satisfy the functional requirements of a de-centralized business ecosystem model. However, our demonstration is scenario-specific and does not yet cover all functionalities needed for the settlement of complex products in a strictly decentralized manner.

In our future work, we will concentrate on these two areas of improvements considering technologies and solutions like, e.g., blockchain and smart contracts, as well as, on the implementation of further actor's roles and estimation whether the proposed business ecosystem model is creating and sharing the desired value.

References

1. Cisco (2013) The Internet of everything for cities. http://www.cisco.com/web/about/ac79/doc s/ps/motm/IoE-Smart-City_PoV.pdf. Accessed 4 Mar 2018
2. James JA (2016) Adverse selection. In: Augier M, Teece JD (eds) The palgrave encyclopedia of strategic management. Palgrave Macmillan, London, pp 1–2
3. Moazed A, Johnson NL (2016) Modern monopolies: what it takes to dominate the 21st century economy. St. Martin's Press
4. Moore JF (1993) Predators and prey: a new ecology of competition. Harvard Bus Rev 71(3):75–86
5. Iansiti M, Levien R (2004) The keystone advantage: what the new dynamics of business ecosystems mean for strategy, innovation, and sustainability. Harvard Business Press
6. Peltoniemi M (2006) Preliminary theoretical framework for the study of business ecosystems. 8(1):10. Emergence Mahwah Lawrence Erlbaum
7. Allee V (2000) Reconfiguring the value network. J Bus Strategy 21(4):36–39
8. Rong K et al (2015) Understanding business ecosystem using a 6C framework in internet-of-things-based sectors. Int J Prod Econ 159:41–55
9. Tian CH et al (2008) BEAM: a framework for business ecosystem analysis and modeling. IBM Syst J 47(1):101–114
10. Turber S et al (2014) Designing business models in the era of internet of things. In: International conference on design science research in information systems. Springer, pp 17–31
11. El Sawy OA, Pereira F (2013) Business modelling in the dynamic digital space: an ecosystem approach. Springer
12. Battistella C et al (2013) Methodology of business ecosystems network analysis: a case study in telecom Italia future centre. Technol Forecast Soc Chang 80(6):1194–1210
13. Westerlund M, Leminen S, Rajahonka M (2014) Designing business models for the internet of things. Technol Innov Manag Rev 4(7):5
14. Adner R (2017) Ecosystem as structure: an actionable construct for strategy. J Manag 43(1):39–58
15. Peffers K et al (2007) A design science research methodology for information systems research. J Manag Inf Syst 24(3):45–77
16. Grefen P (2015) Beyond E-business: towards networked structures. Routledge
17. Radonjic-Simic M, Pfisterer D, Wolff F (2017) Analyzing an ecosystem for complex consumer services. In: ICServ2017—the 5th international conference on serviceology: lecture notes in computer science. Springer International Publishing
18. Pfisterer D, Radonjic-Simic M, Reichwald J (2016) Business model design and architecture for the internet of everything. J Sens Actuator Netw 5(2):7
19. Van Schewick B (2012) Internet architecture and innovation. MIT Press, p 21
20. Yannis Bakos J (1997) Reducing buyer search costs: implications for electronic marketplaces. Manag Sci 43(12):1676–1692
21. Eymann T (2001) Markets without makers—a framework for decentralized economic coordination in multiagent systems. Electronic commerce. Springer, pp 63–74
22. Schmees M (2003) Distributed digital commerce. In: Proceedings of the 5th international conference on electronic commerce. ACM, pp 131–137
23. Serban C et al (2008) The concept of decentralized and secure electronic marketplace. Electron Commer Res 8(1–2):79–101
24. Xia Y, Li Q, Wang L (2010) Research on decentralized E-commerce architecture in P2P environment. In: 2010 international conference on electrical and control engineering (ICECE), pp 2929–2932. IEEE
25. Hausheer D, Stiller B (2005) PeerMart: the technology for a distributed auction-based market for peer-to-peer services. In: 2005 IEEE international conference on communications, ICC 2005, vol 3. IEEE, pp 1583–1587
26. Klems M et al (2018) Desema. https://github.com/markusklems/desema. Accessed 4 Mar 2018

27. BitMarkets White Paper. BitMarkets. https://voluntary.net/bitmarkets/. Accessed 4 Mar 2018
28. OpenMarket. https://github.com/johndpope/openmarket. Accessed 4 Mar 2018
29. Open Bazaar. http://docs.openbazaar.org/. Accessed 4 Mar 2018
30. Searls D (2013) The intention economy: when customers take charge. Harvard Business Press
31. ProjectVRM. Berkman Center for Internet and Society at Harvard University. http://blogs.law. harvard.edu/vrm/projects/. Accessed 4 Mar 2018
32. Kleedorfer F et al (2014) The case for the web of needs. In: 2014 IEEE 16th conference on business informatics (CBI), vol 1, pp 94–101. IEEE
33. Hitz M et al (2016) Generic UIs for requesting complex products within distributed market spaces in the internet of everything. In: Buccafurri F et al (eds) Proceedings of Availability, reliability, and security in information systems: IFIP WG 8.4, 8.9, TC 5 international cross-domain conference, CD-ARES 2016, and workshop on privacy aware machine learning for health data science, PAML 2016, Salzburg, Austria, 31 Aug–2 Sept 2016. Springer International Publishing, Cham, pp 29–44
34. W3C. RDF 1.1 Turtle. http://www.w3.org/TR/turtle/. Accessed 4 Mar 2018
35. X/Open Company Limited. Distributed transaction processing: the XA specification. http://pu bs.opengroup.org/onlinepubs/009680699/toc.pdf. Accessed 4 Mar 2018
36. Mietz R et al (2013) A P2P semantic query framework for the internet of things. PIK Praxis der Informationsverarbeitung und Kommunikation 36(2):73–79
37. Harris S, Seaborne A (2018) SPARQL 1.1 query language. W3C working draft. http://www.w 3.org/TR/sparql11-query/. Accessed 4 Mar 2018
38. WebRTC Project. https://webrtc.org/. Accessed 4 Mar 2018
39. Stoica I et al (2001) Chord: a scalable peer-to-peer lookup service for internet applications. ACM SIGCOMM Comput Commun Rev 31(4):149–160
40. Apache Software Foundation. Apache jena components. https://jena.apache.org/documentatio n. Accessed 4 Mar 2018
41. Xiong L, Liu L (2004) PeerTrust: supporting reputation-based trust for peer-to-peer electronic communities. IEEE Trans Knowl Data Eng 16(7):843–857

Compliance Management in Business Processes

Sridevi Saralaya, Vishwas Saralaya and Rio D'Souza

Abstract Business Process Compliance refers to the act of conformance of a business process with policies, regulations and rules that govern the organization. An imperative requirement of business processes in various fields such as Health care, Insurance, Finance and Online Trade is adherence to a large number of compliance requirements, constraints and quality policies from various sources. Lack of compliance may result in huge compensations and loss of customers and reputation. Compliance issues can be handled either retrospectively i.e. after non-complaint situations are observed or they can be handled proactively i.e. anticipation of possibilities leading to non-compliant circumstances during process execution which may prevent occurrence of deviations and thus save upon compensation effects. Hence compliance management tasks need to be incorporated into each phase of the lifecycle of a business process. In this article we discuss contemporary activities related to lifecycle of compliance management in business processes which involve compliance elicitation, compliance formalization, compliance implementation, compliance verification and compliance improvement based on existing literature. Compliance Monitoring Functionalities (CMFs) which may be used to categorize and also assess existing compliance management approaches and frameworks are also discussed.

S. Saralaya (✉) · R. D'Souza
Department of Computer Science & Engineering, St. Joseph Engineering College, Vamanjoor, Mangaluru, India
e-mail: sridevisaralaya@gmail.com

R. D'Souza
e-mail: riod@ieee.org

V. Saralaya
Department of Microbiology, Kasturba Medical College (Manipal Academy of Higher Education), Mangaluru, India
e-mail: vishwassaralayak@gmail.com

© Springer International Publishing AG, part of Springer Nature 2019
S. Patnaik et al. (eds.), *Digital Business*, Lecture Notes on Data Engineering and Communications Technologies 21, https://doi.org/10.1007/978-3-319-93940-7_3

1 Introduction or Overview

A business process (BP) is a series of activities and/or tasks in an organization that collectively help to attain a business goal which can be in the form of a hardware or software product, a service or something of value to customers. A fundamental necessity for any BP is adherence to a large number of compliance requirements, constraints and specifications from various sources known as *Norms*. Governatori [1] defines "compliance" as: "*An act or process to ensure that business operations, processes and practices are in accordance with prescriptions (often legal) documents*". Compliance is practiced in many disciplines such as insurance, finance, traffic-control, health-care and so on. It is used to ascertain adherence of the procedure/process followed in any field against a set of rules regulating the process. Compliance governance is the practice of placing together the measures, methods and technology by an enterprise to execute, monitor and manage compliance. Managing compliance is a *significant*, *complicated* and *expensive* task to deal with.

It is *significant* as there is growing burden on enterprises from all walks of life to meet guidelines and laws. Compliance requirements originate from diverse sources such as ISO 9000 for quality management,[1] ISO/IEC 27000 for information security,[2] legislation and regulatory bodies (such as Sarbanes-Oxley Act for public accounting firms in US,[3] HIPAA (Health Insurance Portability and Accountability Act[4]), ICMR, which is responsible for regulating bio-medical research in India[5]), Basel III on banking supervision,[6] internal policies and Service-Level Agreements (SLAs) [2]. It is important in every field as absence of compliance may lead to safety mishaps, loss of reputation, money laundering, huge penalties or compensations and further to bankruptcy.

Compliance management is considered *complex* due to two reasons: The primary challenge in compliance management is the alignment of compliance objectives with the business objectives, which are from distinct universe of discourse [3]. Compliance objectives are control objectives which emerge from external sources. As both address separate issues, business process modeling languages cannot be used to model compliance objectives. Business process specifications are descriptive which specify '*How*' activities in the BP are to be executed whereas Contol objectives are based on *norms* and which illustrate '*What*' has to be done to be compliant. There are also possibilities of redundancy, inconsistency and conflicts between the two. Therefore the assimilation of compliance objectives with business objectives should be administered with utmost care.

The second reason to consider managing compliance as a complex pervasive task is because it involves many departments of an organization and spans across

---

[1] http://www.iso.org/iso/home/standards/management-standards/iso9000.htm.

[2] https://www.iso.org/obp/ui/#iso:std:iso-iec:27000:ed-4:v1:en.

[3] https://www.sec.gov/about/laws/soa2002.pdf.

[4] https://www.hhs.gov/sites/default/files/privacysummary.pdf.

[5] http://www.icmr.nic.in/AboutUs/AboutICMR.html.

[6] http://www.bis.org/bcbs/publ/d295.pdf.

multiple processes. The task of managing compliance automatically is not simplistic as multiple perspectives from different stakeholders such as Data, Timing, Resources, Humans and Automated devices are involved in the business process. Often the requirements are imprecise and specified informally. Likewise representing and modelling of rules is not an easy task as well defined standards for the same are not yet available. Conceptual modeling languages used for specifying compliance management functionality should be able to depict all pertinent real-world concepts such as those related to behavioral aspects, men, methods, materials, machines, measurements, environment, business rules and policies, quality control procedure and standards. Business Process Management (BPM) systems emphasize on human resources, work and organization whereas compliance management system focus on rules, policies and regulations. Existing BPM systems lack capabilities to model constraints and requirements.

Compliance management and auditing is a very expensive affair, which not only involves salaries of auditors but also internal costs for the preparation of and assisting audits. Sometimes the compliance implementation cost may be higher than the fine imposed due to non-compliance or loss due to reputation. In the 2008–09 Governance Risk management and Compliance Spending Report, it was estimated that US\$ 32B would be spent by companies only on compliance, governance and risks and greater than US\$ 33B in the year 2009 [4]. To reduce costs enterprises are taking steps at each phase of the BP to make it complaint such as during design time, during execution as well as after execution for auditing and invoking corrective measures.

Conventionally regular audits are conducted to assure compliance. Manual verification by compliance personnel may be time consuming and may fail to deliver the expected level of assurance. The rules are either hard-coded within the business process application or present in the form of implicit knowledge in the minds of domain experts. When the application or the domain expert moves on, the gaps (disparities) have to be filled by the organization. In order to provide incessant updates regarding compliance, a strategic approach is preferred. This transferal from manual reviews to continuous affirmation necessitates automated compliance management systems that provide transparent real-time monitoring of the business process.

This chapter does not propose any new formalism or mechanism to check compliance of business processes. Our objective is to provide information regarding the recent state-of-art developments for handling compliance issues from existing literature and to offer elementary knowledge on requirements for compliance management. Some of the studies which provide a recent survey on business process compliance approaches are by Becker et al. [5] and Fellmann and Zasada [6].

In Sect. 2 we discuss the concepts and terms related to compliance and in Sect. 3 we provide a case study to relate the concepts of compliance. In Sect. 4 we discuss all the phases of compliance management life cycle, in Sect. 5 we discuss functionalities for managing compliance which help in comparing and evaluating existing studies, in Sect. 6 we discuss challenges faced and conclude in Sect. 7.

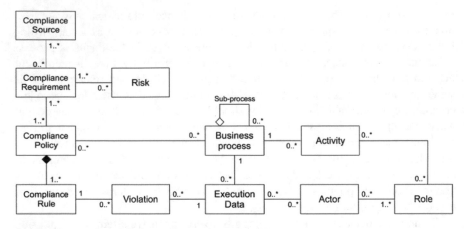

Fig. 1 Conceptual model of compliance governance [7]

2 Compliance Management—Concepts and Terminology

In this section we discuss some of the common concepts and terms related to compliance management with respect to Fig. 1.

BPM—is the acronym for the term *"Business Process Management"*. Business Process Management may be defined as "a discipline to design, model, execute, monitor and optimize business processes so as to increase profitability" rather than a technology or a tool. BPM is not merely a software to manage business issues and humans play a substantial role in management of business process. The main difference between BPM software and workflows is that the BPM software goes further than solely automating work to assisting people with constant process improvement. Business process management (BPM) is a management approach which focuses on synchronizing all features of an enterprise with the needs and requirements of clients. BPM enables enterprises to be more effective, efficient and more competent to change as compared to a functionally engaged, conventional hierarchical management approach.[7] The business process model is usually specified using Business Process Execution Language (BPEL) or Business Process Modelling Notation (BPMN).

Compliance Source—refers to those entities that specify all documents that regulate and provide guidelines for the right conduct of the business process for any particular business domain. A compliance source illustrates a set of rules stated in natural language that defines the way in which a business has to perform. Standards

---

[7]http://www.itinfo.am/eng/business-process-management/.

(such as ISO-9001, CoBIT, CMMI), laws (e.g. HIPAA, SOX) and SLA's are some of the examples of compliance sources.

Compliance Requirements—An organization may have to follow only certain principles from a given set of compliance source documents. The selection of appropriate constraints to be followed constitutes the *Compliance Requirements*. The act of satisfaction of requirements is known as *Complying* with a source.

Compliance Rule—The compliance requirements from various sources have to be identified and extracted into actionable statements termed as *Rules*, which can either be in natural language or some kind of formalism which enables its automated verification.

Compliance Policies—The set of compliance rules which an organization has to adhere to are grouped together to form the internal documents of the organization which also explain how the organization intends to accomplish compliance. These documents form the *Compliance Policies* of an organization. For the evaluation of compliance, knowledge of specific rules and the context in which they are applicable is required for any business process. Hence compliance policies are associated with the business process.

Risk—A risk factor is always involved with violation of compliance rules. For example, in a health care setup, in case the details and reports of patients which are supposed to be confidential are not encrypted while dispatching, would certainly lead to a security risk. The aim of risk management is to understand, detect, prevent and mitigate numerous kinds of risks due to non-compliance in the business process. Mitigation of risk is the real driver for in-house compliance auditing.

Activities—A set of activities when executed in coordination helps to accomplish the goal of the business process. Each activity may be realized by a set of operations or services. These activities may be executed sequentially or in parallel by people or by automated systems.

Actor—An actor may be responsible for execution of every activity in the BP. An actor may have multiple *Roles*. For e.g. An employee of an organization may perform the role of a *Manager* or an *Administrator*. A Microbiologist may play the dual role of *Quality Control Manager* and that of *Certifier* or *Verifier.* If the activities are not performed by the specified *Role*, this leads to a Role or Resource violation.

Business Process Instance—Execution of the business process on concrete data forms an instance of the business process. During each execution, *data* output may be generated by the activities/services/operations. The attainment of compliance is evaluated on the concrete execution of each process instance.

3 Case Study

In order to relate the concepts of compliance we take up a health care example such as the laboratory diagnosis of Dengue virus infection. Dengue is emerging as an important viral disease and presently, afflicts a large portion of humanity around the globe in terms of morbidity and mortality. Laboratory diagnosis of Dengue infection is extremely important due to similarities in clinical features of Dengue with other viral/bacterial/parasitic infections such as Chikungunya, Leptospira and Malaria which are also prevalent in the same geographical areas and therefore pose a problem to physicians in differential diagnosis of the disease. Accurate and early laboratory diagnosis of Dengue is not only important in initiating lifesaving treatment to the patient but also helps in estimation of prevalence and disease control measures Fig. 2.

The laboratory diagnosis of dengue is mainly dependent on appropriate collection, shipment, storage and processing of specimens. During collection of blood for sero-diagnostic studies from the suspected Dengue Fever/Dengue Hemorrhagic Fever cases, universal/standard precautions should be undertaken. Samples required to be collected are:

- When the patient (with symptoms) presents himself at the healthcare facility for the first time or is admitted to the hospital with acute symptoms (acute phase serum, termed as S1).
- At the time of discharge from hospital (convalescent phase sera, S2) or, in case of fatality, at time of death.
- A third serum sample (S3—late convalescent phase serum) may also be collected 3 weeks after collection of acute phase serum S1).

Serological diagnosis of Dengue depends on the demonstration/presence of Antigens of the Dengue virus or the Antibodies produced by the host during infection, against the virus. The basis of paired sera (S1 and S2/S3) testing is to confirm the diagnosis of Dengue virus infection by demonstration of a rise in titer (four-fold increase in levels) of Antibodies against Dengue virus from a basal level in S1 to significantly higher levels in S2 and/or S3. To demonstrate this rise in levels of serological markers, a suitable period of time (7–14 days) is required between the acute phase serum (S1) and the convalescent phase (S2/S3) serum.

The diagnostic process involves the following steps as illustrated by Fig. 3: Blood samples are collected from patients. These samples are barcoded and are preserved between 2 and 8 °C up to 7 days. The test process starts at 8.00 a.m. every day. In order to test for the presence of Dengue virus or Antibodies against the virus in the patient's sample, as a preliminary test, Immuno-chromatographic (Card/Rapid) tests are performed for detection of DNS1 Antigen (Dengue virus non-structural protein) and Dengue IgM and IgG. If any of the three serological markers are detected in this first stage, confirmatory ELISA tests to detect the same markers are conducted. If the preliminary screening tests indicate the patient is positive for Dengue virus infection, the clinician is immediately alerted (critical alert) as life saving measures

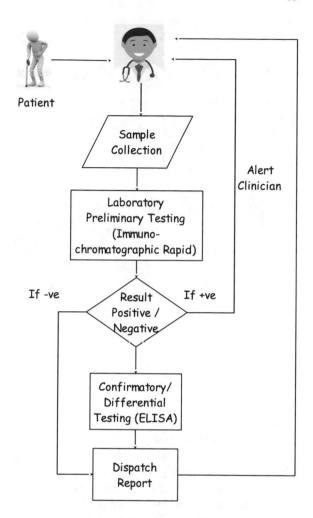

Fig. 2 Strategy to detect dengue virus infection

such as blood/plasma/cells transfusion may be required in the patient and the test process using the confirmatory ELISA's need to be performed for DNS1 Antigen and Dengue IgM/IgG detection. These results have to be validated and verified by a consultant 'Microbiologist'. After verification, the results are entered in the Laboratory Information System (LIS) by the Clerical staff. This transcript of the report has to be verified again by the consultant 'Microbiologist' after which the reports are released to the patient or the clinician through means such as SMS, update on web site, e-mail or by post [8].

Considering the severe morbidity and increased mortality of Dengue, especially that of Dengue Hemorrhagic Fever (DHF), reliability of diagnostic tests conducted is a matter of key concern. The entire process of diagnosis of Dengue is not a simple affair as most of the tests involved in diagnosis (virus isolation; PCR) require costly equipment and expertise which might not be possible or available in rural areas espe-

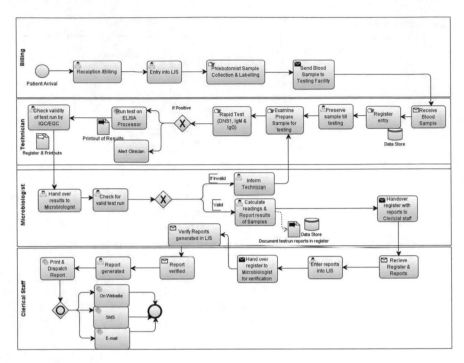

Fig. 3 Elucidation of a health care BP (laboratory diagnosis of dengue viral infection—case study)

cially in developing nations where Dengue virus infection is particularly prevalent. It is further complicated due to the presence of multiple diagnostic kits/reagents with varying sensitivity and specificity. Therefore, strict Quality control measures must be inbuilt into the entire process so that the output of the laboratory is of the highest standards, thereby playing an important role in providing healthcare to the community. All laboratories which undertake diagnostic testing of Dengue virus infections should ensure Quality assurance regarding testing so that they provide accurate, reliable and reproducible results in the shortest period of time. Quality control should include both internal (IQC) and external (EQAS) assurance measures which will constantly monitor and prevent any errors from occurring, thereby providing reliable output at each level of the entire process.

This health care (BP) case is selected for the following reasons: It is a BP run by a single institution to its customers where all the activities related to diagnosis are performed by/on resources belonging to one organization creating an *Orchestration* scenario. The diagnosis BP has to adhere to numerous compliance requirements from various sources for the diagnosis, treatment, prevention and control of Dengue such as World Health Organization (WHO) and Directorate of National Vector Borne Diseases Control Programme—Ministry of Health and Family Welfare, Government of India. This process has demands on various dimensions of compliance such as Resources, Temporal, Control-order and Data output. Resources dimension includes requirements both from human resources (i.e. verification to be done only by *Micro-biologists*) and hardware/software components (such as type of ELISA processors to be utilized, LIS). It also has temporal requirements such as "an activity has to complete within a specified time period" and the process has a specific order in which the activities are to be performed known as control-sequence. It is also important to the healthcare of a huge population involved. It has considerable interaction between its components and compliance to set rules and strict QC over the entire process is also vital to the output.

4   The Compliance Management Lifecycle

In order to address compliance in any BP such as the healthcare example provided in the previous section, it is imperative to find answers to some basic questions [9] such as:

1. *What are we intended to check?*
2. *What languages will we use to model business processes and rules?*
3. *How will we do it i.e. check for compliance?*
4. *When will we check the degree of compliance?*

We try to find answers to the above questions in every phase of the compliance management life-cycle.

Some authors try to separate the concepts of compliance management from the business process by capturing each compliance concept in separate rules [10, 11]. Task of compliance management will have to be incorporated in all phases of the business process life-cycle which involves compliance elicitation, compliance formalization, compliance implementation, compliance verification and compliance improvement. Elicitation refers to the act of identification and extraction of requirements and constraints that need to be satisfied. Formalization refers to precisely formulating the requirements. Implementation and supervising is the enactment of the constraints in the application and their verification. And finally improvement helps to revive compliance based on diagnostic information obtained during execution. In the coming sections we will study in detail each phase of the compliance management lifecycle as depicted in Fig. 4 [10].

4.1   Elicitation

The activities of the elicitation phase helps us to answer the question *"What are we intended to check?"* which involves three main concepts—*(i) Identification of Rules (ii) Categorization of Constraint Properties and (iii) Collection of Metrics.*

Fig. 4 Compliance
management lifecycle [12]

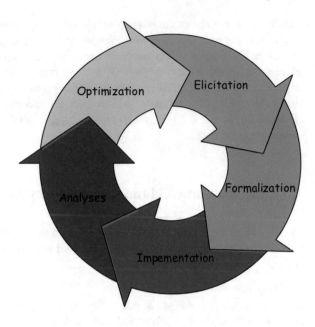

(i) **Identification of Rules**:

This phase involves identification and extraction of requirements and constraints to
be satisfied by the business process from various compliance sources. The requirements newly elicited from the Policies and Standards must be compliant with earlier
existing requirements, which ought to be ensured during elicitation [13]. This is a
very important phase as the requirements stated by the stake holders (end-users,
customers, managers) may not be clear and precise. It is an amalgamation of functional and non-functional requirements of one or more dimensions to be met and
measured. In order to be able to concisely and precisely identify the constraints from
the requirements stated by the users (non-technical or non-experts), the constraints
are extracted into one of the dimensions and stated as Requirements/Rules to be
adhered to. Business rules come from a variety of sources and are spread out across
the organization as depicted in the Fig. 5. Generally a business analyst gathers regulations from various sources. The process of gathering, categorizing and integrating
the rules for supervision and verification is referred to as the Business Rules approach
[14]. The interpretation of all policies/procedures/laws have to be captured by the
Rules. Steinke and Nikolette [15] define a business rule as "*A statement that aims to
influence or guide behavior and information in an organization*". Business Rule as
defined by Ian Graham: "*A business rule is a compact, atomic, well-formed, declarative statement about an aspect of a business that can be expressed in terms that
can be directly related to the business and its collaborator, using simple, unambigu-
ous language that is accessible to all interested parties: business owner, business
analyst, technical architect, customer and so on. This simple language may include
domain-specific jargon.*" [16]. The rules have to emphasize the requirements of the

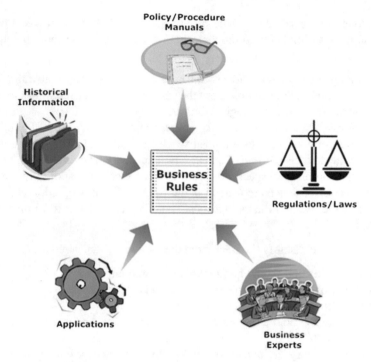

Fig. 5 Sources of business rule [14]

business process and should not feature the technical details of the system under consideration. Every requirement of the business process has to be captured in a corresponding Rule. Studies which emphasize the use of rules as the elements with which compliance can be verified are Alberti et al. [17], Awad et al. [18], Ghose and Koliadis [19], Governatori and Sadiq [20], Hashmi et al. [21] , Mulo et al. [22] and Zhang et al. [23].

Business rules may be classified with respect to their source as [24]:

- *Mandates*: are policies such as the commitment to provide excellent healthcare facilities to the society as per regulations set by the regulatory authorities (For e.g.: Drug Controller of India [DCGI]/Food and Drugs Administration [FDA—US]). Such mandates if not followed may lead to adverse consequences.
- *Policies*: are procedures given by rules and regulations of the company such as objectives, goals and budget statements. Policies are drawn by the health care providers (as per case study) based on the mandate so as to provide health care solutions to the society.
- *Guidelines*: are rules specific to context and environment such as methods, management styles and approaches to be followed for the diagnostic of infectious viral diseases (case study).

Rules can be considered at three different levels of abstraction based on their structural relationship between the elements of the complex business process [24, 25] as:

- *Static or Dynamic Integrity Constraints*—are used to specify constraints on the values/state space. For example from the case study—Every report has to be validated by the same Microbiologist who had verified it.
- *Derivations*—are used to explain derived concepts or conclusions. For example: If the result of both the tests are positive, the patient is confirmed to be infected with Dengue and also differentiated as having primary or re-infections.
- *Reactions*—are used to specify the actions to be taken in response to events i.e. reactive behavior which are usually in the form of Event, Condition and Action. For e.g. If the result of the rapid tests are positive, Clinician is to be immediately notified.

Below we try to identify the rules from the case study discussed as:

1. Sample collection should be followed by Barcoding.
2. Preservation of blood sample between 2 and 8 °C up to 7 days.
3. The test process has to start at 8.00 a.m. every day.
4. If output of Immuno-chromatographic test is positive, confirmatory ELISA test has to be performed.
5. If the preliminary test indicates positive, Clinician has to be immediately notified by phone, SMS or email.
6. The results have to be validated and verified by a consultant Microbiologist.
7. The results are entered in the Laboratory Information System (LIS) by the Clerical staff only after verification by the Microbiologist.
8. The most suitable time interval between collection of S1 and S2/S3 serum is 7–21 days.
9. The duration between the first verification of Report by the Microbiologist and second verification of the transcript after entry in LIS should not be more than 3 h.
10. All details regarding patient details/investigation/report status need to be kept strictly confidential.

(ii) **Categorization of Constraint Properties**

By categorizing the constraint properties to be verified for compliance into various dimensions we are able to address the first question "*What are we intended to check?*" The answer to this question helps us identify the components that have to be observed and their properties to be measured. To ensure conformity of the BP with requirements, the properties of compliance requirement rules to be monitored are categorized into four dimensions—control properties, data properties, temporal properties and resource properties by Cabanillas et al. [9], Elgammal et al. [26] whereas a different categorization of the rules into classes and sub-classes according to their effectiveness is proposed by Hashmi et al. [21]. According to the latter study, categorization is based on the conditions under which the norms/rules come

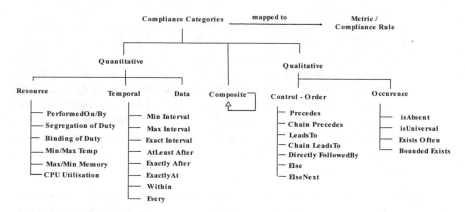

Fig. 6 Categories of compliance

into force i.e. the time period when the rules are effective or applicable until their termination. Below we discuss the two categorizations.

1. **Dimensions of Compliance Constraints (based on Property Specification Patterns)**

Compliance constraints of any business process may be viewed from multiple perspectives as Quantitative and Qualitative Constraints as shown in Fig. 6 [27]. Quantitative properties are those that can be observed and measured whereas Qualitative properties (Control-flow) are those that can only be observed. The foundation for many of these categories is from Dwyer's property specification patterns [28], which have been extended by many researchers over the years [26, 29].

(a) **Quantitative Properties**: Properties such as Data, Temporal and Resource properties which can be measured are quantitative and are explained below:

(i) **Data properties**: Data elements specify details of the activities to be carried out and the information produced by them. The execution of activities in the BP may involve information output produced in the form of data or documents. Data is exchanged between activities, services and operations. The data produced by one activity may be the input for subsequent activities. This data could be modified in whole or in part by the succeeding activities. Compliance rules related to data from the case study:
   - Rule 4—If output of Immuno-chromatographic test is positive, confirmatory ELISA test has to be performed.
   - From workflow description, data output of the Immuno-chromatographic test and the automated ELISA service should be either positive or negative and should never be invalid.

   Data properties are restrictions placed on the output produced by an activity/service. This category verifies the output produced/generated by each service with the expected output during execution. Compliance violations related to data arises when data objects do not satisfy the constraints imposed on them. Cabanillas et al. [30] state three types of problems related to content of data. They are: *insufficient input data* (an activity or service may not receive all the intended data), *insufficient output data* (data output produced by a service may be incomplete) and *prohibited data* (some data such as financial details and medical data may have to be kept confidential and should not be accessible by all operations or actors. Knuplesch [31], Hashmi et al. [32] and Gómez-López et al. [33] deal with data compliance of business process while Schleicher et al. [34] deal with such data restrictions in the cloud environment.

(ii) **Temporal properties**: are used to define time requirements to be satisfied by each service/operation of the business process. Properties that deal with time constraints such as start time of an activity, time span allotted to an activity or deadline to complete fall under this category. Temporal properties which specify exact time when an activity is supposed to start/complete execution are known as *time point* metrics whereas those that specify a duration are known as *time interval* metrics. The temporal concerns of any constraint as stated by Palmirani, et al. [35] fall under three main aspects: (i) the time when a constraint is in force (ii) the time when a constraint is fulfilled and (iii) the time of application of the constraint. Some of sub-patterns under time constraints are:
- *Within*—the time interval within which an event has to occur or an activity has to be completed.
- *Max/MinInterval*—"Event A *min/maxInterval* event B" represents the minimum/maximum duration between occurrence of two events A and B.
- *ExactlyAt*—represents the exact time when an event is supposed to occur.
- *AtLeastAfter*—represents the minimum time after which an event is expected.

Examples of Temporal properties from the case study are:
- Rule 3—The test process has to start at 8.00 a.m. every day.
- Rule 8—The optimal time interval between collection of S1 and S2/S3 serum is 7–21 days.
- Rule 9—The duration between the first verification of Report by the Microbiologist and second verification of the transcript after entry in LIS should not be more than 3 h.

Some of the studies which consider temporal properties are Barbon et al. [36], Zhang et al. [23], Li et al. [37], Governatori et al. [38], Kumar and Barton [39] and Ramezani et al. [10].

(iii) **Resource properties**: are properties which are used to express obligations of resources on which the services are executed or by whom the services are executed. Policies of the business process such as "which service/activity has to be executed on which device or by human actor (role)" falls under this category. In addition, properties such as minimum/maximum hardware required to execute a service such as CPU/memory/temperature range to be maintained come under this property. The sub-patterns under resource constraints are:

- *PerformedOn/PerformedBy*—PerformedOn indicates the node/device on which the service has to be executed. PerformedBy indicates the individual or role responsible for performing the service. (This property is an exception to Quantitative property).
- *Min/MaxMemory*—The min/max free memory required for execution of an activity/service.
- *CPU Utilisation*—Maximum CPU utilisation level.
- *Min/MaxTemp*—The min/max temperature that has to be maintained.

Examples of infrastructure properties from the case study are:

- Rule 2—Specimen/sample can be stored at 2–8 °C up to 7 days.
- Rule 6—The results have to be validated and verified by a consultant Microbiologist.

The studies which consider resource properties are by Cabanillas et al. [40], Nakatumba [41]

(iv) **Composite properties**: Rules that are a combination of more than one category of property form a composite property. Rule 2—"Preservation of blood sample between 2 and 8° C up to 7 days" is a composite property as it is a combination of resource and temporal properties.

(b) **Qualitative Properties**: The properties which can only be observed but cannot be measured numerically are Qualitative properties as given below:

(i) **Control-Order properties**: The workflow of the business process dictates the activities of the process and the sequence in which they have to be executed. Verifying the sequence in which the activities are invoked is very crucial for the business process and is known as control-order constraints. Control-order properties deal with the order of event occurrence during execution of the SBS. The studies which consider control-flow properties are Weidlihh et al. [42] and Ramezani et al. [10]. The three main sub-patterns of control-order constraints are namely *Precedes, LeadsTo* and *Sequence*.

- *Precedes*—"Event A *Precedes* event B" signifies that upon the occurrence of event B, event A must have occurred at least once before event B.
- *LeadsTo*—"Event A *LeadsTo* event B" means once event A happens, event B must occur before the end of the process execution scope.
- *Sequence*—"Event A *Sequence* Event B" requires Event B to be the immediate consequent of antecedent Event A.

Details of control-order sub-patterns are discussed by Elgammal et al. [26]. Examples of control-order properties which are inferred from the workflow of the case study are:

- Collection and Barcoding service *LeadsTo* Preservation service.

- Dispatch Service should be *Preceded* by Report Generation Service.
  The various properties to be observed for each element are derived from the workflow of the case study and the requirements specified as rules.
  The control-order is inherent in the business process modelling languages such as BPMN as they mainly depict the order of execution of activities with time. Hence control-flow is implicit in the BP models.

(c) **Classification based on Effectiveness**: Classification of rules by Hashmi et al. [21] is based on the deontic effects—Obligation, Permission and Prohibition as defined below:

- *Obligation*: "A situation, an act or a course of action to which a bearer is legally bound and if it is not achieved or performed results in violation".
- *Prohibition*: "A situation, an act or a course of action which a bearer should avoid and if it is achieved results in a violation".
- *Permission*: "Something is permitted if the obligation or the prohibition to the contrary does not hold".

The process behavior is limited by the prohibition and obligation constraints. The categorization of the basic three deontic effects and the relationship between these effects, violations and compensations to overcome them is given in Fig. 7.

Prohibitions and Obligations have an inherent association with each other in legal theory and legal relationship. If P is obligatory, then ¬P is prohibited and the prohibition of P is the obligation of ¬P. The two important aspects of obligations are to know the lifespan (validity period) of an obligation and the implications it has on selected activities/services/operations of the BP. Norms or constraints specify the conditions and the duration for which the obligations hold. The lifespan or the duration on the validity of an obligation as specified by the norms indicate a time point or a time interval when an obligation has to be met. An obligation is valid until either it is removed (persistent obligation) or terminated (non-persistent obligation).

- *Persistent Obligation*: is a time interval constraint between two tasks denoted by p and q or on a set of contiguous task such as task q to be performed after task p and before task r.

  - *Achievement* and *Maintenance* obligations are sub-types of persistent obligation.
    Maintenance obligation has to be complied with or adhered to or obeyed for the entire duration by all activities/tasks/operation/services and hence deadlines are not necessary to detect them.
    Achievement obligation on the other hand is complied with if it is achieved or attained at least once during the time interval or duration, hence they are checked/verified for violation at deadlines (which can be even before the start of the validity period). If achievement obligation is fulfilled even before the achievement or active period it is known as "Pre-emptive obligation" otherwise it is a "Non-preemptive obligation". *E.g.* from the case study: Report has to be

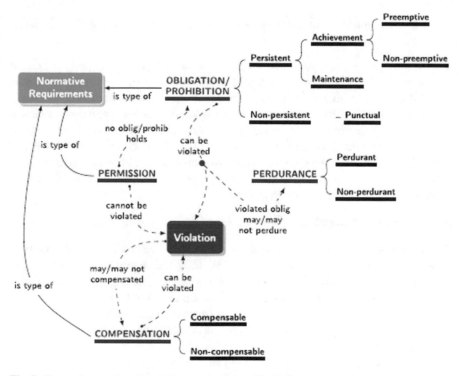

Fig. 7 Normative requirements: classes and relationships [21]

produced within 4 h of collection of the blood sample—is achievement obligation of type non-preemptive as the report cannot be given before the collection of the blood sample.

- *Non-Preemptive Obligation*: is bound by an interval in which the obligation is valid (Report after blood sample is collected)
- *Preemptive e.g.*: Once a blood sample is collected and tested, the report has to be dispatched/given to the Clinician/Patient/Relative.
  (1) The Report must be delivered not later than one day.
  (2) The above clause (1) does not apply if the diagnostic laboratory has previously dispatched the report (either through phone/mail/SMS/hardcopy/website).

- *Non-Persistent Obligation*: has to be followed or fulfilled for a particular instance and known as punctual obligation.

  - *Punctual Obligation*: is on a particular task that has to be performed in one particular instance of a BP. The task has to be immediately performed with no delay when a prequalifying task becomes true (time point constraint). The contents of the obligations for a punctual obligation have to be attained immediately and if not, a violation occurs.

• *Perdurant Obligation*: Even after violation, if the obligation continues to exist, it is termed a Perdurant obligation. Obligations have to be met by certain deadlines and if not met within the deadline, a violation occurs, but the obligation is not yet satisfied and still in effect or force. This triggers compensatory measures/actions or penalties. Therefore if perdurant obligations are not satisfied on time, the original obligation and the associated compensations have to be executed by the BP. For e.g. The report of the blood test has to be dispatched within 4 h, if not accomplished (or delivered) an obligation violation occurs, but the obligation of the diagnostic laboratory to dispatch the report still exists.

(iii) **Collection of Metrics**:

The performance of any business process is normally measured in terms of metrics known as Key Performance Indicators (KPIs) [43], that summarize the attainment of business goals. In order to obtain the status of compliance, every compliance requirement requires its own assessment metric which has to be aligned with the BP rules. Metrics are necessary to assess and evaluate the rules. Similarly metrics also help to assess the compliance status of every individual process instance and also the combination of all instances.

A methodical approach to assess and improve compliance is to define appropriate metrics for the compliance controls. Peter Drucker states, "*If you cannot measure it, you cannot improve it*." Business executives insist on performance metrics to monitor progress towards goals and to evaluate the effectiveness and efficiency of business processes. Metrics are essential for compliance monitoring and are an indispensable part of any business process. Metrics help to determine the status of the business process compliance during monitoring. They are necessary for evaluating compliance and measuring success against objectives. Metrics are performance indicators that illustrate a positive or negative progress and thus provide a snapshot of the status of compliance. The outcome of metric measurement ought to focus on *fault-finding* and evaluating the *root-causes* to resolve violations. Metrics have to be precise and practical in order to diagnose causes for performance anomaly. Hence metrics should be based on SMART criteria which specify five significant characteristics for defining performance indicators which are "Specific, Measurable, Attainable, Realistic and Time-sensitive" [44].

Metrics has been deliberated in various domains by—Frankland [45] consider IT security metrics, Chen et al. [46] discuss metrics such as Availability Time, Response Time, Security and Reliability to verify SLA compliance with the QoS parameters stated by the provider. Mayerl et al. [47] define metrics and the dependencies between them in order to evaluate compliance with SLAs. Hershey et al. [48] define service metrics that assist network operators to manage and monitor complex enterprise systems. Pedrinaci and Domingue [49] discuss methods to define and compute metrics for analysis of Business Processes. Leitner et al. [50] propose metrics for monitoring of applications deployed on Cloud. Sarwar et al. [51] give examples of process metrics in the field of Radiology such as Patient access time and Report turnaround time. Saralaya et al. [52] define metrics for elements at various levels of business process such as Process performance metrics for the business process layer, Service metrics for the service composition and coordination layer and Infrastructure metrics for resource layer.

In the elicitation phase we extract the rules from the requirements, identify the categories to which the rules belong to and finally define apt metris to obtain the values for the rules to verify their adherence to compliance. Figure 8 illustrates the relationship between the Metric, Rule and the Properties to be monitored. Table 1 depicts metrics for compliance requirements pertinent to the diagnostic laboratory case study.

4.2 Formalization (Specification)

Formalism refers to precisely specifying and formulating the requirements. A sound representation of the process model and the norms to be verified is required in order to validate compliance of the business process with the norms. Compliance is an association between the normative specification and process model specification. The normative specification portrays what the business processes has to follow (i.e. compliance rules, policies and regulations) while the process model specification depict the sequence of activities in a business process. A normative document includes the regulatory policies, constraints, obligations and rules to be followed by the organization from various sources. It should be noted that normative statements are only a means to specify the conditions to be fulfilled. They are unable to handle the violations or detect its causes.

Compliance requirements acquired from legal and policy documents identified in the elicitation phase given by various stakeholders are usually abstract and in natural language. Each Rule/Requirement identified during elicitation phase has to be precisely formulated. Compliance constraints are declarative and postulate *what* should be achieved rather than *how* to accomplish them. The requirements in the natural language have to be represented in a structured formal notation as *Compliance Rules* so as to provide tool support for compliance verification.

A formal model is required for illustrating the norms. Similarly the business process has to be represented by a formal model. If the formalism model of the business process and the norms, vary a mechanism to bridge them is required. This requires a collective knowledge of the business process and a suitable formalization

Fig. 8 Relationship between property, rule and metrics

Table 1 Metrics for compliance requirements pertinent to the diagnostic laboratory case study

ID	Compliance rule	Metric	Pattern	Properties
1	Sample collection should be followed by barcoding	M1—order of invoking services	Barcoding sequence sample collection	1. Control-Order—Sequence
2	The specimen may be preserved at 2–8 °C up to one week	M2a—Temp in storage refrigerator not to exceed 2–8 °C M2b—Specimen to be stored only for 7 days	2a—Temperature minim threshold 2 °C, maximum threshold 8 °C 2b—maxInterval to store specimens is 7 days	1. Resource min/maxTemp 2. Temporal—maxInterval
3	Samples are collected and barcoded before the testing process	M3—Barcoding to be done before testing	Sample collection leads to barcoding	1. Control-Order LeadsTo
4	The test process should start at 8:00 AM every day	M4—Testing should start at 8:00 AM each day	Dengue testing has to start at 8:00 a.m.	1.Timing—exactly at k units
5	If output of immuno-chromatographic test is positive, confirmatory ELISA test has to be performed	M5—Data output of immuno-chromatographic test if positive ELISA test has to be executed	Data output of immuno-chromatographic test if positive leads to ELISA test	1. Data condition 2. Control-Order—LeadsTo
6	If the preliminary test indicates positive, clinician has to be immediately notified by phone/SMS or email	M6—If data output of preliminary test is positive notify clinician	Data output of immuno-chromatographic test if positive leads to dispatch and notification of probable positive result to the clinician	1. Data condition 2. Control-Order—LeadsTo
7	The results are entered in the laboratory information system (LIS) by the clerical staff only after verification by the microbiologist	M7—Only after microbiologist verifies the data output, the results are entered into LIS by clerical staff	Entry into LIS by clerical staff has to be preceded by verification by microbiologist	1. Data condition 2. Control-Order—Precedes 3. Resource—PerformedBy
8	Validation of report should be done by the role consultant microbiologist	M8—Only consultant microbiologist is authorized to finalize the report	Only role consultant microbiologist can perform the finalize report operation	1. Resource—PerformedBy

(continued)

Table 1 (continued)

ID	Compliance rule	Metric	Pattern	Properties
9	The optimal time interval between collection of S1 and S2/S3 serum is 7–21 days	M9—An optimal time of 7–21 days has to be maintained between collection of samples S1 and S2/S3	Only after results of S1, S2/S3 samples have to be collected	1. Temporal—Within k units
10	The details of the patient and their reports have to be confidential	M10—Patient's data has to be handled confidentially	Data output and reports have to be confidential	1. Data—Confidentiality
11	The duration between the first verification of report by the microbiologist and second verification of the transcript after entry in LIS should not be more than 30 min	M11—Time gap between first and second verification by the microbiologist of the report not to exceed 30 min	Max time difference between end time of first verification activity and the second verification activity should not exceed 30 min	1. Temporal—Within k units
12	Time taken to generate report after confirmation should not exceed 30 min	M12—Time gap between confirmation and report generation not to exceed 30 min	Max. time difference between start of report generation operation and completion of confirmation operation should not exceed 30 min	1. Temporal—Within k units

language. As this is a difficult task, tools supporting such formalism which hide the technicalities could be helpful to non-technical users. So as to automate the process of compliance verification the requirements have to be formalized into formats that are machine readable, similarly the business process also has to be represented by a machine readable form.

The new aspects introduced by compliance management such as permitted or non-permitted states and operating constraints have to be portrayed in a conceptual model. The selection of an appropriate modelling language for the compliance management framework plays a substantial role as it specifies the formalism to articulate existing and impeding (or forthcoming) business methods and procedures [24].

For modeling the norms/rules formal logical languages such as Temporal Logic (TL) [53], Predicate logic [11], Deontic Logic (DL) and Event Calculus (EC) [54–56] are widely used. A comparative evaluation of temporal logic languages (LTL, CTL) and deontic language (FCL–Formal Contract Language) as building block for the formal specification of compliance requirements have been explained by Elgammal et al. [57]. This answers the second question *What languages can be used to model the rule*.

In literature, temporal logic has been extensively used to formalize hardware and software designs as well as to verify conformance to policies and constraints in various domains. Linear Temporal Logic (LTL) is a modal temporal logic with modalities signifying time. LTL and CTL (Computational Tree Logic) are standards for model checking used to formalize declarative rules. LTL is a logic of traces, an LTL formula is true for a transition system S if and only if it is true for each trace of S. CTL, on the other hand, is a branching-time logic, which handles multiple paths at the same time. The LTL formulae denote properties that will be interpreted on each execution of a program. Each possible execution, can be visualized as a sequence of events or states on a line hence the name "linear time", the satisfiability is checked on the run with no possibility of switching to another run during the checking. On the other hand, the CTL semantics check a formula on all possible runs and will try either all possible runs or only one run when a branch is encountered. LTL may also be addressed as propositional temporal logic (PTL). LTL has been used for formal verification of various computer programs. First-order logic, which is also known as first-order predicate calculus or first-order functional calculus is a group of formal systems expended in computer science, mathematics, linguistics and philosophy. First-order logic quantifies only variables that range over individuals; second-order (an extension of first-order logic) logic, additionally, also quantifies over relations [58].

LTL is a very popular formalism to model a sequence of tasks in a business process workflow and has been effectively employed. On the contrary, Governatori and Hashmi [59] argue that LTL fails to correctly model some aspects of the norms such as the permission constraints. Only structural compliance (i.e. control-order) can be verified by LTL formalism and it also lacks differentiation among constraints such as various types of obligation, violations and their compensations [60]. It is also difficult to obtain data related compliance using temporal operators [32]. Normative reasoning cannot be captured with first-order logic.

As LTL and CTL formalism cannot be easily handled by non-experts, some authors propose graphical notations [61, 62]. Chesani et al. [63] propose an algorithm for transforming the compliance rules which are specified in a graphical language GOSpeL into a declarative language rule based on computational logic. Khaluf et al. [64] propose a language for visually modeling quality constraints. The quality constraint models are transformed into CTL formulas by pattern-based translation. Deontic logic is a formal method that tries to depict the normative concepts such as obligation prohibition and permission [65]. Over the years several variations to the Deontic logic have been proposed.

A BP can be represented using process modelling languages like "Business Process Modeling Notation (BPMN)", "Business Process Execution Language (BPEL)", "Event Process Chains (EPC) XPDL", "UML Activity Diagram" and "Yet Another Workflow Language (YAWL)". BPMN is a standard for modelling business processes whereas BPEL is a standard which can be used for both modelling as well as execution.

Specialized softwares such as ActiveVOS, jBPM, Activiti, Camunda, BonitaOS, TAVERNA help business users to diagrammatically create business processes swiftly. OpenWFE, Freefluo, Syrup, JawFlow. ARIS, intaglioIBPMS, AuraPortal are some of the commercial BPMs. An extensive comparison of the BPM Suites is performed by Meidan et al. [66]. Some of the commercial business process management suites that also incorporate reporting of BP performance are IBMs Business Process Manager, HPs Business Process Monitor, Oracles' Business Process Management Suite, TIBCOs Spotfire and SAPs Business Process Management. This answers the other part of the second question what languages can be used to model the business processes. Little or no support is provided for compliance rules by commercial software systems. The commercial software are declarative i.e. they specify only the control-order for activities to be executed, but with no particulars on how things are to be achieved.

Such process models can be transformed into formal models such as Petri-nets, Workflow-nets (sub-classes of petri-nets) by using translation rules as given by Dijkman et al. [67], Ouyang et al. [68].

The concerns to be addressed by this phase can be summarized as [69]:

- Assist in rightly determining the periphery of compliant behavior.
- Formal semantics to facilitate automated compliance verification.
- Permit specification of all dimensions of compliant behavior i.e. temporal, data, resource and control-flow.
- Allow all possible compliant behavior.
- Allow users to emphasize on restricted behavior.

Thus the formalism phase should facilitate the non-technical personnel to specify required behavior of a compliance requirement without difficulty.

4.3   Implementation

In order to verify that the business process abides with the compliance rules formalized in the previous phase, the formalized rules have to be implemented by suitable means which can be validated during business process execution [12]. This phase helps answer the third question *"How will we check for compliance?"* i.e. How will we measure the compliance requirements/metrics?—Instituting performance measurement procedures help in determining the means to obtain the measure.

Koehler [70] suggests four different patterns for interaction between the business process and rules:

   i. Embedding rules in the business process models as gateways.
  ii. Rules are distinct and referred during gateways.
 iii. Rules referred during gateway decisions may depend on outcome of previous execution instances of the BP.
  iv. Rules dynamically dictate the flow of the business process.

While the first two patterns are frequently observed, the last two are seldom used.

The legal knowledge about the constraints, rules, procedures and policies can be stored and evolved independently of the BP or can be combined with it.

The mantra is to *"Separate the know (domain knowledge) from the flow (sequence of activities)"* [71]. If the Knowledge Base (KB) is separate from the BP, this ensures reusability of BP components and adaptability of the BPs. However the drawback of this approach is that the legal KB may be generic and too large which makes it difficult to maintain and map to the business process. On the other hand the KB can be made more specific to a BP in which case the separation between the KB and the BP may not subsist. The business expert designs the business process by considering the compliance constraints. The compliance expert collaborates with the business expert to elicit, formalize, specify and manage the compliance constraints from multiple sources. The various artifacts of the process models such as activities, service and infrastructure details are stored in the business process repository, whereas the compliance repository maintains and organizes information related to Procedures, Rules, Requirements, Constraints and their sources. A model checker has to be used to enable automatic verification of the business process against the compliance rules. Some of the standard model checkers are: SPIN used for verification of distributed software systems,[8] SMV for hardware circuits[9] and UPPAAL for verification of real-time systems (http://www.docs.uu.se/docs/rtmv/uppaal/index.shtml.).

SPIN is an open source tool that is widely used by industry and academia for formal verification of hardware and software systems. The SPIN model checker verifies the formal business process against the compliance rules. SPIN requires a design model of the system and a compliance rule in the form of a logical property as input and checks whether the property holds for the model. This requires a transformation from the BPEL/BPMN model to a formal representation such as Petrinets or Automata and later to Promela code which is the language used for verification by the SPIN model checker [72, 73]. The response of the model checker is a *true-false* output which indicates the satisfaction of each rule.

UPPAAL has evolved to a commercial product for model checking from an academic research prototype. UPPAAL takes the model of the system in the form of timed automata and the properties to be verified as a subset of Timed Computation Tree Logic (TCTL) query language.

Verification using temporal logic can be based on either Top-Down or a Bottom-up approach [26]. In the bottom-up approach, compliance requirements are formally represented using logical languages such as (LTL or CTL) while the business process modelled using low-level specifications such as BPMN or BPEL should be represented formally using Petri-nets, Finite state automata or process algebraic representation. Verification techniques such as model checking (NuSMV) are used to validate the conformance between the two representations [53, 74, 75, 76].

In Top-Down temporal logic approach, business processes are initially modelled by abstract high level specifications and compliance requirements are modelled using

---

[8]http://www.spinroot.com.

[9]http://www.spinroot.com.

temporal logic. This is followed by verification techniques or model checking. If the process model conforms to the compliance rules, the corresponding BPEL or BPMN business process model may be generated in a semi-automatic manner [77–79].

In-order to obtain metric values and to verify compliance, information has to be extracted from the BP during execution, by the process of monitoring. Information can be obtained from the system under observation in an Invasive or Non-invasive manner. The prime feature to be considered when implementing the monitoring functionality is that, it should be non-invasive with the code of the system being observed i.e. (i) the process of monitoring should be performed parallely during the execution of the BP without affecting its functionality and (ii) with least instrumentation to the BP [56]. Data can be obtained in a non-invasive way during execution, using approaches such as Complex Event Processing (CEP) paradigm or Aspect Oriented Programming (AOP).

Values of metrics defined for compliance rules can be obtained in a non-invasive manner by the process of *Monitoring* based on the notion that events are emitted during execution of the business process. Event processing is one method for tracing and evaluating streams of information (known as *Events*) that are generated during execution of a process and inferring a meaning from them. Complex event processing, (CEP), is event processing which combines data from multiple (several) sources to infer events or patterns that propose more complicated situations. The goal of CEP is to recognise meaningful events and their patterns (such as threats and opportunities) and reciprocate to them instantaneously [80]. CEP engine allows to write queries using Event Query Language (EQL)—a higher level query language similar to SQL. Some of the well-known CEP engines are WSO2 Siddhi, ESPER, Drools Fusion, Apache Flink etc. The studies which utilize CEP paradigm for compliance monitoring are by Weidlich et al. [42], Mulo et al. [22], Saralaya et al. [27], Asim et al. [81], Thullner et al. [82], Barnawi et al. [83], Zahoor et al. [84].

In Aspect-Oriented Programming the program logic is broken down into distinct parts known as *Concerns*. The functions which span multiple points of an application, called '*cross-cutting concerns*' are conceptually distinct from the business logic of the application. Aspects are very well suited for logging, caching, authentication, auditing, security, declarative transactions etc. AOP aims to reduce the amount of "tangled" code i.e. code that has to be repeated at multiple locations of a process. Such requirements cross-cut the system's fundamental operations and are hard to develop and manage using non-AOP programming concepts. Cross-cutting process can be thought of as functionality spanning numerous objects. AOP decouples cross-cutting concerns from the objects that they affect and help to intercept an application which facilitates to monitor the BP in a non-invasive manner. The monitoring logic in the form of *concerns* can be inserted before, after or around the code to be observed. AOP has been implemented as an extension to various programming languages, such as Aspect C++ (C++) and AspectJ (Java). The studies which monitor for compliance using AOP concepts are by Chen et al. [46], Baresi and Guinea [85], Li et al. [37], Kallel et al. [86], Moser et al. [87] and Saralaya et al. [88].

4.4 Supervision

Verifying the compliance of the Business Process with the set of requirements/rules/norms involves:

For every task/action/operation in every instance (trace) of the BP [21]:

- *Determining state corresponding to the task*
- *Determining requirements/rules for each task*
- *Checking for fulfillment of the rules or their violations.*

Various strategies to attain compliance as classified by Sadiq and Governatori [3] are *detective*, *corrective* and *preventative* approaches. Detective approach is an *after-the-fact* approach also known as *Compliance by Detection*. Preventative approach is also known as *Compliance by Design*. Corrective approach tries to mitigate the effect of non-complaint situations after they have occurred.

i. *Compliance by Design* is a preventive approach to managing compliance before process deployment from the initial phases of the business process life-cycle such as the modelling phase. The approach towards Compliance by design can be either Imperative or Declarative. Imperative approach is based on sequence of tasks to be performed in the BP. While the Declarative approach considers the artifacts that are affected during the process. Verification of *Compliance by design* is based on abstract values provided by business experts. The drawback of *Compliance by design* is that some compliance properties such as data output generated during execution and the actual execution time taken by an activity can only be verified during run-time and not design-time. As not much information about the actual data on which a process operates is available at design-time *Compliance by design* should be followed by monitoring and detection of non-compliance during run-time which is *Compliance by detection*. In reality, it should be known that compliant execution cannot be assured by having compliant models. During execution, common problems that occur are handling by unskilled personnel and technical or system failures. It is impractical to discover such errors during design and can only be detected during execution or after it has occurred. Most of the compliance checking approaches focus on design-time compliance verification, some of which are: Sadiq et al. [89], Julisch et al. [90], Awad et al. [18], Sackmann et al. [91], Schumm et al. [92], Elgammal et al. [26] and Goedertier and Vanthienen [93].

ii. *Compliance by detection* may be during or after execution of the business process based on real values and can be classified as (i) Backward Compliance Checking (BCC) or (ii) Forward Compliance Checking (FCC). In Forward Compliance Checking approaches, the control-flow of the business process is verified by the sequences of event occurrence in the event stream during execution. Whereas in Backward Compliance Checking the log is verified after execution of the process instance to match expected pattern.

    a. *Backward Compliance Checking*—is retrospective validation of the business processes which have completed execution. It is also known as *"after-the-*

fact" or *compliance auditing*. The execution traces of the process are examined to check conformance. Approaches such as process mining, probing the event logs (transaction logs, audit trails, message logs, workflow logs) or matching the execution traces with process models such as petri-nets or those based on temporal logic are used for this purpose. The shortcoming of this approach is that it can neither inhibit the occurrence of non-compliant behavior nor alter the actions of the anomalous process execution to avert non-compliance. The studies which adopt the BCC approach are Alberti et al. [17], Chesani et al. [63], Rozinat and Van der Aalst [94], Van der Aalst et al. [95], Saeki and Kaiya et al. [13], Weber et al. [96], Ly et al. [97].

b. *Forward Compliance Checking*—This is a proactive or preventive approach to compliance verification which aims to ensure compliance at design time or run time of the business process. Run-time method of compliance checking verifies that the business process is in accordance with the stated rules, policies or constraints during execution. This method has the advantage that during execution if a non-compliant behavior is identified, the process may either be stopped as it may lead to disastrous situations or there may be possibilities to rectify/overcome the deviations in time to correct the anomalous behavior as explained in the next section. Some of the authors who study run-time compliance monitoring are Ly et al. [98], Mulo et al. [22], Weidlich et al. [42].

4.5 Improvement

Any type of obligation (Persistent, Non-Persistent) may be violated. An occurrence of a violation need not indicate that the execution of the BP has to be stopped (or terminated). Certain types of violations such as that of time deviations (with lesser priority) can be compensated for, so that the BP can continue execution and can still be made compliant by the process of adaptation (replacing services or retry) [85, 99]) or compensatory measures in the form of penalties. Whereas certain types of violations which are of higher priority such as data output, role (resource) or control-order violations need not and should not be compensated and the process has to be terminated as continuation of execution of such BPs may lead to disastrous situations. A compensation is an action or set of activities performed to mitigate the effect of violations/obligations termed as compensation obligation.

If violations to compliance have occurred, it is extremely important to recognize why the violations have occurred i.e. identification of the root-causes to take appropriate measures to improve/optimize the BP known as Adaptation. Violations may be due to snags during BP execution known as *instance-level* violations or due to poorly designed process model known as *model-level* violations [100].

4.5.1 Root-Causes

For each task that is/was violated which was identified in the supervision phase, compensation measures have to be adopted which will result in penalties, monetary loss, loss of goodwill of customers and loss of reputation of the enterprise. Hence it is equally important for the BP analysts to identify the root-causes of violations so that similar problems/errors can be avoided for future instances of the BP. Compliance experts are concerned with every compliance violation and interested in identification of their causes. Whereas the process modeler has to improvise the process models for future instances.

Root Cause Analysis helps to revive compliance based on diagnostic information obtained during execution. The aim of root-cause analysis is to improve product or process quality and should be undertaken in a systematic way so as to be effective. In order to track causalities in any system, the relationships between the effects and their causes can be represented by many well-known techniques and tools such as "Change analysis", "Failure mode and effects analysis (FMEA)", "Fault Tree analysis", "Five whys", "Current Reality Tree", "Ishikawa diagram (Fishbone diagram)", etc. Doggett [101]. Root-cause analysis is a means of understanding why and how a problem/unexpected event occurred. Root cause analysis is based on the principle that problems are better solved by resolving their root-causes when compared to other methods that concentrate on tackling only symptoms of a problem. Through remedial measures, the underlying causes have to be addressed which minimizes the recurrence of the problem. Mdhaffar et al. [102, 103] and Saralaya et al. [27] use fishbone diagrams for root-cause analysis of compliance violations.

The fishbone diagram, well-known as the Ishikawa diagram, was invented by Ishikawa [104] and is a causal diagram that indicates the causes of a particular event. It is a visual method for root-cause analysis to identify the causes of failures. The fishbone diagram is commonly used as a tool for the analysis and prevention of defects in product design and quality defects, i.e., so as to identify potential risk factors causing an effect. Causes are usually clustered into major categories. The categories include:

- Human resources: The people/staff essential for the process.
- Infrastructure requirements: Equipment, tools, computers, network systems, raw materials required to accomplish the process.
- Methodology: How the process is performed and the specific requirements that have to be satisfied for performing it, such as rules and regulations of the land, institutional policies and procedures.
- Measurements: The data which is generated by the process that is used to evaluate the quality of output.
- Environment: The conditions which may include but not limited to, the location, the time, the temperature and/or culture in which the process operates.

Current Reality Tree (CRT) is a complex but powerful method based on representing causal factors in a tree structure developed by Eliahu M. Goldratt, the inventor of the Theory of Constraints (TOC) [105]. The method starts by picking up a list of undesirable factors based on dependencies that subsequently lead to one or more undesirable effects (UDEs) which we refer to as Anomalous behavior. The advantage of CRT is the ease with which we can identify dependencies between effects and root-causes. Elgammal et al. [106] use CRT for identifying and resolving compliance violations.

Some of the studies which try to identify root-causes when monitoring for compliance [27, 42, 107, 108] provide feedback on deviations. Awad et al. [109] try to identify causes for violations of the control-flow category and the data category [110]. The study by Rodriguez et al. [100] helps to assess and improve compliance by identifying where and why compliance violations take place in Service-Oriented Architecture (SOA) enabled business processes.

4.5.2 Improvement/Optimization/Adaptation

The monitored information can be utilized to prevent non-compliance for the current instance or the knowledge can be utilized to avert deviations for future instances. Based on the diagnostic information obtained from the root-cause analysis phase, recovery or adaptation measures can be invoked to enhance compliance. The effect of a violated obligation on the BP as well as other obligations has to be considered in order to take compensatory measures. For e.g. If a temporal violation occurs, i.e. a service/operation does not complete at the expected time, then the adaptation strategy has to consider the effect of the time delay on succeeding operations as well as to the total completion time of the BP. Once the amount of delay is calculated, the services/operations succeeding the violated/deviated service can be replaced by services with shorter execution time. Studies by Saralaya et al. [99], Ismail et al. [111], Angarita et al. [112], Aschoff and Zisman [113] devise mechanisms to mitigate such delays.

Violations of other categories such as control-flow, data output or resource/role violation, adaptation measures are not suggested as they may be fatal and lead to erroneous output. A control-flow violation indicates incorrect invocation sequence of activities or operations. A data violation denotes wrong output from a service operation. A resource violation reveals either a forbidden role performing an operation or violation of resource constraints such as inappropriate temperature range or ignoring minimum memory requirements.

5 Compliance Monitoring Functionalities (CMFs)

The effectiveness of any compliance monitoring framework may be evaluated based on the functionalities specified by Ly et al. [114, 115]. This evaluation framework postulates ten Compliance Monitoring Functionalities (CMFs) categorized into three sets namely, Modelling requirements (CMF 1–3), Execution requirements (CMF 4–6) and User requirements (CMF 7–10) which can be used to categorize and also assess existing compliance management approaches and frameworks. We briefly discuss the ten CMFs below:

1. **Modelling Requirements**: The monitoring framework should permit the specification of compliance requirements to be observed from various perspectives such as control-flow, timing, data and resources.

CMF 1: Specifies that the model should be able to express and monitor quantitative and qualitative time related constraints. *After* and *Before* patterns are qualitative constraints whereas the time metrics which enforce an interval between two events referring to latencies, delay and deadlines are quantitative.
CMF 2: This functionality deals with the data output produced by a service/operation/process.
CMF 3: This functionality states that the verification model should not only check for control-flow and data related properties but also have the capacity to verify resource constraints such as roles and agents.

2. **Execution Requirements**: The set of constraints under this category deal with events, data and conditions available during execution of a process.

CMF 4: Stipulates that the model should support atomic and non-atomic activities. Atomic activities are related with a time-point (e.g. An activity should start at 8.00 A.M.) whereas non-atomic activities are related with time-duration (e.g. an activity should be completed within 4 h).
CMF 5: Support for activity life-cycle events i.e. capability to associate every completion event with the corresponding start event i.e. correlation between events.
CMF 6: Should support constraints of multiple-category which implies the framework should be able to deal with composite constraints made up of more than one category.

3. **User Requirements**: The framework should be capable to not-only detect non-compliance but also assist in analysing causes for non-compliance.

CMF 7: Should not only be capable to continuously monitor and reactively detect but also be able to notify non-compliance and propose recovery actions.
CMF 8: Support for proactive monitoring to avoid possible violations.
CMF 9: Provision to identify root-causes of violations.
CMF 10: Quantify the degree of adherence to compliance requirements.

The performance of any compliance monitoring and analysis framework may be evaluated in terms of (i) *Degree of Compliance* for every instance (ii) *Level of Compliance* for multiple instances.

Degree of compliance acts as an indicator to assess the level of adherence to the compliance requirements at any position during the execution of a business process instance [116]. The metric that can be used to calculate the degree of compliance at any time t during the execution of a process is given by Eq. (1):

$$1 - \frac{\sum_i weight_i \# viol_i(t)}{\# Rules(t) \sum_i weight_i} \tag{1}$$

where i ranges over the constraints rules. For each constraint i, $weight_i$ denotes the assigned weight and $\# viol_i(t)$ denotes the number of rule violations until time t; $\# Rules(t)$ denotes the total number of rules encountered until time t. At any point of time t, the current state a process instance might be is: *compliant* (all compliance rules satisfied till time instant t), *violated* (most weighted data/resource/control-order compliance rules are violated and beyond repair), *possibly violated* (less weighted temporal rules such as end time of an operation or service violated which can be overcome through adaptation).

Metric known as Key Compliance Indicator (KCI) may be used to verify the compliance status of all executed BP instances as given below by—Eq. (2):

$$KCI_{CompInst} = \frac{|Compliant Instances|}{|All Instances|} \tag{2}$$

Governatori and Shek [60] define a process as compliant "*if and only if all traces are compliant (all obligations have been fulfilled or if violated they have been compensated)*". Table 2 presents a summary and evaluation of prominent studies.

6 Challenges

Some of the challenges faced during managing compliance are addressed by Boella et al. [71], Abdullah et al. [118] and Hashmi et al. [119]. Boella et al. [71] propose the general challenges that have to be fulfilled by softwares used for legal knowledge management:

1. *Open Norms*: Norms cover a wide range of scenarios and unanticipated imminent developments. But, the BP and automated computer systems are able to handle only specific situations. Thus software systems should be equipped with the capability to interpret and translate norms into executable BP.
2. *Communication among disciplines*: Compliance is a field which involves contribution of skilled personnel from multiple facets such as legal experts, compliance officers, BP designers, senior managers and operational managers who have their specific views expressed in domain-specific languages. The duties of several departments can be connected by the *Three lines of Defense Model* as given by [120]. The first line of defense must be placed in the business which should incorporate controls in the BP to diminish risks from non-compliance of regulatory requirements. The specialists who set norms such as compliance officers and risk managers form the second line of defense. The external and internal auditors who verify adherence and effectiveness of norms form the third line of control.

Table 2 Comparison of compliance monitoring functionalities

Approach	CMF 1 time	CMF 2 data	CMF 3 resources	CMF 4 non-atomic	CMF 5 lifecycle	CMF 6 multi-instance	CMF 7 reactive mangt.	CMF 8 proactive mangt.	CMF 9 root-cause	CMF 10 compliance degree
Barnawi et al. [83]	Y	Y	H	Y	Y	Y	–	–	–	Y
Mulo et al. [22]	Y	Y	–	Y	Y	–	–	–	–	–
Elgammal et al. [106]	Y	–	Y	–	–	Y	P	–	Y	–
Thullner et al. [82]	Y	Y	–	Y	Y	Y	–	–	–	–
Barbon et al. [36]	Y	Y	–	Y	Y	Y	–	–	–	–
Li et al. [37]	Y	Y	–	Y	–	–	Y	–	–	–
Weidlich et al. [42]	Y	–	–	–	–	–	–	–	Y	–
Baresi and Guinea [85]	Y	Y	–	Y	Y	Y	Y	–	–	–
Awad et al. [117]	Y	Y	–	–	Y	–	–	–	Y	–
Turetken et al. [29]	Y	Y	Y	Y	–	Y	–	–	–	Y
Saralaya et al. [27]	Y	Y	H/A	Y	Y	Y	Y	Y	Y	Y

Y Supported, *–* Not Supported, *H* Human Resources, *A* Automated Resource, *P* Partially Supported

3. *Dynamics*: The business process and also the compliance requirements are frequently updated. Hence in order to support the dynamic environment, compliance management requires:

   - The ability to swiftly modify the business processes
   - The ability to swiftly change the legal model
   - The ability to swiftly evaluate which business processes and activities are affected by the changing norms
   - The ability to swiftly update the business model to incorporate the new norms
   - The ability to record and document decision making and changes incorporated to oblige the audit trail.

As financial BP sectors such as loan landing, insurance etc. profoundly rely on computer software for managing compliance, the organizations have a huge burden to incorporate compliance features more rapidly than the support provided by traditional software development lifecycle.

Abdullah et al. [118] discuss the many challenges faced by compliance management practitioners, such as Frequent Changes in Regulations, Lack of Compliance Culture, Lack of Perception of Compliance as Value-Addition, Lack of Efficient Risk Management, Lack of Resources, Inadequate Reporting/Documentation, Inadequate Monitoring and Lack of IT Support/Tools. Some of the challenges as discussed by Hashmi et al. [119] are:

(1) *Norm Extraction and Elicitation*: The constraints/norms come from a variety of sources and are generally in natural language. It is a complex task to extricate the requirements from the legal documents and to see that they are not misinterpreted. Hence it is challenging to extricate and automate the requirements from the legal documents consisting of technical terms and legal jargon.

(2) *Formalism of Norms*: The efficiency of a compliance approach rests on the formalism approach adopted to represent the norms. Failure to effectively represent the norms results in inefficient handling of compliance. Even though a large number of formal languages exist to formally represent compliance requirements, each have their own limitations. For e.g. Event Calculus is unable to represent the effects of a task/operation and first order logic is unable to handle timing operators.

(3) *Multi-Jurisdictional Requirements*: Some rules may cross organizational or national boundaries and are interpreted in unique various ways across boundaries. The identification and automation of such constraints is challenging so that inconsistencies between different interpretations are eliminated.

7 Conclusion

Compliance is a practice of identifying whether or not a Business process adheres to or fulfills a set of obligations. Norms enforce restrictions on the way the activities of the BP have to be performed and impose penalties when deviation in behavior is observed. Ensuring compliance with regulations and policies is increasingly becoming significant to organizations as adherence to rules and regulations not only stipulates the quality of the organizations and the value added to the services they provide but also helps in diminishing the risks and losses that would be incurred due to compensations, litigations and penalties. Compliance not only deals with how the activities/operations are executed in a business process but also addresses the artefacts and the effect of actions on the surrounding environment. In this article we have tried to provide some basic knowledge of managing compliance through the different phases of compliance lifecycle.

References

1. Governatori G (2005) Representing business contracts in RuleML. Int J Coop Inf Syst 14(02n03):181–216
2. Bianco P, Lewis GA, Merson P (2008) Service level agreements in service-oriented architecture environments (No. CMU/SEI-2008-TN-021). Carnegie-Mellon University, Pittsburgh Pa Software Engineering Institute
3. Sadiq S, Governatori G (2010) Managing regulatory compliance in business processes. In: Handbook on business process management, vol 2. Springer, Berlin, Heidelberg, pp 159–175
4. Hagerty J, Hackbush J, Gaughan D, Jacobson S (2008) The governance, risk management, and compliance spending report, 2008–2009: inside the $32 B GRC Market. AMR Research, Boston, USA, 25 Mar
5. Becker J, Delfmann P, Eggert M, Schwittay S (2012) Generalizability and applicability of model-based business process compliance-checking approaches—a state-of-the-art analysis and research roadmap. Bus Res 5(2):221–247
6. Fellmann M, Zasada A (2014) State-of-the-art of business process compliance approaches. In 22st European conference on information systems (ECIS 2014), Tel Aviv, Israel, June 9–11. http://aisel.aisnet.org/ecis2014/proceedings/track06/8
7. Silveira P, Rodríguez C, Birukou A, Casati F, Daniel F, D'Andrea V, Worledge C, Taheri Z (2012) Aiding compliance governance in service-based business processes. In: Handbook of research on service-oriented systems and non-functional properties: future directions. IGI Global, pp 524–548
8. World Health Organization, Special Programme for Research, Training in Tropical Diseases, World Health Organization. Department of Control of Neglected Tropical Diseases, World Health Organization. Epidemic and Pandemic Alert, 2009. *Dengue: guidelines for diagnosis, treatment, prevention and control.* World Health Organization
9. Cabanillas Macías C, Resinas Arias de Reyna M, Ruiz Cortés A (2010) Hints on how to face business process compliance. *III Taller De Procesos De Negocio E Ingeniería De Servicios, PNIS2010, Valencia, España*
10. Ramezani E, Fahland D, van der Aalst W (2012) Where did i misbehave? Diagnostic information in compliance checking. Bus Process Manag 262–278
11. Van der Aalst W, Van Hee K, Van der Werf JM, Kumar A, Verdonk M (2011) Conceptual model for online auditing. Decis Support Syst 50(3):636–647
12. Ramezani E, Fahland D, van der Werf JM, Mattheis P (2011) Separating compliance management and business process management. In: International conference on business process management. Springer, Berlin, Heidelberg, pp 459–464
13. Saeki M, Kaiya H (2008) Supporting the elicitation of requirements compliant with regulations. In: Advanced information systems engineering. Springer, Berlin, Heidelberg, pp 228–242

14. Davis J (2009) Open source SOA. Manning Publications Co
15. Steinke G, Nickolette C (2003) Business rules as the basis of an organization's information systems. Ind Manag Data Syst 103(1):52–63
16. Graham I (2007) Business rules management and service oriented architecture: a pattern language. Wiley
17. Alberti M, Chesani F, Gavanelli M, Lamma E, Mello P, Montali M, Torroni P (2008) Expressing and verifying business contracts with abductive logic programming. Int J Electron Commer 12(4):9–38
18. Awad A, Decker G, Weske M (2008) Efficient compliance checking using BPMN-Q and temporal logic. In: BPM, vol 8, pp 326–341
19. Ghose A, Koliadis G (2007) Auditing business process compliance. In: Service-oriented computing (ICSOC 2007), pp 169–180
20. Governatori G, Sadiq S (2008) The journey to business process compliance
21. Hashmi M, Governatori G, Wynn MT (2016) Normative requirements for regulatory compliance: an abstract formal framework. Inf Syst Front 18(3):429–455
22. Mulo E, Zdun U, Dustdar S (2009) Monitoring web service event trails for business compliance. In: 2009 IEEE international conference on service-oriented computing and applications (SOCA). IEEE, pp 1–8
23. Zhang P, Leung H, Li W, Li X (2013) Web services property sequence chart monitor: a tool chain for monitoring BPEL-based web service composition with scenario-based specifications. IET Softw 7(4):222–248
24. Zur Muehlen M, Indulska M, Kamp G (2007) Business process and business rule modeling languages for compliance management: a representational analysis. In: Tutorials, posters, panels and industrial contributions at the 26th international conference on conceptual modeling, vol 83. Australian Computer Society, Inc, pp 127–132
25. Wagner G (2005) Rule modeling and markup. In: Reasoning web. Springer, Berlin, Heidelberg, pp 251–274
26. Elgammal A, Turetken O, van den Heuvel WJ, Papazoglou M (2016) Formalizing and appling compliance patterns for business process compliance. Softw Syst Model 15(1):119–146
27. Saralaya S, D'Souza R, Saralaya V (in press) An event-driven dynamic cross-layer business process compliance monitoring and analysis framework. Int J Bus Inf Syst. http://www.inderscience.com/info/ingeneral/forthcoming.php?jcode=ijbis
28. Dwyer MB, Avrunin GS, Corbett JC (1998) Property specification patterns for finite-state verification. In: Proceedings of the second workshop on formal methods in software practice. ACM, pp 7–15
29. Turetken O, Elgammal A, Van den Heuvel W, Papazoglou M (2012) Capturing compliance requirements: a pattern-based approach. IEEE Softw 29(3):28–36
30. Cabanillas Macías C, Resinas Arias de Reyna M, Ruiz Cortés A (2010) On the identification of data-related compliance problems in business processes. Vi Jornadas Científico-Técnicas En Servicios Web Y Soa, JSWEB 2010, Valencia, Spain
31. Knuplesch D, Ly LT, Rinderle-Ma S, Pfeifer H, Dadam P (2010) On enabling data-aware compliance checking of business process models. ER 10:332–346
32. Hashmi M, Governatori G, Wynn MT (2012) Business process data compliance. In: International workshop on rules and rule markup languages for the semantic web. Springer, Berlin, Heidelberg, pp 32–46
33. Gómez-López MT, Gasca RM, Pérez-Álvarez JM (2015) Compliance validation and diagnosis of business data constraints in business processes at runtime. Inf Syst 48:26–43
34. Schleicher D, Fehling C, Grohe S, Leymann F, Nowak A, Schneider P, Schumm D (2011) Compliance domains: a means to model data-restrictions in cloud environments. In: 2011 15th IEEE international enterprise distributed object computing conference (EDOC). IEEE, pp 257–266
35. Palmirani M, Governatori G, Contissa G (2011) Modelling temporal legal rules. In: Proceedings of the 13th international conference on artificial intelligence and law. ACM, pp 131–135

36. Barbon F, Traverso P, Pistore M, Trainotti M (2006) Run-time monitoring of instances and classes of web service compositions. In: International conference on web services, 2006 (ICWS'06). IEEE, pp 63–71
37. Li B, Ji S, Liao L, Qiu D, Sun M (2013) Monitoring web services for conformance. In: 2013 IEEE 7th international symposium on service oriented system engineering (SOSE). IEEE, pp 92–102
38. Governatori G, Hulstijn J, Riveret R, Rotolo A (2007) Characterising deadlines in temporal modal defeasible logic
39. Kumar A, Barton RR (2017) Controlled violation of temporal process constraints–models, algorithms and results. Inf Syst 64:410–424
40. Cabanillas C, Resinas M, Ruiz-Cortés A (2011) Defining and analysing resource assignments in business processes with ral. In: International conference on service-oriented computing. Springer, Berlin, Heidelberg, pp 477–486
41. Nakatumba J (2013) Resource-aware business process management: analysis and support. PhD thesis, Eindhoven University of Technology. https://doi.org/10.6100/ir760115
42. Weidlich M, Ziekow H, Mendling J, Günther O, Weske M, Desai N (2011) Event-based monitoring of process execution violations. In: BPM, vol 11, pp 182–198
43. Popova V, Sharpanskykh A (2010) Modeling organizational performance indicators. Inf Syst 35(4):505–527
44. Shahin A, Mahbod MA (2007) Prioritization of key performance indicators: an integration of analytical hierarchy process and goal setting. Int J Product Perform Manag 56(3):226–240
45. Frankland J (2008) IT security metrics: implementation and standards compliance. Netw Secur 2008(6):6–9
46. Chen C, Li L, Wei J (2007) AOP based trustable SLA compliance monitoring for web services. In Seventh international conference on quality software, 2007 (QSIC'07). IEEE, pp 225–230
47. Mayerl C, Huner KM, Gaspar JU, Momm C, Abeck S (2007) Definition of metric dependencies for monitoring the impact of quality of services on quality of processes. In: 2nd IEEE/IFIP international workshop on business-driven IT management, 2007 (BDIM'07). IEEE, pp 1–10
48. Hershey P, Runyon D, Wang Y (2007) Metrics for end-to-end monitoring and management of enterprise systems. In: Military communications conference, 2007 (MILCOM 2007). IEEE, pp 1–7
49. Pedrinaci C, Domingue J (2009) Ontology-based metrics computation for business process analysis. In: Proceedings of the 4th international workshop on semantic business process management. ACM, pp 43–50
50. Leitner P, Inzinger C, Hummer W, Satzger B, Dustdar S (2012) Application-level performance monitoring of cloud services based on the complex event processing paradigm. In: 2012 5th IEEE international conference on service-oriented computing and applications (SOCA). IEEE, pp 1–8
51. Sarwar A, Boland G, Monks A, Kruskal JB (2015) Metrics for radiologists in the era of value-based health care delivery. Radiographics 35(3):866–876
52. Saralaya S, D'Souza R, Saralaya V (2016) Metrics for monitoring a hierarchical service-based system. Int J Appl Eng Res 11(6):3897–3904
53. Giblin C, Liu AY, Müller S, Pfitzmann B, Zhou X (2005) Regulations expressed as logical models (REALM). In: JURIX, pp 37–48
54. Hashmi M, Governatori G, Wynn MT (2014) Modeling obligations with event-calculus. In: International workshop on rules and rule markup languages for the semantic web. Springer, Cham, pp 296–310
55. Kowalski R, Sergot M (1989) A logic-based calculus of events. In: Foundations of knowledge base management. Springer, Berlin, Heidelberg, pp 23–55
56. Spanoudakis G, Mahbub K (2006) Non-intrusive monitoring of service-based systems. Int J Coop Inf Syst 15(03):325–358
57. Elgammal A, Turetken O, van den Heuvel WJ, Papazoglou M (2010a) On the formal specification of regulatory compliance: a comparative analysis. In: International conference on service-oriented computing. Springer, Berlin, Heidelberg, pp 27–38

58. Hinman PG (2007) Fundamentals of mathematical logic
59. Governatori G, Hashmi M (2015) No time for compliance. In: 2015 IEEE 19th international enterprise distributed object computing conference (EDOC). IEEE, pp 9–18
60. Governatori G, Shek S (2012) Rule based business process compliance. In: RuleML (2)
61. Awad A (2007) BPMN-Q: a language to query business processes. In: EMISA, vol 119, pp 115–128
62. Van Der Aalst WM, Pesic M (2006) DecSerFlow: towards a truly declarative service flow language. In: International workshop on web services and formal methods. Springer, Berlin, Heidelberg, pp 1–23
63. Chesani F, Mello P, Montali M, Storari S (2007) Testing careflow process execution conformance by translating a graphical language to computational logic. In: AIME, vol 4594, pp 479–488
64. Khaluf L, Gerth C, Engels G (2011) Pattern-based modeling and formalizing of business process quality constraints. In: Advanced information systems engineering. Springer, Berlin, Heidelberg, pp 521–535
65. Governatori G, Milosevic Z, Sadiq S (2006) Compliance checking between business processes and business contracts. In: 10th IEEE international enterprise distributed object computing conference, 2006 (EDOC'06). IEEE, pp 221–232
66. Meidan A, García-García JA, Escalona MJ, Ramos I (2017) A survey on business processes management suites. Comput Stand Interfaces 51:71–86
67. Dijkman RM, Dumas M, Ouyang C (2008) Semantics and analysis of business process models in BPMN. Inf Softw Technol 50(12):1281–1294
68. Ouyang C, Dumas M, Breutel S, ter Hofstede A (2006) Translating standard process models to BPEL. In: International conference on advanced information systems engineering. Springer, Berlin, Heidelberg, pp 417–432
69. Ramezani E (2017) Understanding non-compliance. PhD thesis, Eindhoven University of Technology. https://pure.tue.nl/ws/files/52365079/20170116_Ramezani_Taghiabadi.pdf
70. Koehler J (2011) The process-rule continuum-Can bpmn & sbvr cope with the challenge?. In 2011 IEEE 13th Conference on Commerce and Enterprise Computing (CEC), pp. 302–309. IEEE, 2011
71. Boella G, Janssen M, Hulstijn J, Humphreys L, Van Der Torre L (2013) Managing legal interpretation in regulatory compliance. In: Proceedings of the fourteenth international conference on artificial intelligence and law. ACM, pp 23–32
72. Holzmann GJ (1997) The model checker SPIN. IEEE Trans Softw Eng 23(5):279–295
73. Holzmann G (2003) Spin model checker, the: primer and reference manual. Addison-Wesley Professional
74. Halle S, Villemaire R, Cherkaoui O (2009) Specifying and validating data-aware temporal web service properties. IEEE Trans Softw Eng 35(5):669–683
75. Eshuis R (2006) Symbolic model checking of UML activity diagrams. ACM Trans Softw Eng and Methodol (TOSEM) 15(1):1–38
76. Wang HJ, Zhao JL (2011) Constraint-centric workflow change analytics. Decis Support Syst 51(3):562–575
77. Abouzaid F, Mullins J (2008) A calculus for generation, verification and refinement of BPEL specifications. Electronic Notes in Theoretical Computer Science 200(3):43–65
78. Awad A, Goré R, Thomson J, Weidlich M (2011) An iterative approach for business process template synthesis from compliance rules. In: Advanced information systems engineering. Springer, Berlin, Heidelberg, pp 406–421
79. Yu J, Han YB, Han J, Jin Y, Falcarin P, Morisio M (2008) Synthesizing service composition models on the basis of temporal business rules. J Comput Sci Technol 23(6):885–894
80. Luckham D (2002) The power of events, vol 204. Addison-Wesley, Reading
81. Asim M, Llewellyn-Jones D, Lempereur B, Zhou B, Shi Q, Merabti M (2013) Event driven monitoring of composite services. In: 2013 international conference on social computing (SocialCom). IEEE, pp 550–557

82. Thullner R, Rozsnyai S, Schiefer J, Obweger H, Suntinger M (2011) Proactive business process compliance monitoring with event-based systems. In: 2011 15th IEEE international enterprise distributed object computing conference workshops (EDOCW). IEEE, pp 429–437
83. Barnawi A, Awad A, Elgammal A, Elshawi R, Almalaise A, Sakr S (2016) An anti-pattern-based runtime business process compliance monitoring framework. Int J Adv Comput Sci Appl 7(2)
84. Zahoor E, Perrin O, Godart C (2011) An event-based reasoning approach to web services monitoring. In: 2011 IEEE international conference on web services (ICWS). IEEE, pp 628–635
85. Baresi L, Guinea S (2011) Self-supervising BPEL processes. IEEE Trans Softw Eng 37(2):247–263
86. Kallel S, Charfi A, Dinkelaker T, Mezini M, Jmaiel M (2009) Specifying and monitoring temporal properties in web services compositions. In: Seventh IEEE European conference on web services, 2009 (ECOWS'09). IEEE, pp 148–157
87. Moser O, Rosenberg F, Dustdar S (2008) Non-intrusive monitoring and service adaptation for WS-BPEL. In: Proceedings of the 17th international conference on world wide web. ACM, pp 815–824
88. Saralaya S, D'Souza R, Saralaya V (2015) Cross layer property verification with property sequence charts. In: 2015 international conference on soft-computing and networks security (ICSNS). IEEE, pp 1–7
89. Sadiq S, Governatori G, Namiri K (2007) Modeling control objectives for business process compliance. Bus Process Manag 149–164
90. Julisch K, Suter C, Woitalla T, Zimmermann O (2011) Compliance by design–bridging the chasm between auditors and IT architects. Comput Secur 30(6):410–426
91. Sackmann S, Kähmer M, Gilliot M, Lowis L (2008) A classification model for automating compliance. In: 2008 10th IEEE conference on e-commerce technology and the fifth IEEE conference on enterprise computing, e-commerce and e-services. IEEE, pp 79–86
92. Schumm D, Turetken O, Kokash N, Elgammal A, Leymann F, Van Den Heuvel WJ (2010) Business process compliance through reusable units of compliant processes. In: International conference on web engineering. Springer, Berlin, Heidelberg, pp 325–337
93. Goedertier S, Vanthienen J (2006) Designing compliant business processes with obligations and permissions. In: Business process management workshops. Springer, Berlin, Heidelberg, pp 5–14
94. Rozinat A, Van der Aalst WM (2008) Conformance checking of processes based on monitoring real behavior. Inf Syst 33(1):64–95
95. Van der Aalst WM, De Beer HT, van Dongen BF (2005) Process mining and verification of properties: an approach based on temporal logic. In OTM confederated international conferences on the move to meaningful internet systems. Springer, Berlin, Heidelberg, pp 130–147
96. Weber I, Governatori G, Hoffmann J (2008) Approximate compliance checking for annotated process models
97. Ly LT, Rinderle-Ma S, Göser K, Dadam P (2012) On enabling integrated process compliance with semantic constraints in process management systems. Inf Syst Front 14(2):195–219
98. Ly LT, Rinderle-Ma S, Knuplesch D, Dadam P (2011) Monitoring business process compliance using compliance rule graphs. In: OTM confederated international conferences on the move to meaningful internet systems. Springer, Berlin, Heidelberg, pp 82–99
99. Saralaya S, D'Souza R, Saralaya V (in press) Temporal impact analysis and adaptation for service-based systems. Int J Inf Commun Technol. http://www.inderscience.com/info/ingene ral/forthcoming.php?jcode=ijict
100. Rodríguez C, Schleicher D, Daniel F, Casati F, Leymann F, Wagner S (2013) SOA-enabled compliance management: instrumenting, assessing, and analyzing service-based business processes. SOCA 7(4):275–292
101. Doggett AM (2005) Root cause analysis: a framework for tool selection. Qual Manag J 12(4):34–45
102. Mdhaffar A, Halima RB, Jmaiel M, Freisleben B (2014) CEP4Cloud: complex event processing for self-healing clouds. In: 2014 IEEE 23rd international WETICE conference (WETICE). IEEE, pp 62–67

103. Mdhaffar A, Rodriguez IB, Charfi K, Abid L, Freisleben B (2017) CEP4HFP: complex event processing for heart failure prediction. IEEE Trans NanoBiosci
104. Ishikawa K (1982) Guide to quality control (No. TS156. I3713 1994.)
105. Dettmer HW (1997) Goldratt's theory of constraints: a systems approach to continuous improvement. ASQ Quality Press
106. Elgammal A, Turetken O, Van Den Heuvel WJ (2012) Using patterns for the analysis and resolution of compliance violations. Int J Coop Inf Syst 21(01):31–54
107. Elgammal A, Turetken O, van den Heuvel WJ, Papazoglou M (2010b) Root-cause analysis of design-time compliance violations on the basis of property patterns. In: Service-oriented computing, pp 17–31
108. Taghiabadi ER, Fahland D, van Dongen BF, van der Aalst WM (2013) Diagnostic information for compliance checking of temporal compliance requirements. In: International conference on advanced information systems engineering. Springer, Berlin, Heidelberg, pp 304–320
109. Awad A, Smirnov S, Weske M (2009) Towards resolving compliance violations in business process models. GRCIS. ceur-ws.org
110. Awad A, Weidlich M, Weske M (2009) Specification, verification and explanation of violation for data aware compliance rules. In: Service-oriented computing, pp 500–515
111. Ismail A, Yan J, Shen J (2013) Incremental service level agreements violation handling with time impact analysis. J Syst Softw 86(6):1530–1544
112. Angarita R, Cardinale Y, Rukoz M (2014) Reliable composite web services execution: towards a dynamic recovery decision. Electronic Notes in Theoretical Computer Science 302:5–28
113. Aschoff RR, Zisman A (2012) Proactive adaptation of service composition. In: 2012 ICSE workshop on software engineering for adaptive and self-managing systems (SEAMS). IEEE, pp 1–10
114. Ly LT, Maggi FM, Montali M, Rinderle-Ma S, van der Aalst WM (2013) A framework for the systematic comparison and evaluation of compliance monitoring approaches. In: 2013 17th IEEE international enterprise distributed object computing conference (EDOC). IEEE, pp 7–16
115. Ly LT, Maggi FM, Montali M, Rinderle-Ma S, van der Aalst WM (2015) Compliance monitoring in business processes: functionalities, application, and tool-support. Inf Syst 54:209–234
116. Maggi FM, Montali M, van der Aalst WM (2012) An operational decision support framework for monitoring business constraints. In: International conference on fundamental approaches to software engineering (FASE), vol 12, pp 146–162
117. Awad A, Weske M (2009) Visualization of compliance violation in business process models. In: Business process management workshops. Springer, pp 182–193
118. Abdullah NS, Sadiq S, Indulska M (2010) Information systems research: aligning to industry challenges in management of regulatory compliance. Inf Syst Res 1:1–2010
119. Hashmi M, Governatori G, Lam HP, Wynn MT (2017) Are we done with business process compliance: state of the art and challenges ahead. Knowl Inf Syst 1–55
120. Doughty K (2011) Guest editorial: the three lines of defence related to risk governance. ISACA J 5:6

Part II
Cloud Computing

Sustainable Cloud Computing Realization for Different Applications: A Manifesto

Sukhpal Singh Gill and Rajkumar Buyya

Abstract In cloud computing, an application design plays an important role and the efficient structure of an application can increase the energy-efficiency and sustainability of cloud datacenters. To make the infrastructure eco-friendly, energy-efficient and sustainable, there is a need for innovative applications. In this chapter, we comprehensively analyze the challenges in sustainable cloud computing and review the current developments for different applications. We propose a taxonomy of application management for sustainable cloud computing and identified research challenges. We also map the existing related studies to the taxonomy in order to identify current search gaps in the area of application management for sustainable cloud computing. Furthermore, we propose open research challenges for sustainable cloud computing based on the observations.

Keywords Cloud computing · Sustainable computing · Cloud application
Sustainability · Energy efficiency

1 Introduction

Cloud resources are generally not only shared by a number of cloud users but are also reallocated dynamically based on the changing demand of cloud users. Sustainable use of the computing resources is required to improve the energy-efficiency of cloud datacenters [1]. To satisfy the future demand of application users, cloud computing needs an advanced sustainability theory for efficient management of resources. The technologies with green cloud computing are required for effective management of

S. S. Gill (✉) · R. Buyya
Cloud Computing and Distributed Systems (CLOUDS) Laboratory,
School of Computing and Information Systems,
The University of Melbourne, Melbourne, Australia
e-mail: sukhpal.gill@unimelb.edu.au

R. Buyya
e-mail: rbuyya@unimelb.edu.au

© Springer International Publishing AG, part of Springer Nature 2019
S. Patnaik et al. (eds.), *Digital Business*, Lecture Notes on Data Engineering
and Communications Technologies 21, https://doi.org/10.1007/978-3-319-93940-7_4

all the resources (including cooling systems, networks, memory, storage and servers) in a holistic manner [2]. Presently, cloud computing paradigm supports an extensive variety of applications, but the usage of cloud services is growing swiftly with the current development of the Internet of Things (IoT) based applications. To fulfill the dynamic requirements of user applications, the next generation of cloud computing should be sustainable and energy-efficient. Currently, service providers are facing problems to ensure the sustainability of their cloud services [3]. Moreover, the sustainability of cloud services is also affected by usage of the large number of Cloud DataCenters (CDCs) to fulfill the demand of cloud users.

The various types of applications are executing on cloud infrastructure such as data-intensive or compute-intensive. There is a need to execute the applications in a concurrent manner to improve the performance of cloud computing systems [4]. Initially, Quality of Service (QoS) parameters for each cloud application is essential to be recognized based on the user requirements for provisioning of resources [5]. To improve the energy efficiency and sustainability of cloud datacenters, effective application modeling approach is used to design the cloud application [6, 7]. Green Information and Communications Technology (ICT)-based innovative applications are required to make the environment eco-friendly and sustainable infrastructure.

The rest of the chapter is organized as follows. In Sect. 2, we present the architecture for sustainable cloud computing for application management. After that, we discuss the existing related studies in Sect. 3. Based on existing research work, we propose a taxonomy of application management for sustainable cloud computing and map the existing research works to the proposed taxonomy in Sect. 4. In Sect. 5, we analyze research gaps and present some promising directions towards future research in this area. Finally, we summarize the findings and conclude the chapter in Sect. 6.

2 Sustainable Cloud Computing: Architecture

Figure 1 shows the layered architecture for sustainable cloud computing, which provides a holistic management of cloud computing resources (cooling systems, networks, memory, storage and servers), to make more sustainable and energy-efficient cloud services. There are three main services in cloud computing: Infrastructure as a Service (IaaS), Platform as a Service(PaaS) and Software as a Service (SaaS). An application manager is deployed at SaaS layer to manage the user applications such as smart home, ICT, healthcare, wearable devices, agriculture, aircraft, astronomy, education, artificial intelligence etc. At PaaS layer, Service manager controls the important aspects of the system. *IT device manager* manages all the devices attached to cloud datacenter. *Application model* defines the type of application for effective scheduling of different applications. *Application Scheduler* manages the user applications from the application model, finds the QoS requirements for each application for their effective execution and transfers the information about the quality of service of an application to the *Resource/VM manager*. *Remote CDC manager* handles the migration of VMs and workloads between local and remote cloud datacenters

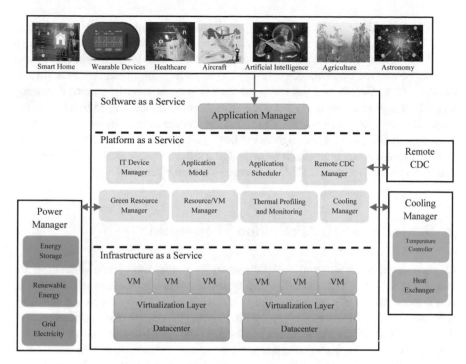

Fig. 1 Architecture for sustainable cloud computing

for effective utilization of energy. *Resource/VM manager* uses physical or virtual machines for scheduling of provisioned resources and executes the workloads based on their QoS requirements. *Green resource manager* manages the electricity coming from power manager and it prefers renewable energy as compared to grid electricity to enable sustainable cloud environment. It also manages the energy consumption of cloud datacenters to ensure sustainability of cloud services. The thermal sensors are used to monitor the value of temperature and *Thermal profiling and monitoring* technique analyze the variations of the temperature of the cloud datacenter. The *Cooling manager* controls the temperature of the cloud datacenter at the infrastructure level.

IaaS layer contains the details about virtual and physical machines of cloud datacenters. For an efficient execution of workloads, migrations of Virtual Machine (VM) is performed at virtualization layer for load balancing. The variation of the temperature of different VMs running at various cores is monitored proactively by the temperature-aware scheduler. The Cooling manager is deployed to maintain the required temperature to ensure the smooth functioning of computing resources of a cloud datacentre. If the temperature of CDC is more than its threshold value, then thermal alerts will be generated, and *Heat Exchanger* takes a required action to control the temperature of CDC with minimal effect on its performance. Further, the temperature is controlled by District heating management using water economizer, outside air economizer and chiller plant. Moreover, renewable energy resources and fossil fuels (grid electricity) generates power, which is controlled by Power Manager. The renewable energy is highly preferred instead of grid energy, which helps to enable energy-efficient and sustainable cloud environment. Further, grid energy can be used to execute deadline-oriented workloads, which improves the reliability of cloud services. Generally, solar and wind are two main sources of renewable energy and this energy is stored using batteries. The thermal sensor is used to monitor the value of temperature to generate an alert if the temperature is more than its value of threshold and pass the message to the heat controller for further action. Remote CDC is a cloud datacenter, which is located at a different place. VMs and workloads can be migrated to a remote CDC to balance the load effectively.

3   State-of-the-Art: Application Management

Application management plays a vital role in improving the performance of cloud computing systems. There are different types of cloud applications which need to be managed effectively from application modeling to its execution. An application should follow an efficient modeling approach during the design of an application. Further, green ICT-based innovative applications are required to make the infrastructure more sustainable. This section discusses the existing related studies in sustainable cloud computing for application management. Hazen et al. [8] analyzed supply chain functions' sustainability for big data analytics and identifies the factors affecting the performance of economic measures to improve data-driven process. They suggested that business people can use data analytics tools to improve the operational and strategic capabilities of the business. Furthermore, this research work analyzed the effect of data analytics on the sustainability of supply chain functions. Bifulco et al. [9] explored and analyzed the importance of ICT-based smart sustainable cities to improve environmental conditions while delivering the services with minimum consumption of energy. Further, Consumption Based Lifecycle (CBL) is presented to calculate the amount of consumption of power for daily household activities. Moreover, Dynamic Voltage and Frequency Scaling (DVFS) technique-based energy management mechanism is recommended to design a smart sustainable city, which can improve energy efficiency during the execution of ICT based cloud applications.

Rehman et al. [10] developed a Big Data Based Business (BDBB) model to manage the whole cloud applications using sustainability-aware approach to reduce the impact of carbon footprints on the environment. BDBB model manages the data by providing the following advantages: (1) improving resource utilization, which reduces cost, (2) establishing trust between provider and user and (3) improving data security and privacy preservation during data sharing. Cottrill and Derrible [11] recommended that big data can be used to develop sustainability indicators for transporting applications to create more efficient and sustainable transport system, which may also improve the GPS to trace the exact location of traffic congestion. Yekini et al. [12] introduced a new way to provide technical and vocational education using an autonomic cloud computing-based system, which delivers education resources to staff and student.

Maksimovic [13] developed an IoT based energy efficient model (Green-IoT), which uses big data analytics to develop more sustainable, smarter and secure cities. Further, energy-efficient IoT devices are considered for the network to assess the knowledge, which is used to deliver more sustainable services by making effective decisions during execution of IoT based applications. Bradley et al. [14] developed a cost-effective IoT based Parallel Design (IoT-PD) model for an efficient management of big data for cloud applications. Further, IoT-PD model generates the sustainable value using machine learning method to reduce its impact on the environment, which improves energy efficiency and sustainability of cloud services. Moreover, Perera and Zaslavsky [15] suggested utilizing energy-efficient IoT devices to increase the sustainability of services and developed a Trading-Based Value Creation (TBVC) model for effective utilization of IoT devices to execute real-time applications. Zuo et al. [16] proposed an IoT based Energy Efficient (IoT-EE) model to estimate the power consumption by cloud datacenters to execute user's workloads and analyze the energy consumption. Further, an objective function is developed to assess the delivering cloud service and IoT-EE model can be used in industry for designing and making of a product.

Bossche et al. [17] presented a Data Mining Based Model (DMBM) for text mining application, which uses unstructured data to extract useful information by removing extra information. Further, DMBM forms sustainable clusters using available resources efficiently. Cappiello et al. [18] developed a Green Computing Based Model (GCBM) to develop an application with lesser carbon emissions and maximum energy efficiency. Further, the reliability of a service is improved by ranking the application components based on their importance. Moreover, this research work also suggests two ways to choose a driven site for CO_2: (1) based on the availability of resources and CO_2 emissions and (2) based on the possible delay time. One limitation of this research work is that, authors did not suggest any action such as high cost associated to selected locations. Xia et al. [19] developed a Semantic Information services Architecture (SIA) for management of cloud services for matchmaking, retrieval and advertisement for electrical and electronic equipment. Further, the lifecycle of cloud providers is described using an ontology-based technique to analyze sustainability factors related to the semantic applications. Zgheib et al. [20] listed the advantages of IoT applications while delivering healthcare as a cloud service and developed a model to manage healthcare applications efficiently using message-oriented middleware. Further, exchanging of data is performed between different IoT devices using semantic web ontology language (OWL) messages. Gupta et al. [21] developed an IoT based Security-Aware Framework (IoT-SAF) to design sustainable and secure healthcare centres by analyzing different security aspects of IoT applications. The value of different parameters of health are managed using embedded sensors of the equipment and the gathered information is transferred to centralized cloud datacenter. Further, secure and faster data transmission in IoT-SAF is provided using XML web-based services.

NoviFlow [22] proposed a Green-Software Defined Network (G-SDN) model to make network infrastructure efficient and sustainable by offering sustainable solutions to reduce carbon emissions. Further, G-SDN model improves economic incentives by increasing energy efficiency of cloud datacenters. NoviFlow [22] identified that existing models are focusing on QoS and cost to provide energy-efficient cloud services. Pesch et al. [23] developed a Thermal-Aware Scheduling (TAS) technique for energy-efficient management of cloud resources to execute user applications, which improves the sustainability and energy efficiency of cloud datacenters. Experimental results show that TAS reduces the consumption of energy up to 40%. Chen et al. [24] presented a Cloud and Big Data (CBD) based model called *Smart Clothing* to use wearable devices to monitor the health status of patients. Experimental results show that CBD works efficiently for the long duration as compared to existing models [22, 23]. Waga and Rabah [25] proposed a sustainable Cloud-Based Framework (CBF) to manage agriculture data such as rainfall, wind speed and temperature to improve the use of climate and land in an efficient manner, which helps farmers to produce a profitable crop.

For desktop resource virtualization, Park et al. [26] proposed a Cloud-Based Clustering Simulator (CBCS) to choose the cluster for sustainable and energy-efficient cloud services. Based on network infrastructure and execution time, CBCS utilizes resources effectively without considering other resources such as processor, memory and storage. To improve the resource utilization and reduce the product-service cost for industrial product service system, Ding et al. [27] developed the Sustainability-Aware Resource Management (SARM) framework. Further, SARM helps to increase the satisfaction of users. Gmach et al. [28] presented a Power Profiling Technique (PPT) for CDCs, which increases the energy-efficiency and manages to decrease the carbon emissions and water consumption. Further, PPT analyzed the effect of demand and supply of energy on the environment. Moreover, PPT executes the workloads by fulfilling its QoS requirements while delivering the sustainable and energy-efficient cloud service.

Islam et al. [29] proposed a cloud-based Water-Aware Workload Scheduling (WAWS) framework to improve the water efficiency for cooling cloud datacenters during execution of cloud workloads. Further, water consumption is optimized by exploiting the spatio-temporal varieties of water efficiency. Dabbagh et al. [30] proposed an Energy-Efficient Technique (EET), which decreases the monthly expenditures of CDCs by postponing the execution of non-urgent workloads, which also helps urgent-workloads to achieve their deadline. Moreover, the control policy for prediction of demands of computing resources (storage, power memory etc.) of CDCs is designed for incoming requests. Further, EET is validated by using real traces of a Google CDCs. Garg et al. [31] developed an Environment Conscious Workload Scheduling (ECWS) technique for execution of High Performance Computing (HPC) applications on resources to improve the sustainability of cloud datacenters. Cheng et al. [32] developed Workload Placement and Migration Framework (WPMF) to improve throughput and sustainability of cloud datacenter for efficient execution of cloud workloads. WPMF executes transactional workloads on available resources and the workloads are migrated to different datacenters with the maximum value of energy efficiency.

Chen et al. [33] developed renewable energy sources based an Energy-Efficient Workload Management (EEWM) technique for sustainable cloud datacenter. Moreover, the temperature of CDC is controlled in EEWM using spatio-temporal diversity-based cooling facilities. Sehgal et al. [34] developed a smart human security framework based on IoT and Fog Computing (IoT-FC), which protects humans form an accident using the concept of wearable and pervasive computing. The secure communication between fog layer and IoT devices is managed by the cloud layer. Further, at fog and cloud level, the data is protected from external attacks. Khosravi and Buyya [35] proposed a Gaussian Mixture model based Short-term Prediction (GMSP) model to predict the renewable energy level to run datacenters and helps to take decisions for migration of VMs from one datacenter site to another. Further, GMSP can also perform online VM migration and experimental results show that GMSP can predict up to 15 min ahead. Desthieux et al. [36] proposed a Cloud-based Decision Support System (CDSS), which has high computing performance to control three-dimensional digital urban data that aids to provide an environmental analysis. Further, an analysis module of CDSS improves decision-making process to assess solar energy potential for installing new solar projects.

Table 1 describes the summary of these related techniques and their comparison based on the focus of study along with open research issues.

4 Taxonomy of Application Management for Sustainable Cloud Computing

Application management plays an important role to improve the sustainability and energy-efficiency of cloud services. The components of the application management taxonomy are: (i) QoS parameter, (ii) application parameters, (iii) application scheduling, (iv) application category, (v) application domain, (vi) resource administration, (vii) coordination and (viii) application model as shown in Fig. 2. Table 2 shows the comparison of existing techniques based on the proposed taxonomy of application management.

4.1 QoS Parameter

There are different QoS requirements for different user applications. Literature reported that three different types of QoS parameters for sustainable computing [2, 41, 42]: (i) cost related (utility and energy), (ii) time related (response time, throughput, availability and delay or latency) and (iii) others (correctness, easiness, resource utilization, privacy, security, reliability, robustness or scalability and inter-

Table 1 Comparison of existing techniques and open research challenges

Year	Technique	Focus of study	Open issues
2018	GMSP [35]	Prediction of level of renewable energy	Due to overloading of cloud resources, a huge number of workloads are waiting for their execution
	CDSS [36]	Solar energy potential management	The service delay is increasing by switching of resources between low and high scaling modes
2017	IoT-PD [14]	IoT devices	Secure communication is required to exchange information
	TAS [23]	Energy utilization	Heterogeneous workloads are not considered
	WAWS [29]	Water efficiency	Need to improve reliability of applications
	EET [30]	Storage efficiency	Over-utilization of resources affects throughput
	SARM [27]	Industrial product service	The response time is increased by switching of resources between low and high scaling modes
2016	IoT-SAF [21]	Web service security	The secure communication of IoT devices is required
	CBL [9]	ICT based smart cities	To improve energy efficiency, power is to be managed automatically
	CBD [24]	Smart clothing	Execution time is more, which reduces customer satisfaction
	EEWM [33]	Cooling management	The manipulation of supply voltage can save energy consumption by reducing the processor frequency
	BDBB [10]	Server utilization	Energy utilization is lesser, which leads to wastage of energy
2015	SIA [19]	Web ontology language (OWL)	There is need to improve retrieval speed in SIA
	IoT-FC [34]	Wearable and pervasive computing	The reliability of the storage component is affected by putting servers in sleeping mode or turning on/off
2014	TBVC [15]	Trading based value	The large consumption of energy by IoT devices increases carbon footprints
	GCBM [18]	CO_2 emissions	Under-utilization of resources increases cost of CDC
	CBF [25]	Agriculture	Energy consumption is more, which increases carbon footprints
	CBCS [26]	Desktop resource virtualization	The workload execution affects the resource utilization
	WPMF [32]	Heterogenous workloads	Due to speed gap between processor and main memory, a huge amount of clock speed is lost while waiting for the incoming data

(continued)

Table 1 (continued)

Year	Technique	Focus of study	Open issues
2013	DMBM [17]	Text mining	Response time is larger, which reduces customer satisfaction
2012	G-SDN [22]	Software defined network	Underloading of resources affects resource utilization
2011	ECWS [31]	HPC applications	Reserve resources are in advance increases cost
2010	PPT [28]	Greenhouse gas emissions	SLA violation can be reduced by using autonomic resource management technique

operability). "*Energy* is the amount of electricity consumed by a resource or resource set to finish the application's execution". A *Utility* is the effectiveness of an application while using practically after the development like banking application has high utility. *Response time* is defined as the amount of time consumed by a specific application to respond with the desired output. "*Throughput* is the ratio of the total number of tasks of an application to the amount of time required to execute the tasks". *Availability* is the amount of time (hours) a specific application will be available for use per day. *Delay or Latency* is a defined as a delay before the transfer of user request for processing. *Correctness* is defined as the degree to which the cloud service will be provided accurately to the cloud customers. *Easiness* is defined as the amount of efforts are required to use a specific application to complete a task. "*Resource Utilization* is a ratio of an execution time of a workload executed on a particular resource to the total uptime of that resource". *Privacy* is a QoS parameter through which user and provider can store their information privately using authorization and authentication.

Security is the capability of the computing system to protect from malicious attacks. *Reliability* is the capability of an application to sustain and produce correct results in case of occurrence of faults such as network, hardware or software related faults. *Robustness or Scalability* is the maximum number of users, which can access the application without degradation of performance. *Interoperability* is the degree to which an application can be ported to other platforms. Other QoS requirements of cloud computing systems can be network bandwidth etc.

4.2 Application Parameters

There are different types of application parameter, which can help to measure the status of an application [43, 44]. There are four main types of application parameters: (a) budget, (b) deadline, (c) capacity and (d) performance. *Budget* is the amount of cost that user wants to spend for the execution of their task and measured in dollars ($). *Deadline* is the maximum time limit allowed to execute the user application and measured in seconds. *Capacity* is defined as the capability of an application to

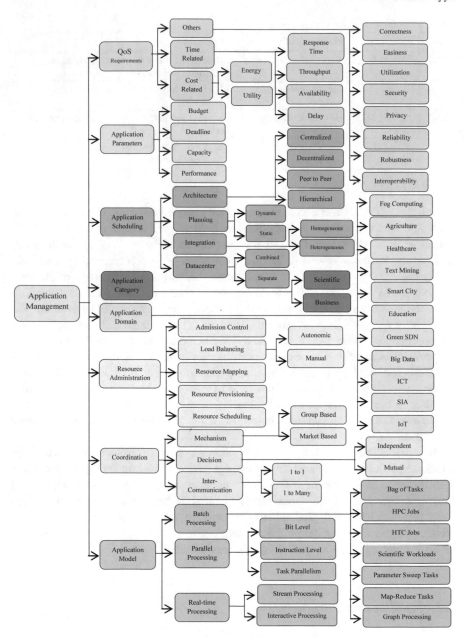

Fig. 2 Taxonomy of application management for sustainable cloud computing

execute user tasks using available resources such as power infrastructure, IT devices etc. *Performance* is a parameter, which is used to measure the status of an application during an unexpected spike in traffic.

4.3 Application Scheduling

There are four components of an application scheduling as studied from literature [1, 2, 45, 46]: (a) architecture, (b) planning, (c) integration and (d) datacenter. The *architecture* of cloud computing system is a vital component and four different types of architectures are identified from literature: hierarchical, peer to peer, decentralized and centralized. *Peer-to-peer* is networking architecture in which each cloud node, has the same capabilities and responsibilities to execute user task. There is two levels of schedulers in the *hierarchical* architecture: lower and higher level. The lower-level scheduler assigns resources to every small task, which is controlled by a higher-level scheduler to reduce the complexity of resource allocation. There is no mutual coordination in *decentralized* architectures and it executes the tasks by allocating resources independently. In *centralized* architectures, all the tasks that are needed to be executed are managed by a central controller and scheduled resources are used to execute these tasks using coordination mechanism. *Planning* of a cloud application is done using two types of resource scheduling: static and dynamic. In *static* resource scheduling, a required resource is mapped to user workload based on their QoS requirements, while *dynamic* resource scheduling maps and executes the user workloads using provisioned resources. To find out the result of different execution units, an *integration* function is used by the scheduler, which can be *combined* or *separate*. The literature reported two different types of *datacenters*: heterogeneous and homogeneous. A CDC with the same configuration of computing resources such as processors, storage, networking and operating systems to process user applications is called a *homogenous* datacenter, while a CDC is a combination of different configurations of computing resources is called a *heterogeneous* datacenter.

4.4 Application Category

The cloud computing paradigm is a very effective platform, which can handle the increasing complexity of user applications. As studied form literature [3, 39, 40], two types of user applications are identified: scientific and business. The *Business* application is a platform, which enables the execution of business functions. For example: banking systems and online shopping websites. *Scientific* applications contain the real-world activities, which needs high computing capacity to execute user requests. For example: oil exploration, aircraft design and fuel efficiency, weather prediction and climate modeling, flight control system etc.

Table 2 Review of state-of-the-art based on taxonomy

Technique	Application domain	Application category	QoS parameter	Application scheduling				Application parameter	Resource administration	Coordination			Application model		
				Architecture	Planning	Integration	Datacenter			Mechanism	Decision	Inter communication	Batch processing	Realtime processing	Parallel processing
GMSP [35]	Smart city	Business	Energy	Centralized	Dynamic	Separate	Homogeneous	Capacity	Resource mapping	Group based	Independent	1 to many	Map-Reduce tasks	Bit level	Stream processing
CDSS [36]	ICT	Scientific	Cost	Peer to peer	Dynamic	Combined	Heterogenous	Performance	Resource mapping	Market based	Independent	1 to 1	Bag of tasks	Bit level	Interactive processing
IoT-PD [14]	IoT	Business	Time	Centralized	Static	Separate	Heterogenous	Deadline	Admission control	Group based	Mutual	1 to many	Map-Reduce Tasks	Task Parallelism	Interactive Processing
TAS [23]	Smart city	Scientific	Delay	Decentralized	Dynamic	Separate	Homogeneous	Deadline	Load balancing	Market based	Independent	1 to Many	HPC Jobs	Instruction Level	Stream Processing
WAWS [29]	Smart city	Business	Availability	Hierarchical	Static	Combined	Heterogenous	Capacity	Resource Selection	Group Based	Mutual	1 to 1	HTC Jobs	Bit Level	Interactive Processing
EET [30]	Big data	Business	Throughput	Hierarchical	Dynamic	Separate	Homogeneous	Performance	Admission Control	Market Based	Independent	1 to Many	Bag of Tasks	Instruction Level	Stream Processing
SARM [27]	Smart city	Scientific	Response time	Centralized	Static	Separate	Heterogenous	Deadline	Load Balancing	Group Based	Mutual	1 to 1	HTC Jobs	Bit Level	Interactive Processing
IoT-SAF [21]	Healthcare	Business	Interoperability	Peer to peer	Static	Combined	Homogeneous	Performance	Load Balancing	Group Based	Independent	1 to Many	HPC Jobs	Bit Level	Interactive Processing
CBL [9]	ICT	Scientific	Cost and robustness	Decentralized	Static	Combined	Heterogenous	Budget	Resource Selection	Market Based	Independent	1 to 1	Bag of Tasks	Task Parallelism	Interactive Processing
CBD [24]	Healthcare	Business	Time and reliability	Centralized	Dynamic	Separate	Homogeneous	Capacity	Resource Provisioning	Group Based	Mutual	1 to Many	Scientific Workloads	Instruction Level	Interactive Processing
EEWM [33]	ICT	Scientific	Energy and privacy	Peer to Peer	Dynamic	Combined	Heterogenous	Deadline	Resource Selection	Market Based	Independent	1 to Many	Scientific Workloads	Instruction Level	Stream Processing
BDBB [10]	Big data	Scientific	Security	Hierarchical	Dynamic	Combined	Homogeneous	Capacity	Load Balancing	Group Based	Independent	1 to 1	HPC Jobs	Instruction Level	Interactive Processing

(continued)

Table 2 (continued)

Technique	Application domain	Application category	QoS parameter	Application scheduling				Application parameter	Resource administration	Coordination			Application model		
				Architecture	Planning	Integration	Datacenter			Mechanism	Decision	Inter communication	Batch processing	Realtime processing	Parallel processing
SIA [19]	Semantic Service	Business	Resource utilization	Centralized	Static	Separate	Heterogenous	Performance	Resource Selection	Market Based	Mutual	1 to 1	Parameter Sweep Tasks	Task Parallelism	Interactive Processing
IoT-FC [34]	Fog computing	Scientific	Cost and easiness	Hierarchical	Static	Combined	Homogeneous	Budget	Admission Control	Group Based	Independent	1 to Many	Map-Reduce Tasks	Instruction Level	Interactive Processing
TBVC [15]	IoT	Business	Time and correctness	Hierarchical	Static	Separate	Heterogenous	Performance	Load Balancing	Group Based	Independent	1 to Many	Bag of Tasks	Task Parallelism	Stream Processing
GCBM [18]	Smart city	Scientific	Response time	Decentralized	Dynamic	Separate	Heterogenous	Budget	Load Balancing	Market Based	Mutual	1 to Many	Map-Reduce Tasks	Instruction Level	Interactive Processing
CBF [25]	Agriculture	Scientific	Availability	Decentralized	Dynamic	Combined	Homogeneous	Performance	Resource Selection	Group Based	Mutual	1 to 1	Map-Reduce Tasks	Bit Level	Interactive Processing
CBCS [26]	Big data	Business	Throughput	Peer to Peer	Static	Separate	Homogeneous	Performance	Load Balancing	Market Based	Mutual	1 to 1	Graph Processing	Bit Level	Stream Processing
WPMF [32]	Big data	Scientific	Cost	Centralized	Dynamic	Separate	Heterogenous	Budget	Resource Provisioning	Group Based	Independent	1 to Many	Map-Reduce Tasks	Instruction Level	Interactive Processing
DMBM [17]	Text mining	Business	Time	Hierarchical	Static	Combined	Heterogenous	Deadline	Resource Selection	Group Based	Mutual	1 to 1	Graph Processing	Task Parallelism	Interactive Processing
G-SDN [22]	Green-SDN	Scientific	Energy and utilization	Decentralized	Dynamic	Combined	Homogeneous	Capacity	Resource Provisioning	Market Based	Independent	1 to Many	HPC Jobs	Instruction Level	Stream Processing
ECWS [31]	Smart city	Business	Time and cost	Centralized	Static	Separate	Homogeneous	Performance	Resource Provisioning	Group Based	Mutual	1 to 1	HPC Jobs	Instruction Level	Interactive Processing
PPT [28]	ICT	Scientific	Energy and robustness	Peer to peer	Static	Combined	Heterogenous	Budget	Resource Selection	Market Based	Independent	1 to Many	Parameter Sweep Tasks	Bit Level	Interactive Processing

4.5   Application Domain

In order to satisfy different kinds of customers, the applications are created for an extensive range of domains to make cloud services more sustainable and to improve customer satisfaction [2, 37, 38]. To develop sustainable smart cities, *ICT based applications* are designed, which reduces energy consumption of cloud datacenters. The use of Internet-based smart devices is growing exponentially, and it is important to adopt *IoT based technology* for secure, fast and reliable communication, which delivers the sustainable cloud services. To improve the consistency of that knowledge among all the cloud users, *Semantic* web ontology language exchanges the information among smart devices, IoT devices etc. Cloud user can access the unstructured data and retrieve the required information using text mining-based cloud applications. Further, to improve the process of matchmaking and retrieval of data, applications are designed to provide *Semantic information services*. Moreover, *Fog computing* based models such as driverless car and human security models are developed using IoT devices. For example, people can use wearable devices to protect themselves from any future accident, which improves the sustainability of IoT devices. To reduce the carbon emissions and improve the communication in terms of sustainability, *Green-Software Defined Network (G-SDN)* is designed, which improve the network lifetime of IoT devices. ICT based application and IoT devices are used in an integrated manner to design smart cities, which can control energy consumption and provide a way for effective use of resources in an efficient manner. Other important application domains of sustainable cloud computing are agriculture, education and healthcare. *Agriculture* applications aid to manage agriculture data such as rainfall, wind speed and temperature to improve the use of climate and land in an efficient manner, which helps farmers to produce a profitable crop. Students and staff can access available open education resources using *education*-based applications. Further, wearable devices can be used to measure the health status of patients using IoT based *healthcare* applications.

4.6   Resource Administration

Resource Administration consists of five main components [47, 48]: (1) load balancing, (2) admission control, (3) resource provisioning, (4) resource scheduling and (5) resource monitoring. *Load balancing* is a process to effectively distribute the workload over all the available resources to maintain the performance of a computing system and load balancing can be *autonomic* or *manual*. *Admission control* mechanism is used to ensure that sufficient resources are available to provide failover protection by reserving resources in advance. The process of identifying adequate resources from resource pool based on application requirement is called *Resource provisioning* [8], whereas mapping of provisioned resources through resource provisioning for application execution [10] is called *resource scheduling*. The main aim of effective resource management is to schedule the provisioned cloud resources for execution of an application, so that application can execute with maximum resource utilization [11]. *Resource monitoring* is a process to measure the value of QoS parameters during workload execution.

4.7 Coordination

Coordination is a process through which different applications can communicate to achieve a common goal [1, 41, 42] and three components of coordination are: (1) mechanism, (2) decision and (3) intercommunication protocol. In the cloud, there are two types of *mechanism*: market-based and group-based. In *market-based* mechanism, the concept of Service Level Agreement (SLA) based negotiation is used to deliver the resources to different applications. In *group-based* mechanism, resources are shared within the groups formed based on the same QoS requirements of an application. Literature reported that there are two different types of *decisions*: independent and mutual. Resource scheduler schedules the resources independently for workload execution in the *independent* decisions scheme without focusing on resource utilization. The concept of mutual coordination is used in the *mutual* decisions scheme, to make the coordination between high-level and low-level scheduler for execution of all the tasks of user application. The components of an application are interacting with each other using two types of intercommunication protocol: one to many and one to one. Based on negotiated SLA, one consumer is getting with one provider in *one to one* protocol, while one cloud provider offers service to more than one user in *one to many* intercommunication protocol.

4.8 Application Model

There are three types of *application models* as identified from literature [47, 48]: (1) real-time processing, (2) batch processing and (3) parallel processing. *Real-time processing* is a processing of data which requires continuous input, process and output of data and it processes data in a short span of time. For example: processing of data at radar systems and bank ATMs. There are two types of real-time processing: (1) stream processing and (2) interactive processing. *Stream processing* is a processing of small-sized data (in Kilobytes) generated continuously by thousands of data sources (geospatial services, social networks, mobile or web applications etc.), which typically send data records simultaneously. In *interactiveprocessing*, the workloads can be executed anytime but its execution of a workload must be finished before their desired deadline. *Batchprocessing* is a type of data processing (large batches of data) which is needed to run all time i.e. 24 × 7 like Internet services, delay torrent etc. to execute user workloads.

There are seven types of batch processing: (1) bag of tasks, (2) HPC jobs, (3) HTC jobs, (4) scientific workloads, (5) parameter sweep tasks, (6) map-reduce tasks and (7) graph processing. *Bag-of-tasks* refers to the jobs that are parallel among which there are no dependencies, for example: video coding and encoding. *High-Performance Computing (HPC) jobs* in which single computer is used to solve large problems in business, engineering or science such as advanced application programs, which need to be executed efficiently, reliably and quickly. *High-Throughput Computing (HTC) jobs* in which large number of computing resources are running to finish the execution of a computational task. In *scientific workloads*, real workload activities can be simulated like weather prediction, flight control system etc. which requires high processing capacity to process it. *Parameter sweep tasks* are identical in their nature and differ only by the specific parameters used to execute them. *Map-reduce tasks* split the input data-set into independent chunks and parallel way of execution, which is used to execute the mapped tasks. Further, the outputs of the maps are sorted and used as an input to the reduce tasks. *Graph processing* involves the process of analyzing, storing and processing graphs to produce effective outputs. *Parallel processing* is an operation in which job is divided into small independent parts and execute all simultaneously on different processing nodes to improve the speed of an application. There are three types of parallel processing: (1) bit-level, (2) instruction level and (3) task parallelism. *Bit-level* parallel processing divides the job into the number of bits for execution, while instruction level parallel processing executed different instructions of a specific job. In *task parallelism*, a large task is divided into little tasks and execute them parallelly.

5   Gap Analysis and Future Research Directions

Researchers have done a large amount of research work in the area of application management for sustainable cloud computing but there are some research issues still pending to address. Following are research gaps as identified from the literature [2–7, 37–49] as discussed in Sect. 3.

5.1   *Application Modeling*

The future cloud applications should be developed with three-tiered architectures, which has three different layers: (i) database, (ii) application processing and (iii) user interface. To improve the reliability, simplicity and performance of an application, the functionality of every tier must be independent to execute at heterogeneous resources. The components of applications should follow loosely coupled design to decrease dependency among them, so that application can be migrated from one CDC to another without degrading its efficiency of their execution. Furthermore, data security should be provided to protect a data of e-commerce applications from unauthorized users. The future application of cloud computing can be smart home, smart city, smart dust, smart lipstick etc.

5.2 Resource Management

Resource management is the organized method of scheduling of available resources to the required customer's workloads over the Internet. Applications should be executed by allocating virtual resources in optimized manner and workload should be executed with minimum cost and time. Effective resource management in the virtual environment can improve resource utilization and user satisfaction. There is a problem of under-provisioning and over-provisioning of resources in existing resource allocation techniques. To overcome this problem, a QoS-aware resource management technique is required for efficient execution of cloud applications.

5.3 Energy Efficiency

To provide a reliable cloud service, it is required to identify that how the occurrences of failures affect the energy efficiency of cloud computing system. Moreover, it is necessary to save the checkpoints with the minimum overhead after predicting an occurrence of the failure. Therefore, workloads or VMs can be migrated to more reliable servers, which can save the energy consumption and time. Further, consolidation of multiple independent instances (web service or email) of an application can improve the energy efficiency, which improves the sustainability and availability of cloud service.

5.4 Reliability and Fault Tolerance

The prominent cloud providers such as Google, Facebook, Amazon and Microsoft are providing highly available cloud computing services using thousands of servers, which consists of multiple resources such as processors, network cards, storage devices and disk drives. With the growing adoption of cloud, CDCs are rapidly expanding their sizes and increasing the complexity of the systems, which increases the resource failures. The failure can be SLA violation, data corruption and loss and premature termination of an application's execution, which can degrade the performance of cloud service and affect the business. For next-generation clouds to be reliable, there is a need to identify the failures (hardware, service, software or resource), their causes and manages them to improve their reliability. To solve this problem, a model and system is required that introduces replication of services and their coordination to enable reliable delivery of cloud services in a cost-efficient manner for an execution of cloud application.

5.5 Security and Privacy

Real cloud failure traces can be used to perform the empirical or statistical analysis about failures to test the performance in terms of the security of the system. Security during migration of virtual machines from one CDC to another is also a significant issue because a state of a VM can be hijacked during its migration. To solve this problem, there is a need for encrypted data transfer to stop user account hijacking, which can provide a secure communication between user and provider. To improve the sustainability and reliability of cloud service to next level, homomorphic encryption methods can be used to provide security against malicious attacks like denial of service, password crack, data leakage, DNS spoofing and eavesdropping. Further, it is required to understand and address the causes of security threats such as VM level attacks, authentication and authorization and network-attack surface for efficient detection and prevention from cyber-attacks. Moreover, data leakage prevention applications can be used to secure data, which also improves the sustainability and reliability of cloud-based applications.

5.6 Scalability

The unplanned downtime can violate the SLA and affects the business of cloud providers. To solve this problem, a cloud computing system should incorporate dynamic scalability to fulfill the changing demand of user applications without the violation of SLA, which helps to improve the sustainability of cloud services during peak load.

5.7 Latency

Virtualization overhead and resource contention are two main problems in computing systems, which increases the response time. A reliability-aware computing system can minimize the problems for real-time applications such as video broadcast and video conference, which can reduce latency while transferring data to improve the sustainability of computing systems.

5.8 Data Management

Computing systems are also facing a challenge of data synchronization because data is stored geographically, which overloads the cloud service. To solve this problem, rapid elasticity can be used to find the overloaded cloud service and it adds new instances to handle the execution of current applications. Further, there is a need for an efficient data backup to recover the data in case of server downtime.

5.9 Auditing

To maintain health stability of the cloud service, there is a need for periodic auditing by third parties, which can improve the sustainability, reliability and protection of computing system for future cloud-based applications.

5.10 New Technological Developments

Cloud computing paradigm utilizes the Internet to provide on-demand services to cloud users and emerged as a backbone of a modern economy. Recent technological developments such as the Internet of Things, fog computing, software-defined clouds, big data and smart city are creating new research areas for cloud computing. There is need of the re-evaluation of existing application models of cloud computing to address research issues such as energy efficiency, sustainability, privacy and reliability.

5.10.1 IoT and Big Data for Smart Cities

The emerging big data and Internet of Things (IoT) applications such as smart cities, healthcare services etc. are increasing, which needs fast data processing to improve the sustainability of computing systems. However, these applications are facing large delay and response time because computing systems need to transfer data to the cloud and then cloud to an application, which affect its performance. Fog computing is a solution to reduce the latency, in which cloud is extended to the edge of the network. IoT environment using fog-assisted cloud computing for processing of data to make smarter decisions in a permitted time period. The data collected from different IoT devices have a large variety and volume (also known as Big Data), which also needs fog servers with high processing power. As a result of regular capturing and collection of datasets, they grow with the velocity of 250 MB/min or more. The continuous exchange of data in IoT environments is using for efficient decision making and real-time analytics for smart cities.

5.10.2   Software Defined Clouds

Software Defined Network (SDN) based software defined clouds can be used to provide secure communication during the execution of a user applications. Transport Layer Security (TLS)/Secure Sockets Layer (SSL) encryption techniques can also be used to provide secure communications between the controller(s) and OpenFlow switches, the configuration is very complex, and many vendors do not provide support of TLS in their OpenFlow switches by default. SDN security is critical since threats can degrade the availability, performance and sustainability of the network.

5.10.3   Fog Computing

The Fog computing paradigm offers a virtualized intermediate layer to provide data, computation, storage, and networking services between cloud datacenters and end users. The megatrend of Internet of Things (IoT) based real-time applications such as health monitoring, disaster management and traffic management requires lesser response time and latency to process user tasks. Therefore, fog computing is a solution to improve performance, in which cloud is extended to the edge of the network. Further, following open issues and challenges are required to be addressed to realize the full potential of fog-based application management for sustainable cloud computing.

- *Trade-off between Security and Reliability*: It is very difficult to incorporate security protocols in fog computing due to its distributed environment. One of the main security issues is calling authentication at different levels of fog devices. Trusted execution environment and Public-Key Infrastructure based on authentication solutions can provide a security to fog computing. To reduce authentication cost, rogue devices can be detected using measurement-based methods. Reliability is one of the main issues since fog computing is comprises of large number of geographically distributed devices. Reliability protocols for sensor networks can be used in case of failure of a cloud application, service platform, network and individual sensors. Reading of sensors can be affected by noise; concept of redundancy can be used to solve the problem of information accuracy.
- *Trade-off between Resource Utilization and QoS*: Fog devices have additional compute and storage power, but it is not possible for these devices to provide the resource capacity of cloud. Therefore, efficient resource management technique is required to process the application requests in a timely manner. To solve this problem, resource usage of user application should be predicted accurately in advance which can utilize resources efficiently. Moreover, existing resource management techniques in fog computing considers execution time only. In addition; required technique needs to consider the basic features of cloud computing in order to optimize the important QoS parameters such as execution time, energy consumption and network usage.

- *Trade-off between Latency and Power Consumption*: Fog environment consist of large number of fog devices in a distributed manner and computation may be consuming more energy than centralized cloud environment, hence it is an important research issue. Existing research reported that fog devices are more capable to reduce latency as compared to cloud by experiencing a little larger energy consumption during the execution of cloud-based applications. In fog computing system, trade-off between delay and power consumption is an open research area.

6 Summary and Conclusions

In this chapter, the survey of recent developments in application management for sustainable cloud computing has been presented. We identified the focus of study of existing techniques and proposed open research challenges. Based on the identified important open issues and focus of study, a taxonomy of application management for sustainable cloud computing has also been presented. Our taxonomy categorizes and investigates the existing research works based on their techniques towards addressing the research challenges. Moreover, we proposed some promising research directions based on the analysis, that can be pursued in the future.

Acknowledgements One of the authors, Dr. Sukhpal Singh Gill [Post Doctorate Research Fellow], gratefully acknowledges the Cloud Computing and Distributed Systems (CLOUDS) Laboratory, School of Computing and Information Systems, The University of Melbourne, Australia, for awarding him the Fellowship to carry out this research work. We thank SaraKardani Moghaddam and ShashikantIlager for their comments on improving the chapter.

References

1. Mastelic T, Oleksiak A, Claussen H, Brandic I, Pierson J, Vasilakos AV (2015) Cloud computing: survey on energy efficiency. ACM Comput Surv (CSUR) 47(2):1–36
2. Gill SS, Buyya R (2017) A taxonomy and future directions for sustainable cloud computing: 360 degree view, pp 1–35. arXiv:1712.02899
3. Ficco M, Rak M (2016) Economic denial of sustainability mitigation in cloud computing. Organizational innovation and change. Springer, Cham, pp 229–238
4. Li X, Jiang X, Garraghan P, Wu Z (2018) Holistic energy and failure aware workload scheduling in cloud datacenters. Future Gener Comput Syst 78:887–900
5. Gill SS, Chana I, Buyya R (2017) IoT based agriculture as a cloud and big data service: the beginning of digital India. J Organ End User Comput (JOEUC) 29(4):1–23
6. Singh S, Chana I (2016) QoS-aware autonomic resource management in cloud computing: a systematic review. ACM Comput Surv (CSUR) 48(3):1–46
7. Abbasi Z (2014) Sustainable cloud computing. PhD thesis, Arizona State University
8. Hazen BT, Skipper JB, Ezell JD, Boone CA (2016) Big data and predictive analytics for supply chain sustainability: a theory-driven research agenda. Comput Ind Eng 101:592–598
9. Bifulco F, Tregua M, Amitrano CC, D'Auria A (2016) ICT and sustainability in smart cities management. Int J Public Sect Manag 29(2):132–147

10. Rehman MH, Chang V, Batool A, Wah TY (2016) Big data reduction framework for value creation in sustainable enterprises. Int J Inf Manag 36(6):917–928
11. Cottrill CD, Derrible S (2015) Leveraging big data for the development of transport sustainability indicators. J Urban Technol 22(1):45–64
12. Yekini NA, Inyang-Udoh U, Doherty F (2016) Open educational resources (OER) for sustainable development using autonomic cloud computing system. Int J Eng Manuf 6(6):60–68
13. Maksimovic M (2017) The role of green internet of things (G-IoT) and big data in making cities smarter, safer and more sustainable. Int J Comput Digit Syst 6(4):175–184
14. Bradley R, Jawahir IS, Murrell N, Whitney J (2017) Parallel design of a product and internet of things (IoT) architecture to minimize the cost of utilizing big data (BD) for sustainable value creation. Procedia CIRP 61:58–62
15. Perera C, Zaslavsky A (2014) Improve the sustainability of internet of things through trading-based value creation. In: 2014 World forum on internet of things (WF-IoT). IEEE, pp 135–140
16. Zuo Y, Tao F, Nee AYC (2017) An internet of things and cloud-based approach for energy consumption evaluation and analysis for a product. Int J Comput Integr Manuf 1–12
17. Bossche RVD, Vanmechelen K, Broeckhove J (2013) Online cost-efficient scheduling of deadline-constrained workloads on hybrid clouds. Future Gener Comput Syst 29(4):973–985
18. Cappiello C, Melia P, Pernici B, Plebani P, Vitali M (2014) Sustainable choices for cloud applications: a focus on CO_2 emissions. In: 2nd international conference on ICT for sustainability (ICT4S), pp 352–358
19. Xia K, Gao L, Wang L, Li W, Chao K (2015) A semantic information services framework for sustainable WEEE management toward cloud-based remanufacturing. J Manuf Sci Eng 137(6):1–11
20. Zgheib R, Conchon E, Bastide R (2017) Engineering IoT healthcare applications: towards a semantic data driven sustainable architecture. In: eHealth 360°. Springer International Publishing, pp 407–418
21. Gupta PK, Maharaj BT, Malekian R (2017) A novel and secure IoT based cloud centric architecture to perform predictive analysis of users activities in sustainable health centres. Multimed Tools Appl 76(18):18489–18512
22. NoviFlow Inc. (2012) Green SDN: software defined networking in sustainable network solutions, pp 1–7
23. Pesch D, Rea S, Galdiz JIT, Zavrel V, Hensen JLM, Grimes D, O'Sullivan B (2017) ICT-Energy concepts for energy efficiency and sustainability. In: Globally optimised energy-efficient datacenters, pp 1–248
24. Chen M, Ma Y, Song J, Lai C, Hu B (2016) Smart clothing: connecting human with clouds and big data for sustainable health monitoring. Mob Netw Appl 21(5):825–845
25. Waga D, Rabah K (2014) Environmental conditions' big data management and cloud computing analytics for sustainable agriculture. World J Comput Appl Technol 2(3):73–81
26. Park JH, Kim HW, Jeong YS (2014) Efficiency sustainability resource visual simulator for clustered desktop virtualization based on cloud infrastructure. Sustainability 6(11):8079–8091
27. Ding K, Jiang P, Zheng M (2017) Environmental and economic sustainability-aware resource service scheduling for industrial product service systems. J Intell Manuf 28(6):1303–1316
28. Gmach D, Chen Y, Shah A, Rolia J, Bash C, Christian T, Sharma R (2010) Profiling sustainability of datacenters. In: IEEE international symposium on sustainable systems and technology (ISSST), pp 1–6
29. Islam MA, Ren S, Quan G, Shakir MZ, Vasilakos AV (2017) Water-constrained geographic load balancing in datacenters. IEEE Trans Cloud Comput 5(2):208–220
30. Dabbagh M, Hamdaoui B, Rayes A, Guizani M (2017) Shaving datacenter power demand peaks through energy storage and workload shifting control. IEEE Trans Cloud Comput 1–14
31. Garg SK, Yeo CH, Anandasivam A, Buyya R (2011) Environment-conscious scheduling of HPC applications on distributed cloud-oriented datacenters. J Parallel Distrib Comput 71(6):732–749
32. Cheng D, Jiang C, Zhou X (2014) Heterogeneity-aware workload placement and migration in distributed sustainable datacenters. In: IEEE 28th international parallel and distributed processing symposium, pp 307–316

33. Chen T, Zhang Y, Wang X, Giannakis GB (2016) Robust workload and energy management for sustainable datacenters. IEEE J Sel Areas Commun 34(3):651–664
34. Sehgal VK, Patrick A, Soni A, Rajput L (2015) Smart human security framework using internet of things, cloud and fog computing. In: Intelligent distributed computing. Springer, Cham, pp 251–263
35. Khosravi A, Buyya R (2018) Short-term prediction model to maximize renewable energy usage in cloud datacenters. In: Sustainable cloud and energy services. Springer, Cham, pp 203–218
36. Desthieux G, Carneiro C, Susini A, Abdennadher N, Boulmier A, Dubois A, Camponovo R (2018) Solar cadaster of Geneva: a decision support system for sustainable energy management. In: From science to society. Springer, Cham, pp 129–137
37. Moghaddam FA, Lago P, Grosso P (2015) Energy-efficient networking solutions in cloud-based environments: a systematic literature review. ACM Comput Surv (CSUR) 47(4):1–35
38. Subirats J, Guitart J (2015) Assessing and forecasting energy efficiency on cloud computing platforms. Future Gener Comput Syst 45:70–94
39. Fiandrino C, Kliazovich D, Bouvry P, Zomaya A (2017) Performance and energy efficiency metrics for communication systems of cloud computing datacenters. IEEE Trans Cloud Comput 5(4):738–750
40. Singh S, Chana I (2016) A survey on resource scheduling in cloud computing: issues and challenges. J Grid Comput 14(2):217–264
41. Singh S, Chana I (2015) Q-aware: quality of service based cloud resource provisioning. Comput Electr Eng 47:138–160
42. Singh S, Chana I (2016) Cloud resource provisioning: survey, status and future research directions. Knowl Inf Syst 49(3):1005–1069
43. Uddin M, Rahman AA (2012) Energy efficiency and low carbon enabler green IT framework for datacenters considering green metrics. Renew Sustain Energy Rev 16(6):4078–4094
44. Singh S, Chana I (2016) Resource provisioning and scheduling in clouds: QoS perspective. J Supercomput 72(3):926–960
45. Kramers A, Höjer M, Lövehagen N, Wangel J (2014) Smart sustainable cities–exploring ICT solutions for reduced energy use in cities. Environ Model Softw 56:52–62
46. Gill SS, Buyya R (2018) Resource provisioning based scheduling framework for execution of heterogeneous and clustered workloads in clouds: from fundamental to autonomic offering. J Grid Comput 1–33. https://doi.org/10.1007/s10723-017-9424-0
47. Singh S, Chana I, Singh M, Buyya R (2016) SOCCER: self-optimization of energy-efficient cloud resources. Clust Comput 19(4):1787–1800
48. Singh S, Chana I (2013) Consistency verification and quality assurance (CVQA) traceability framework for SaaS. In: IEEE 3rd international advance computing conference (IACC), pp 1–6
49. Wang T, Li Y, Wang G, Cao J, Bhuiyan MZA, Jia W (2017) Sustainable and efficient data collection from WSNs to cloud. IEEE Trans Sustain Comput 1–12. https://doi.org/10.1109/tsusc.2017.2690301

Auction Based Scheme for Resource Allotment in Cloud Computing

R. Bhan, A. Singh, R. Pamula and P. Faruki

Abstract In cloud computing resource allotment is one of the most demanding areas. Resources are attempt through fixed price model by the cloud provider and users, which is not much efficient and justified scenario. The optimal processing cost for each task by using resource bidding procedures which consider the impact of cost on long-term trade. Most of the existing resource allocation techniques focus on static task based allocation. The dynamic resource bidding price model based on auction is efficient and achieves optimal cost computation. The proposed dynamic model namely Double Auction Procurement Game for Resource Allocation (DAP-GRA) uses winner determination scheme for cost computation to achieve optimal resource allocation for tasks. The technique takes into account requirements of both users and Cloud Service Providers (CSPs) and calculates the final cost, based on the trade information. Results show that proposed schemes/mechanisms outperform other existing schemes/mechanism.

Keywords Cloud computing · Auction · Resource

R. Bhan (✉)
Department of Computer Science and Engineering, National Institute of Technology, Hamirpur, Hamirpur, India
e-mail: ratibhan@gmail.com

A. Singh
Department of Computer Science and Engineering, National Institute of Technology, Jalandhar, Jalandhar, India
e-mail: avtarz@gmail.com

R. Pamula
Department of Computer Science and Engineering, Indian Institute of Technology-ISM, Dhanbad, Dhanbad, India
e-mail: rajendrapamula@gmail.com

P. Faruki
Department of Information Technology, Sir BPTI Engineering College, Bhavnagar, India
e-mail: parvezfaruki.kg@gmail.com

© Springer International Publishing AG, part of Springer Nature 2019   119
S. Patnaik et al. (eds.), *Digital Business*, Lecture Notes on Data Engineering
and Communications Technologies 21, https://doi.org/10.1007/978-3-319-93940-7_5

1   Introduction

Cloud computing help us to solve the purpose of sharing resources, information, and software via the internet. Cloud computing and grid have several similarities in relations of characteristics and tasks. Together these models are used to improve the utilization of resources available thereby reduce the total cost. Cloud computing has various types of resources having different cost with respect to the service provider. The growth in the scope and several cloud users ask for healthy and effective cost models for mid and large size cloud vendors focused on market share. These days majority of pricing strategies for cloud resources are sort of usage based pricing, in which customers pay a static per unit price to access cloud resources. The pricing strategy to select cloud is difficult due to a large number of user's requirement and heterogeneous demand.

Auctions are generally referring products having low standard price rates, at a specific time their rate disturbs demand and supply. To resolve resource acquirement situation auction is a great tool. Everywhere in the global there exists infinite variety of virtual auction techniques. The case of auction model have mutual platform where user and provider act individually and freely. This work classifies them in two ways: the list of commodities and the list of participants such as venders and customers. The former indicates how several ranges of commodities like service etc. stay traded in an auction having similar time. The final indication which side competition occurs for the applicants.

Resources are available to cloud service providers and the user can only get those resources through bidding. The user can bid for resources because there are many service providers. Service providers are pricing the resources as per there term and conditions to gain more profit from the users. Therefore, there is a need to implement various pricing models to the cloud users.

The last section depicted the comparative study of fixed price model and dynamic pricing model using proposed algorithm. Twenty-five participants took place for biding and only four customers are capable to get resources form each supplier. Fixed pricing model fail to serve any customer by last supplier but dynamic pricing model perform better to reduce the total cost of resources for all customers served by every supplier. At last result show the bidding process for each type of resources where dynamic pricing model able to earn more revenue than fixed pricing model.

2   Related Work

The complexity of cloud computing is more complex as compared to grid computing proposed by Samimi et al. [1]. In a cloud environment, providers deal with services, whereas customers accept resources that track their tasks on the criteria of QoS and pricing requests. Due to uncertainty of the resource allocation in heterogeneous cloud computing paradigm. Marcos et al. [2] and Rodrigo et al. [3] proposed the resource allocation in optimal way is also a challenging task. Numerous researcher discovered that combine both model to allocate resource to find better solutions for market mechanism and economic models [4–9].

In heterogeneous cloud, resource allocated in system centric and user centric way. The performance of the system is optimized in such a way overall throughput which is the traditional way to find resources allocation in system centric method. The user centric method, focus on providing higher limit utilization for users based on QoS needs. There are two techniques for resource allocation in the economic model between various competing entities. One way of resource allocation is pricing and non-pricing on the basis of the economy model. The technique based market is a favorable method to bargain with difficulty while filling budget constraint. In recent time an explicitly combinatorial auction based scheme is evolved that have the ability to optimize allotment package of multiple commodities proposed by Zaman and Grosu [10].

Zhao et al. [11] suggested a cloud storage systems (e-commerce) based on auctions for resource pooling. Based on vender's requests, customer (truthfulness) usages a minimal cost task to control the shared instruction and reduce the expense cost of the resource sharing.

Wu et al. [12] presented self-organizer cloud for resource allotment using the automatic and scalable technique. The scheme support resource effectiveness between customer and vendor. Assumed customer bids planner for resource chooses on the basis of maximum price purchase and determines trader's resource with minimum cost to deliver the customer.

Khethavath et al. [13] presented distributed cloud architecture using effective resource discovery and ideal resource allotment. Praveen et al. [14] suggested in a distributed cloud based on game theory method to resource provisioning. Based on vendor's requests, customer estimates the corresponding service (resource) values. The vendor request for highest purchase resource service is chosen a winner then eligible for incentive cost.

Chard et al. [15] proposed Social Cloud (storage service) based on posted price in social networks. Based on archives vendor along with post price offer the vendor chose a customer based on SLA for the cloud application. The payment by customer take place through the bank to the vendor. The payment take place through bank between customer and vender.

Chard et al. [16] presented Social Cloud (storage service) based on reverse vickrey auction for resource sharing in the social cloud. Based on Vendors request the auction administrator chose customer with minimum cost as the winner. The expense engaged with payment procedure rules of reverse vickrey auction.

Khan [17] suggested resource sharing services (task) based framework for developing a market to share resources in the social cloud like a web portal. The broker calculates the score utility for the customer. The winner are choose and selected on the basis of vendor with top score utility. The expense involved with the cost procedure rules of reverse vickrey auction.

2.1   A Taxonomy of Market Technical Aspects

First, let us categories certain classic market techniques mention by Buyya's taxonomy [18].

i.   Posted Price

Commodity market internal shows the vender present special discount such as discounted products or fascinate customers with free gifts. The price posted suggestion model permit suppliers promote in an open way such an offerings are unequal along with particular conditions. In everyday life this type of shared commodity bazaar prototype is an extension.

ii.   Commodity Bazaar

The phrase bazaar possible commodity marketplace, where the vender's group cost for products and customers pay the price for it. An example farmer selling potatoes at Rs. 5 in a commodity bazaar for 1 kg; purchase 2 kg of these potatoes have to pay Rs. 10 in commodity market. The cost stated by the vender and change over period of time or not changing i.e. flat cost based on the stability of stock and supply. The sum in general proportion toward the amount of products. The commodity market prototype to a great extends direct and widespread; hence it viewed as start point to calculate the resource market home.

iii.   Bargaining (Nash Bargaining)

In various countries cost of products are not displayed at all as an alternative, shopkeeper and customer negotiate for cost with mutual agreement. Generally, vender begin with higher cost and customer with lower price. The trading fails after agreeing on fixed final payment. The negotiating process is upsetting when it comes to practical, when vender has to limit the problematic cost based on stock and supply.

iv.   Proportional Allotment

In some cases where demand goes beyond supply. The answer is to reduce quantity of allotment and increase the cost price. The resources are shared in proportion at a given cost to the customers. For illustration, the vender has 100 kg of pulses and customer A payment Rs. 60 and customer B payment Rs. 40. After payment the vender gives customer A 70 kg and buyer B 30 kg of pulses. It looked as a change in commodity bazaar, where vender provide cost control and customers get together to distribute the resource in a sensible way.

v.   Tendering

Public achievement by a government office is envisioned, for instance construction of a bridge. The detail of the essential product is state by government first. The enterprises then inspect and take decision regarding the bid or disregard it. The government gathers the bids, generally selects the low cost appropriate one and tender is given to the chosen enterprise have low cost. The enterprise received whole price after completion of work. This process is known as tendering and focus is to give agreeable cost for quality product and service.

Table 1 Market features technical aspects

	Vender:Customer	Price leader
Posted price	1:N	Vender
Commodity bazaar	1:N	Vender
Bargaining	1:1	both
Proportional allotment	1:N	Customer
Tendering	N:1	Customer
Auction	Vary	Auctioneer

Table 2 Type of Auction

Commodities		#Venders:#Customers	
		M:1or 1:N	M:N
	One	Single-sided auctions	Double-sided auctions
	Many	Combinatorial	

vi. **Auction**

A situation where many customers approach for one product that is not shareable. The solution modify the cost so that single customer needs to give that cost. The auction prototype fits in similar conditions to change the cost by challenge biddings between the customers. The process tender viewed alternative sort of auctions where the competition is among the venders. There are various auction sorting techniques considered in the next part. The auction prototype is a complex approach for choosing the cost in conditions of equality and optimality as a result, it is the most favorable technique to design a calculating resources market.

The upper mention market technical aspects are outline in Table 1.

2.2 Auction Theory

The market system is a sort of auction where cost negotiation is not at all predetermined neither attained, but is described by the execution of economic bidding. Normally, procedure is an auction permitting brokers to specify their concern either on one or many resources by using these suggestions to describe equally an allotment of resources and pay to brokers in group [19].

Auction permits many customers to bargain with one vendor via offering broker bids or alternative way. The broker groups the procedure of the auction and act as a judge. Negotiation carries on up to the time for achieving a single definite cost. Thus, auction corrects demand and supply created on the economic cost bid between venders and customers. The auction is often based on two approaches, summarizes in Table 2 namely single side auctions and double side actions.

2.2.1 Single-Sided Auctions

i. Dutch Auction

The cost importance goes low from highest to lowest in the Dutch auctions. The method is (1) broker states a highest primary cost and begins a timepiece that point to the reducing cost, (2) a customer pressing a key at a random time to end the timepiece, and (3) auction ends straightaway. The first customer pushes a key becomes the winner and must buy acquisition the commodity at that cost. For example flower bazaar in the Amsterdam is proper to limit the winner slowly possible.

ii. English Auction

The reader may be envision English auction unbiased using term auction since auction is broadly applied at the same time in traditional stocks and auctions at present time based on internet. There is one vender and the race happens between multiple customers in the English auctions. The technique is (1) the broker initial states the first cost of the commodity, (2) then customer increase cost by bidding (3) at last bidding process stopped when no additional bid by broker or reach specified fix period. The last bidder with the maximum cost suits the winner and needs to purchase the commodity at that cost.

iii. Japanese Auction

The cost rises by broker not buy customer in the Japanese auctions. The method is (1) the broker groups the first cost and begins to increase it at the time, (2) in every time a customer must state whether state they continue auction or not (3) auction completes when single customer left. The remained end customer turn into the winner and need to acquisition the commodity near to close cost. The customer left behind cannot come back to the same auction.

iv. Sealed Bid Auctions

The bid is kept secret from different customers. The biddings are not open to the public in sealed tender bargains inverse of auctions that are discussed above in three ways. The process is (1) customers present their "sealed" biddings straight to the broker, (2) these biddings are related with a broker, and (3) customer with the maximum bid turns out to be the winner and has to acquisition the commodity. The cost at which the winner acquisition the commodity is decided independently in many directions: (a) first cost bargains the winner's unpaid like personal offer, (b) next cost bargains the winner's unpaid like the another maximum offer and (c) kth cost bargains the winner's unpaid like kth maximum offer.

Cost bargaining also possible in English auction with grouping like Yahoo sell-off on internet. (1) where customer states his booking cost to a delegate broker, (2) delegate broker then saves bidding above the earlier maximum bid until the booking cost is touched, and (3) auction ends at established time frame. The maximum competitor turn out winner plus cost pays just for one entity above the next topmost applicant. The observe sort of auction are grouped in English auctions and next for sealed applicant auction.

v. **Reverse Auctions**

The above discussed four parts has different circumstances having one vender and various customers. The alternative one has studied: auctions with several venders and one customer. This type of family auctions is known as reverse auctions. An instance, Public finding having single customer such as local administration office and many venders like construction firms challenging each other to deliver the lowest cost for a commodity like bridge. The conversation on the non-reverse auctions is functional to the reverse bargain by opposing the costs and exchange the expression vender in lieu of customer, maximum in lieu of minimum and etc., lacking cost overview. So less unit's debate resolve no reverse bargain anymore.

2.2.2   Double-Sided Auctions

There are numerous customers and multiple venders in same period for one type of commodities. This determining is known as double-sided auctions. A classic instance stock bazaar, where several people are sharing resources to purchase or sell an enterprise's stock. Remark that double-sided bargain challenge in two ways for rivalry come at same time frame: one is customers and other is venders. The internet auctions are different in this case alike Yahoo bargains where vendors indirectly compete with another.

The two ways auctions method represented in double sided bargains: the continuous double auction (CDA) and the periodic double auction (PDA, recognized as stock clearing sell-off). An applicant place orders individually the CDA and the PDA at any time and several times as needed. All demand comprises a cost (buy aimed at maximum or vend aimed at minimum) and a quantity (buy aimed positive or vend aimed negative). The broker registers the arriving orders on the stock record.

The CDA and PDA have variance among when the trading happens in the same period. The demand arises fast in CDA, the broker tries to equalizer it versus the orders placed on the demand book. For instance, received order for 10 items buy are same versus accurate end demand for 4 items and additional vend demand for 6 items, as extended as the buying cost go beyond the vending cost. In circumstances of the partial equalizer, the unsold items are placed on the stock record.

The PDA assessment, the broker is unable to counter balance the order within period of time; vice versa simply place on the stock record. So, at certain predefined period of time, efforts counter balance the orders approximately likely. The Itayose1 procedure frequently utilized for the counter balance, via which ordered are ranked with supportive cost and meet quantity demand decision facts.

2.2.3 Combinatorial Auctions

It is the generic procedure of bargain, which permit sun like categories product to trade in the same time, not like the single side bargains debated in upper parts. Over all it is important for the brokers (customers and/or venders) whose assessments count on the group of commodities they cover, rather than the commodity itself. For instance, while using a computing machine, necessity to secure not the single computing machine but also its operating system that are linked to internet. Combinatorial auctions requirement in real global, consist of the bargains for, energy, shipping routes, corporate finding and radio range [19].

In easy way, study group brokers $N = \{1, \ldots,\}$ and group commodities $S = \{1, \ldots, m\}$. Allow $v = (v_1, \ldots, v_n)$ signify the valuation purposes of the dissimilar brokers, where valuation for every $i \in N$, $v_i : 2S \to \mathbb{R}$. There is a hypothesis that there are no expansion. In distinction from others, it is declared that a broker's valuation be determined by only on the group of commodities wins; particularly, do not recognize such a broker who always take care about other brokers allotments and payments. The combinatorial auctions broker purposes estimate have non preservative. Non-preservative are of two kinds: substitutability and complementarities.

Substitutability signifies shared value of numerous commodities is lower in count of their individual prices. For instance, suppose purchase a car having two option one is Honda and other is Swift. Since unable to drive these car at same time for individual is not possible, satisfying individual is lower than double as much as consider one; vice versa, Limitation of two cars alternates with all other. Whenever two commodities are exact alternates, assessment of their joint equals to the price for any single, that circumstance these things are viewed as multiple objects having one type of commodity.

Complementarities signifies shared value of several commodities is higher in count of their individual prices. For instance, suppose commanded to travel check-in (New Delhi) check-out (Bangalore) transit thru Bombay; requirement a fight ticket for the individual and merely not any of these else unable to complete task. A business development is considerable when worth of corporation is produced not for all job alone but with mutual result of the entire enterprise.

Combinatorial auctions have simple broker's policy advantage. In situation with no combinatorial bargain for instance, a broker request to organize a package of various commodities in singly manner have threat of incompletion. So the broker job is complex and risky. Using same auctions in the arrange form, a broker necessities to only express requirement for a package to the broker. Here no threat is yet finished as the allotment is guaranteed in the direction win-or-lose.

Basically, broker complexity reduces from the combinatorial auction and an alternative shifting this one to the buyer. Practically, auctions are significant topic for individual in current field of computer science and economics [19].

3 Market-Based Resource Management Model

Numerous research schemes are motivated on swapping of market for cloud and grid infrastructures. Between them, the maximum outstanding, which are associated with this work, are Mandi [20], Bellagio [21], SORMA [22], GridEcon [23], Ocean Exchange [24], Tycoon [25], SCDA [26, 27], CatNet [28, 29], Nimrod\G [30–32], Dynamic VM provisioning [33], Cloud Market [34], E commerce [11], Self-Organizer Cloud [12–14], Social Cloud (Posted Price) [15], Social Cloud (Reverse Vickrey Auction) [16] and Social Cloud (Web portal) [17] are describe in Table 3. Several existing schemes (namely Bellagio [21] and Tycoon [25]) have limited pricing and bargaining strategies. Auction allow only one kind of auction and retained at a fixed period.

A further generic market structural design like CatNet, GridEcon, SORMA and Ocean Exchange also provision individual one or two market prototypes like combinatorial auctions and mutual negotiation. In SORMA, bidding performed automatically to share an auction or negotiation with its supplier that may see large delays for customers who need resources on urgent basis. The GridEcon project proposes research in a feasible business prototype for trading facilities in the open bazaar. This model and Ocean Exchange allows only commodity market prototype, while CatNetcon firms only the negotiating and contract prototypes.

In earlier work, the option of market model is defined by the market itself. In Mandi system [20], the option of pricing and bargaining model is left to the customer and suppliers. This is essential as the selection of the market prototype and pricing (fixed, adjustable) alter from applicant to applicant depending on the usefulness gained. Tycoon [25] by Lai et al. proposed resource allocation assign on market based scheme built on proportional share, where resources are distributed in the proportion to the sum of the cost of customer uses.

It goals to permit customers to distinguish the value of their tasks in a cluster computing paradigm. Auction Share engaged by tycoon process to assign resources immediately and reliably. The customers are user brokers, the venders are servers in the cluster, the auctioneer is a procedure executing on the server, and the commodities to utilities resources as CPU cycles. Remark that there are many free auctioneers on each server. (1) The vender itself indexes there service site facility in trading process, (2) the customers bid for the resource on a retailer side, (3) the retailer limits share of the resource for the customers, and (4) the customers use the resource. The enterprise cloud paradigm is not supportive for Tycoon's strategy since its divide up single side auction prototype reasons an exposure threat for the customers who need a group of many resources.

Buyya et al. [30] proposed Nimrod/G a bargaining built grid scheduler expanded from Nimrod [32] constructed on the uppermost of Globus Toolkit [33]. Its objective to support researchers who run a limit package study on heterogeneous resources that are gather in the limited time along with economic constraint. Its services a commodity market prototype with the constraints of cost and time. The customer is end user and vendor act as resource supplier. The market is Nimrod/G elements and resources

Table 3 Market based resource management model

Name	Auction policy/Model	Centralized	Pricing	Customer/Vendor	Type of market and monitoring	Year
Mandi	Commodity bazaar, Single Side Auction, Double Side Auction	Yes	Static and dynamic pricing like spot cost	Discover, newcomer or offer in an auction, buying product in bazaar	Flexible	2011
Bellagio	Periodical and Combinatorial	Yes	K-pricing	Bidding	Job monitoring	2004
Sorma	Combinatorial	No	K-pricing	Bidding	Job monitoring	2007
GridEcon	Commodity bazaar, Double side	Yes	Static, k-pricing	Bidding	Job monitoring	2007
Ocean Exchange	Tendering/Bargaining	No	Static	Discover and negotiation	Job monitoring	2003
Nimrod\G	Commodity Market	No	–	–	Job monitoring	2002
Tycoon	Proportional sharing	No	–	–	Job monitoring	2005
SCDA	Double Side	Yes	K-pricing	Bidding	Job monitoring	2007
CatNet	Bargaining	No	Static and Dynamic	Bidding	Job monitoring	2005
Dynamic VM Provisioning	Combinatorial, Double Side	Yes	K-pricing	Bidding	Job monitoring	2013
Cloud Market	Double Side	Yes	K-pricing	Bidding	Job monitoring	2014
e-commerce	online reverse auction	Yes	Static	Bidding	Cloud Storage, customer Truthfulness	2015
Self-Organizer Cloud	Double Auction	Yes	Static and dynamic pricing like spot cost	Bidding	Resource effectiveness (Resources)	2016
distributed cloud architecture	Reserve Auction, kademlia protocol	Yes	Static and dynamic pricing like spot cost	Discover and negotiation	Resources as CPU, memory, bandwidth and QoS (i.e., a latency)	2013 2014
Social Cloud	Posted Price	Yes	Static	Bidding	Storage Services	2010
Social Cloud	Reverse Vickrey Auction	Yes	K-pricing	Bidding	Cloud Storage, customer truthfulness	2012
Social Network (Web portal)	Multi attribute Reverse Vickrey Auction	Yes	K-pricing	Bidding	Task Execution Services, customer truthfulness	2014

used to gain more customers. (1) The customer needs scheduler using the parametric engine to rearrange resource position which meets its cost and time limit, in trading process (2) the scheduler bargain with the resource suppliers and chooses lowest cost that satisfies bid, (3) customers message to particular task resource through scheduler. It is designed particularly for the parallel applications on the grid paradigm for this reason not suitable for business organization in the cloud paradigm. Where different applications are running in combination not using the same ones.

4 Objective and Design of Proposed Algorithm

The goal of resource allocation technique is to investigate optimal processing cost of each task that is achievable, using resource bidding. The following section presents the details of architecture for resource allocation in cloud computing.

4.1 Architecture of Cloud

The proposed algorithm "Double Auction Procurement Game for Resource Allocation" (DAPGRA) in the cloud is depicted in Fig. 1. The issue of resource allocation is resolved in the algorithm by using auction based resource allocation method and using a dynamic pricing prototype for evaluating the resources. Procurement auction is recognized as a reverse auction in which function of broker and customer is reversed. A customer requests for resources and supplier, typically bid designed for providing facilities to customers. The auction win leader has to deliver services to customers. The architecture contains several cloud resource providers and many user's resources to finish their task. The users submit their needs to the auctioneer. There are many types of resources offered to cloud resource providers. The user demands any quantity of resources from each particular sorts in categorize of a package. After receiving requests, the auctioneer sends messages to cloud providers and begin bidding. The bid considered to be completed only when sufficient resources are provided to the bidder or overbooking. Overbooking signifies that the cloud provider distinguishes the completion of the auction method, resources are allowed from any other user who is presently using resources, else go for the present auction procedure. Next bid providers, begin customers mapping and suppliers later calculating the minimum bid and maximum resource demanding user. For minimum bid provider have sufficient resources, assigned to the main customer. Else, check for the result resource provision.

4.1.1 Entities of Algorithm

User: User entity defines about requirement of resources to fulfill their needs to complete tasks. Then it goes to auctioneer or broker for increasing in value resource demand where it mentions a number of resources essential for each kind, time of resource necessity and its budget. The user informed when assigned specific resource by cloud resource supplier. When needs are fulfilled, the user pays the total valuation by the auctioneer.

Auctioneer or Broker: Auctioneer entity is the intermediate of entire auction method. The liability is to assign specific resource provider to the user. The broker finalized the charge needed to be given by the client. The customer raises needs and necessity to the broker by sending information for bidding to cloud resource

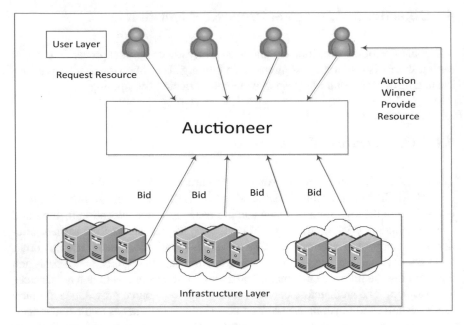

Fig. 1 Architecture of cloud

suppliers. The complete auction procedure is organized and achieved by broker. Auctioneer is answerable for checking auction winners based on the query.

Cloud Resource Provider: The entity discusses resources available to them. The resources namely CPU, Memory, storage etc., are available with the cloud provider. The supplier needs to trade these resources to earn maximum profits. They look up for the user to supplier services. For the provider, take part in auction procedure when the auctioneer directs messages. When the supplier has sufficient resources to take part and bid for resources. Once wins the auction bid, offer services to other customers and earn gains. The provider participates in several auction processes allowing to obtainable resources without any limit. The only circumstance when commitment is done then they have to provide services. Figure 2 depicts the relationship between different participants of the algorithm.

4.1.2 DAPGRA Parameters

- Let there be a N number of users and n represents a user number, $N = \{1, 2, 3, \ldots, n\}$.
- Consider the M number of Cloud resource providers and m represents a cloud resource provider number. $M = \{1, 2, 3, \ldots, m\}$.
- Total number of resource types are K and $K = \{1, 2, 3, \ldots, k\}$.

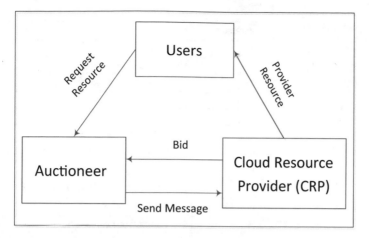

Fig. 2 Relationship between entities

- Each user asks for the different amount of each resource type. Every cloud provider has also different amount of each resource type.
- The user requirement vector is represented as
- $A_n = \{a_{1n}, a_{2n}, a_{3n}, \ldots, a_{tn}, bn\}$ where tn represents the time for which user require resources on per hour basis and bn is the budget of the user.
- When cloud resource provider bid, it has to advertise resources available represented by vector $Qm = \{q1m, q2m, q3m, \ldots, qkm\}$ and price for each resource type as $Pm = \{p1m, p2m, p3m, \ldots, pkm\}$.
- Weight vector W of resources representing different weight for each resource are represented as $W = \{w1, w2, \ldots, k\}$.

Table 4 shows the complete list with various parameters used in the algorithm model for calculating the auction winner and price calculation are described.

4.1.3 Winner Determination

The user request resources from auctioneer then go to cloud resource suppliers for bidding. After receiving bids from the supplier, auctioneer calculates winners of the auction. Auctioneer start calculating the total weight of resources requested by the user which are denoted by Xbt as described in Eq. 1.

$$Xbt_i = \sum_{k=1}^{n} (a_i w_i) \tag{1}$$

Then calculate per resource bid value of user as given by Eq. 2.

Table 4 Parameters of algorithm

Parameter	Description
N	Count of users
n	User count at current
M	Count of Cloud Resource Suppliers
m	Provider Number
K	Total number of different resources
A	User requirement vector
Q	Quantities available with provider vector
P	Provider bid price vector
t	Time for which user requires resources
b	Budget of the user
Xb	Per resource bid value of user
Xbt	Time multiplied by per resource bid value
u	Total bid of provider
Mb	Per resource bid value of provider
W	Weighted vector of resources
w	Weight of each resource
wt	Weighted total of resources
Up	Price paid by user
Pw	Price by ultimate winner provider
Pd	Price by defaulter provider

$$Xb_n = \frac{b_n}{Xb_t} \tag{2}$$

Next per resource bid value are multiplied by the time for which users require resources as to give more preference to users who require resources for a long time which ultimately generate a profit for the provider as explained in Eq. 3.

$$Xb_{tn} = Xb_n \times t_n \tag{3}$$

The users who require resources for more time that get higher value and the list is sorted in descending direction signifies that a user with greater bid density is allocated in priority. Maximum value user comes in first place. In the similar fashion, provider per resource bid density is calculated and then the list are sorted in ascending order using Eq. 4.

$$Mb_m = \frac{\sum_{i=1}^{k}(p_i)}{K} \tag{4}$$

In the sorted list, cloud supplier start with the lowest bidder in the auction. If enough resources for the first user, the mapping is done by the first user and the corresponding cloud supplier. Otherwise, check for the next cloud supplier. Iterate through all suppliers until user's needs are satisfied. The same process is followed for all users. The pay price of the user is described in the next section.

4.1.4 Price Calculation

The user is mapped to the supplier. Then users have to pay prices, which is calculated in Eq. 5.

$$U p_n = \left(\frac{b_i}{\left(\sum_{i=1}^{k} (a_i \times t_n) \right)} + M b_m \right) \times t_n \times \sum_{i=1}^{k} a_i \tag{5}$$

The total price calculates by the auctioneer are communicated to the user which have to pay at the end of completion of tasks to the individual supplier.

4.1.5 Execution Steps

   I. The section explain the task execution steps for the cloud provider is as follows:
  II. All cloud providers initially register themselves with the broker or auctioneer so that in the near future they send bid in the auction.
 III. Users at a given point of time send their requirements to the auctioneer. In the requirement message, they recollect their budget and time for which require resources.
  IV. The auctioneer collect the list of all the user's requirements and send the messages to all registered providers for bidding.
   V. Auctioneer activate the timer. Providers who bid within the time limit are considered for the current auction process.
  VI. Cloud provider after receiving a message from auctioneer check his resources. If resources available or overbooking, would bid for each type of resource and recollect the quantity of resources available.
 VII. Auctioneer after completing the time, stop the bidding process. Later on, if any provider bids, are not considered for the auction process.
VIII. Auctioneer start calculating process. Initially, calculate the total weight of requirements of each user according to Eq. 1. Then it finds per resource bidding value as mentioned in Eq. 2.
  IX. The preference is given to users who require resources for a longer duration, above calculated value are multiplied by the time for which resources are required using Eq. 3.
   X. The sorted user's list prepared are in decreasing order. The way users require resources for longer durations are placed in the starting.

XI.   The similar approach, using Eq. 4, per resource value are calculated for cloud
       providers.
XII.   The list is sorted in increasing numbers as a provider with minimum bidding
       come at the starting position.
XIII. Mapping of users and cloud providers are done and is calculated using Eq. 5.
       If the first cloud provider has enough available resources, the user is mapped
       otherwise check for next provider. After satisfying the needs of one user if left
       with enough resources, is compete for next user resource otherwise removed
       from the list.
XIV.   All user's needs are satisfied. The provider initiates virtual machine to fulfill
       a user's need.
XV.   When all the needs of users are satisfied, the auction process is closed. The
       sequence order of the execution flow of the DAPGRA algorithm depicts pay
       price.

4.2 Proposed Resource Allocation Algorithm

The Fig. 3 shows the proposed Algorithm of each resource provided for bidding at
given time.
   Input: Registered Provider_List
   Output: Auction winners and price matrix
   Above algorithm is also described with the help of flow chart in Appendix 1.

4.3 Simulation Parameters

The proposed auction based resource allotment algorithm implemented in the
CloudSim toolkit by impending the framework. Initially auction procedure requests
for the number of customer applicants and cloud suppliers participants in the auction.
Assume four sorts of VM (VM1, VM2, VM3, and VM4) namely small, medium,
large, and huge are created to provide resources to the users. The parameters of different VM were initialized accordingly so user requests resources in package form.
The range of resources has also been specified and based on these VM parameters
are set up.
   Four specifications are measured namely, processor speed, storage, memory size,
and bandwidth. The assortment of memory size is (256, 512, 1024, 2048) MB,
storage (2000, 40000) MB, bandwidth (150, 1500) b/s and processor sort is (320,
1200). Creation of VM and scenario of twenty-five participant's resources from the
auctioneer and cloud suppliers are ten at present the auction signifies they bid for
the current auction process. All market-based resource allocation model, enclosed
a resource attribute, type, and quantity. In the scenario, VMs are recognized to be

Algorithm	Double Auction Procurement Game Resource Allocation Algorithm (DAPGRA)
Step 1	Set i, j, temp ← 0
Step 2	Users demand resources to the broker and notify their financial plan and time for which resources are essential.
Step 3	Broker sends a message to cloud resource provider for bidding with an activated timer.

Step 4	Cloud resource suppliers check the availability or overbooking of their resources.
	If sufficient resources available
	    Begin
	      Bid for current auction
	      recollect total available resources
	  End
	    Else
	    Begin
	    Do not bid for current auction
	  End

Step 5	Auctioneer close bidding process and deactivate the timer.

Step 6	Find auction winners
	Calculate resource bid value per user according to equation 1.1 and equation 1.4.
	Multiply the time to resource bid value per user according to equation 1.3.
	Sort users list in descending direction and provider in higher order and map customers and suppliers.
	Set n ← 1
	for i = 1 to M
	for j ← 1 to k
	if (R[m][j]) <A[n][j])
	Begin
	i++;
	break;
	End
	End for
	temp ← i;
	break;
	End for
	Map nth user with temp provider.
	Decrement provider resources
	m ← temp
	for j ← 1 to k
	R[m][j] = R[m][j] − A[n][j]
	End for
	Set n ← n + 1 and go to step e)

Step 7	Calculate price according to equation 1.5.
Step 8	Notify Cloud resource provider rough auction results.

Step 9	Cloud resource provider begin completing user task
	 If not sufficient resource available
	  Begin
	  Notify auctioneer
	  Assign user task to next provider in the list and go to step 6(e)
	    Provider pay price according to equation 1.5
	  End
	  Else if
	  Begin
	    Start VM completion of the task.
	    Complete user task and end VM.
	    User pay price.
	  End

Step 10	Close auction process.

Fig. 3 Proposed algorithm of auction process

Table 5 List of parameters for broker

Parameters	Descriptions
Broker	Broker id
VM	The features of demanded VMs
CPU	The need of computer processor's speed and power, which is measured in millions of instructions per second (MIPS)
Memory	The demand quantity of ram in MB
Storage	The requirement quantity of storage in MB
BW	The need number of bandwidth in bits per second
Quantity	The number of requested VMs
Bid	The amount of money that a customer pay for the requested items in Cost/MI
Time	The interval that a customer occupies the resources

Table 6 List of parameters for cloud providers

Parameters	Descriptions
Cloud provider	Cloud provider id
VM	The features of the current VM
CPU	The presented computer processor's speed and power, which is measured in millions of instructions per second (MIPS)
Memory	The presented quantity of ram in megabytes (MBs)
Storage	The presented quantity of storage in MBs
BW	The presented quantity of bandwidth in bits per second
Quantity	The count of presented VMs
Bid	The amount of money that a customer sell for presented items in Cost/MI

a resource type, and the VM features are study as the aspects, which cover the following:

1. Computer processor's quickness and power, which is roughly measured in Millions of Instructions Per Second (MIPS).
2. Memory that signified the total of ram in Mega Bytes (MBs).
3. Storage that represented the volume of storage in Mega Bytes (MBs).
4. Bandwidth that signified the quantity of bandwidth in Bits per Second (B/S).

The lists of limitations for brokers and cloud suppliers are shown in Table 5 and Table 6 respectively. The used attribute's ranges according to the ranges.

The range of processor quickness are recognize to be (250, 1200) MIPS, the memory range (256, 512, 1024, 2048) MB, the storage volume limits (2000, 40000) MB, the bandwidth in among (150, 1500) B/S, the bid limits Rs (0.014–0.1023)

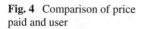

Fig. 4 Comparison of price paid and user

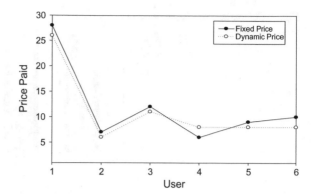

cost/MI (bids were offered in the units of cost per Million of Instructions), and the time interval (5, 60) min. Four sorts of VMs were measured in the model, but the count rise without limit. Further, the machine is scalable because the count of brokers and cloud suppliers are improved without limit.

5 Result and Discussion

The scenario for a resource allotment, a supplier had twenty-five participants. Supplier one is capable to meet allocates resources among four customers. Likewise, others are able to meet allocate resources among various providers, but the last supplier has failed in the auction to serve any customer with allocated resources. It occurs because of very high bidding importance or less number of resources obtainable. The present auction method is successful because it has been capable to satisfy user's needs.

The price for several scenarios is calculated by using both fixed price model and dynamic pricing model using the designed algorithm. The results are depicted in Fig. 4. The graph shows that as compared to fixed price model, in most of the cases dynamic model provides benefit by lowering the total resource cost for the users. In a few cases, the results of the fixed price model are better. The graph shows that dynamic pricing model is beneficial as it is cost saving for the user. Figure 5 depicts the price bid by four providers for each resource type and the trade price which is the price of actual selling of resources. Trade price is generally more than the bidding value of the provider due to which they earn more revenue in the dynamic pricing model.

Fig. 5 Comparison of bid price and trade price

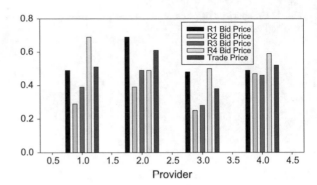

6 Conclusion

In cloud computing resource allotment is one of the most challenging areas. The resources are shared at different costs which depend on the bidding of individual customers and cloud service providers, which is not much efficient and justified assumption. To resolve the resource allotment issue, auction based algorithms extend benefits equally to the customer and cloud resource provider. The technique helps to achieve the dynamic prizing prototype built on an economic framework. Proposed auction based process has been observed that proficiently assigns usable resources of the cloud supplier and fulfills the customer needs. The proposed auction based algorithm executed using the CloudSim and price has been calculated dynamically. The result indicate that proposed algorithm are valuable for cloud providers as they make more profits and fulfills the customers for their optimal demands and resources allotment. The further implementation of auction model based on dynamic resource bidding is important to reduce computational cost and achieve efficient algorithms for Resource Allotment in Cloud Computing.

Appendix: Flow Chart of the Proposed Auction Algorithm

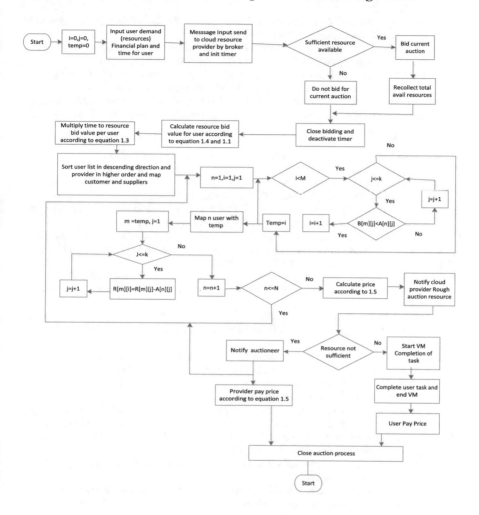

References

1. Samimi P, Teimouri Y, Mukhtar M (2011) A combinatorial double auction resource allocation model in cloud computing. In: IEEE symposium on computers and informatics, pp 634–639
2. Marcos DA, Alexandre C, Buyya R (2010) A cost benefit analysis of using cloud computing to extend the capacity of clusters. Clust Comput 13(3):335–347
3. Rodrigo NC, Ranjan R, Beloglazov A, De Rose CAF, Buyya R (2011) CloudSim: a toolkit for modeling and simulation of cloud computing environments and evaluation of resource provisioning algorithms. J Softw Pract Exp Arch 41(1):23–50
4. Fujiwara I, Aida K, Ono I (2009) Market-based resource allocation for distributed computing. IPSJ SIG technical report, vol 2009-HPC-121, No 34

5. Li C, Li LY (2012) Optimal resource provisioning for cloud computing. J Supercomput 62(2):989–1022
6. Sun M, Zang T, Xu X, Wang R (2013) Consumer-centered cloud services selection using AHP. In: International conference on service sciences, pp 1–6
7. Zhu Q, Agrawal G (2012) Resource provisioning with budget constraints for adaptive applications in cloud environments. IEEE Trans Serv Comput 5(4):497–511
8. Du L (2012) Pricing and resource allocation in a cloud computing market. In: IEEE/ACM international symposium on cluster, cloud and grid computing, pp 817–813
9. Bonacquisto P, Modica GD, Petralia G, Tomarchio O (2014) A strategy to optimize resource allocation in auction-based cloud markets. In: IEEE international conference on services computing, pp 339–346
10. Zaman S, Grosu D (2013) A combinatorial auction-based mechanism for dynamic VM provisioning and allocation in clouds. IEEE Trans Cloud Comput 1(2):129–141
11. Zhao J, Chu X, Liu H, Leung YW, Li Z (2015) Online procurement auctions for resource pooling in client-assisted cloud storage systems. In: Proceedings of the IEEE conference on computing communication (INFOCOM), Hong Kong, pp 576–584
12. Wu X, Liu M, Dou W, Gao L, Yu S (2016) A scalable and automatic mechanism for resource allocation in self-organizing cloud. Peer-to-Peer Netw Appl 9(1):28–41
13. Khethavath P, Thomas J, Chan-Tin E, Liu H (2013) Introducing a distributed cloud architecture with efficient resource discovery and optimal resource allocation. In: Proceedings of the 9th world congress on services (SERVICES), pp 386–392
14. Praveen K, Thomas J, Liu H (2014) Game theoretic approach to resource provisioning in a distributed cloud. In: Proceedings of the international conference on data science and engineering (ICDSE), pp 51–56
15. Chard K, Caton S, Rana OF, Bubendorfer K (2010) Social cloud: cloud computing in social networks. In: Proceedings of the IEEE 3rd international conference on cloud computing (CLOUD), pp 99–106
16. Chard K, Bubendorfer K, Caton S, Rana OF (2012) Social cloud computing: a vision for socially motivated resource sharing. IEEE Trans Serv Comput 5(4):551–563
17. Khan MA (2014) A service framework for emerging markets. In: Proceedings of the 21st international conference on telecommunication (ICT), pp 272–276
18. Buyya R, Abramson D, Giddy J, Stockinger H (2002) Economic models for resource management and scheduling in Grid computing. Concurr Comput: Pract Exp 14:1507–1542
19. Shoham Y, Leyton-Brown K (2009) Multibroker systems: algorithmic, game-theoretic, and logical foundations. Cambridge University Press
20. Garg SK, Vecchiola C, Buyya R (2011) Mandi: a market exchange for trading utility and cloud computing services. Springer
21. AuYoung A, Chun B, Snoeren A, Vahdat A (2004) Resource allocation in federated distributed computing infrastructures. In: Proceedings of the 1st workshop on operating system and architectural support for the on-demand IT infrastructure, NV, USA
22. Neumann D, Stoesser J, Anandasivam A, Borissov N (2007) Sorma-building an open grid market for grid resource allocation. In: Proceedings of the 4th international workshop on grid economics and business models, France
23. Altmann J, Courcoubetis C, Darlington J, Cohen J (2007) GridEcon-the economic-enhanced next generation internet. In: Proceedings of the 4th international workshop on grid economics and business models, Rennes, France
24. Padala P, Harrison C, Pelfort N, Jansen E, Frank M, Chokkareddy C (2003) OCEAN: the open computation exchange and arbitration network, a market approach to meta computing. In: Proceedings of the 2nd international symposium on parallel and distributed computing, Ljubljana, Slovenia
25. Lai K, Rasmusson L, Adar E, Zhang L, Huberman BA (2005) Tycoon: an implementation of a distributed, market-based resource allocation system. Multiagent Grid Syst 1:169–182
26. Tan Z, Gurd JR (2007) Market-based grid resource allocation using a stable continuous double auction. In: 8th IEEE/ACM International Conference on Grid Computing, pp 283–290

27. Zhu T, Gurd JR (2007) Market-based grid resource allocation using a stable continuous double auction (PhD thesis)
28. Schnizler B, Neumann D, Veit D, Reinicke M, Streitberger W (2005) Theoretical and computational basis for CATNETS—annual report year 1
29. Veit D, Buss G, Schnizler B, Neumann D (2007) Theoretical and computational basis for CATNETS—annual report year 3
30. Buyya R, Abramson D, Giddy J (2000) Nimrod/G: an architecture for a resource management and scheduling system in a global computational grid. Computer 1:283–289
31. Abramson D, Foster I, Giddy J, Lewis A, Sosic R, Sutherst R, White N (1997) The Nimrod computational workbench: a case study in desktop metacomputing. Aust Comput Sci Commun 19(1997):17–26
32. Foster I, Kesselman C (1997) Globus: a metacomputing infrastructure toolkit. Int J Supercomput Appl 11:115–128
33. Bonacquisto P, Modica GD, Petralia G, Tomarchio O (2014) A strategy to optimize resource allocation in auction-based cloud markets. In: IEEE international conference on services computing, pp 339–346
34. Zaman S, Grosu D (2013) A combinatorial auction-based mechanism for dynamic VM provisioning and allocation in clouds. IEEE Trans Cloud Comput 1(2):129–141

M-Cloud Computing Based Agriculture Management System

Vinay Kumar Jain, Shishir Kumar and Prabhat K. Mahanti

Abstract Modernization in agriculture sector is one of the major challenging problems in India. Currently, Indian farmers faced many problems in agriculture domain such as lack of irrigation infrastructure, market infrastructure and transport infrastructure along the presence of a chain of middlemen through whom most agricultural commodities must circulate before finally reaching consumers etc. One of the possible solutions for improvement is by using mobile applications help in gathering information from farmers such location-based information and environmental. This chapter presents a framework to solve the problems in agriculture by using Mobile based Cloud computing platform which makes smart farmers and increases the productivity. This framework promotes a fast development of agricultural modernization, realize smart agriculture and effectively solve the problems concerning agriculture, countryside, and farmers.

Keywords Cloud computing · Distributed system · Agriculture
Mobile computing · Bigdata

1 Introduction

Agriculture is an important sector in India with more than 70% of the Indian population living in rural areas and earns its live hood by agriculture and allied means of income. The area faces significant difficulties of upgrading creation in a circumstance of decreasing regular assets essential for generation [1]. This sector will have

V. K. Jain (✉) · S. Kumar
Jaypee University of Engineering and Technology, Guna, India
e-mail: vinay2588@gmail.com

S. Kumar
e-mail: dr.shishir@yahoo.com

P. K. Mahanti
University of New Brunswick, Saint John, Canada
e-mail: pkmahanti@yahoo.co.in

© Springer International Publishing AG, part of Springer Nature 2019
S. Patnaik et al. (eds.), *Digital Business*, Lecture Notes on Data Engineering
and Communications Technologies 21, https://doi.org/10.1007/978-3-319-93940-7_6

characteristics like Spatiality, Complexity, Enormousness, Dynamics and number of factors that affect production and quality and will be more as it is different for each crop. The environment of Indian farmland is an extremely confused nature and includes numerous sorts of elements from the environment to human, from environment to monetary, from topography to society and so on. The method for collecting data usually incurs the substantial costs and technologies [2]. The growing demand for agricultural products, however, also offers opportunities for producers to sustain and improve their livelihoods. But due to the wide gap between what the technologist gets in the experimental farm and what a farmer gets on his farm are different and it creates a wide gap between the "best-practices" and the common run of farmers.

Data and correspondence innovations (ICT) assume a key part in tending to different difficulties and elevating the jobs of the country agriculturists. It offers a chance to present new exercises, new administrations, and applications into rustic regions or to upgrade existing administrations.

The role of Information Technology (IT) to develop agricultural education, research, and extension to improve the quality of life in the rural area is well established in India. It can help an average Indian farmer to get relevant information regarding agro-inputs, crop production technologies, agro-processing, market support, agro-finance and management of farm agri-business etc. [3]. Currently, Indian farmers are looked with narrowing overall revenues expenses of numerous sources of info like composts and fuel have expanded, while item costs have remained genuinely steady or even dropped [4]. Expanded globalization and market deregulation increment weights on numerous smallholder farmers in creating nations. So as to utilize full impact of these worldwide changes, strategies of rural items, showcasing, valuing, and exchange must be returned to them.

Mobile phone infiltration has been becoming quickly even in the remote rustic regions. The phenomenal speed of selection of cell phone innovation has raised the general assumptions about its potential commitments to spreading of imaginative cultivating innovation. Mobiles based technology provides low-cost communication system which can help in resolving farmer's problems and provide a medium for educating farmers.

Big data is expected to play significant roles in agriculture by utilizing M-cloud computing arm. It will provide weather forecasting, real-time optimization, real-time information resources for farmers, intelligent irrigation recommendations, price monitoring, logistic management, decision making on inventories and budgets. With the availability of modern ICT based technologies, we have an opportunity in agriculture to make more informed decisions based on the data.

Researchers have applied big data analytics to agricultural data and weather records, revealing how climate variation impacts their crop yields.

Rapid developments in the Internet of Things (IOT), Big data technologies and Cloud Computing are propelling the phenomenon called Smart Farming. Both Big Data and Smart Farming are relatively new concepts, so it is expected that knowledge about their applications and their implications for research and development is not widely spread. This chapter provides brief reviews of new concepts and technologies used in smart agriculture management system.

2 Problem in Agriculture

Over the years, Indian rural farmers depend on home-grown practices or local knowledge for improved farming system. These practices refer to skills and experience gained through old tradition and practice over many generations and not much helped in improving agricultural yield. The major problem where ground level improvement is needed is the emergence of new crops, detection of animal diseases, plant weeds, pests, quality of fertilizers etc. The second problem was to avoid middlemen who maintained a stronghold on the market scene. India rural farmers are much depended on middleman (broker) because they provide essential resources like quick credit, non-bureaucratic and quick payment for goods [5]. They stay basic for items that require time, stockpiling, space and vitality contributions, for instance for items that must be dried, put away, transported, handled and bundled before circulation. As a rule, these wares are sold and purchased a few times, including an incentive at each progression, before achieving the buyer. The innovation and back to play out these capacities are for the most part past the span of low-pay agriculturists and are left to agents who have the assets. Some other important problems are also faced by Indian farmers are:

2.1 Linkage Problems

The lack of a close working relationship between national agricultural research and extension organizations, and with various classes of farmers and homestead associations, is a standout amongst the most troublesome institutional issues defying services of farming in numerous creating countries. Linkage systems are utilized to channel data amongst gatherings and to arrange required errands during the time spent getting pertinent advances to agriculturists.

2.2 Weather Information

Weather and climate are the biggest risk factors which impacting farming performance and management. Extreme weather and climate events such as floods, droughts, or temperature impact on decline agricultural production. Factors, for example, unreasonable precipitation and huge change in temperature add to the powerlessness of individual homesteads, and on entire country groups [6].

2.3 Transportation and Infrastructure

In urban areas, foundation and transportation are typically all around grew, yet not in remote zones where poor framework and high expenses of transportation are obstructions for potential market contestants, prompting a less focused market condition.

2.4 Agriculture Marketing

The most common problems faced by small Indian farmers are the lack of market information on prices and factors which influencing the market prices.

2.5 Credit Facilities

For credit facilities, Indian farmers have a tendency to depend on relatives, merchants, and brokers. Scarcely any ranchers obtain from formal budgetary establishments like banks and government societies but a large number of farmers lack due to their illiteracy.

2.6 Storage of Food Grains

Storage of food grains is a big problem. Almost 10% of our gather goes squander each year without legitimate storerooms [7].

3 Cloud Computing

Computing Processing is being changed to a model comprising of administrations that are commoditized and conveyed in a way like customary utilities, for example, water, power, gas, and communication. In such a model, clients get to administrations in light of their prerequisites without respect to where the administrations are facilitated or how they are conveyed. The term Cloud Computing is moderately new however the idea driving the expression has been around for quite a while [5]. Such Cloud applications as programming as-a-benefit, for instance, are not especially new [6] and over-the-net applications have been accessible for a long time. "Hotmail" or Web hosting sites are well-known Cloud Computing applications par excellence.

While there are numerous meanings of Cloud Computing [7], the most straightforward one characterizes Cloud Computing as getting PC administrations/assets from the Internet as opposed to from neighborhood singular stages. Cloud is a known allegory for depicting the Internet. Platforms in Cloud Computing are [4]:

- Software-as-a-Service (SaaS)—software application services obtained from the Internet.
- Platform-as-a-Service (PaaS)—the user utilizes the Internet as a computing platform, rather than having his own individual, localized platform.
- Infrastructure-as-a-Service (IaaS)—a computing infrastructure based on the Internet rather than local servers.

Cloud computing is a significant alternative in today's agricultural perspective. The technology gives the farmers and analyst the opportunity to quickly access various application platforms and resources through the web pages on-demand.

Cloud computing with mobile phones provides much more ease for the user because of low cost and high performance at remote locations. Mobile based farming generation administration framework will give far reaching apparatuses to organization, creation, and deals. Agriculturists will have the capacity to gather, store and investigate information about on-cultivate activities, trim plantings and yields by means of mobile phones [4].

4 Agriculture with M-Cloud Computing

At the most theoretical level, technology is the application of knowledge for practical purposes. By and large, innovation is utilized to enhance the human condition, the indigenous habitat, or to complete other financial exercises. Technology can be classified into two major categories [8]:

(i) *Material technology*

It is exemplified in a mechanical item, for example, devices, hardware, enhanced plant assortments or cross breeds, agrochemicals, enhanced types of creatures and immunizations.

(ii) *Knowledge-based technology*

It refers to management skills and technical knowledge which help farmers to successfully grow a crop or produce animal products.

Mobile agriculture computing delivers low-cost timely information that helps Indian farmers to understand and analyze market prices, facilitates trade, weather reports and credit facilities. Figure 1 illustrated an M-cloud computing based agricultural system. M-cloud Computing also reduces transaction time, travel, and costs by bridging distances and allowing for a more effective use of time.

Utilizing this communications help to promote social networks and communities to progress in health, safety, employment, recreation, and other areas. Major roles and methods which can improve agriculture growth are illustrated in Table 1. One of the major improvement is it increases the levels of community participation, facilitating

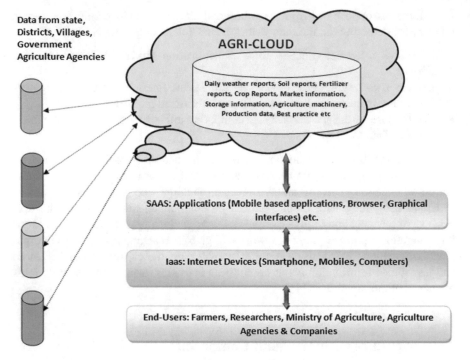

Data from state,
Districts, Villages,
Government
Agriculture Agencies

Fig. 1 M-cloud computing based agricultural system

Table 1 Role and methods in agriculture using M-Cloud Computing

Roles	Methods
Data Collection	Application based data collection from every geographic region
Education and Awareness	Information such as good practices, crop varieties or pest and disease management using images or videos will be given to farmers using mobile phones
Market Information	Periodically provide information on market for helping farmers in decision making
Transport Facilities	GPS-based Mobile services
Disease outbreak	Precaution steps can be given using mobiles phones

an informed decision-making process, particularly greater participation from rural women.

Some are the example illustrated which can lead to the rural farmer communities in improvement in their agriculture production.

Fig. 2 An illustration of
multiple crop diseases
present in Farm Cloud

4.1 Crop Disease Identification Management

Cloud computing can we use by mobile like a farmer could take a photo of a yield
with his telephone and transfer it to a database where a specialist could survey the
development of the harvest in view of its shading and different properties. Individuals
could give their own perusing on temperature and stickiness and be a substitute for
sensor information if none is accessible. Figure 2 present a scenario where multiple
crop disease is send using the mobile phone from different fields.

4.2 Smarter Agricultural Practices

Capturing the empirical knowledge of current farmers in the form of data will make
farming more reliable and aid new entrants to the industry.

(a) *Visualization of the environment*

This involves sensors based technology to measure factors such as humidity, temperature or carbon dioxide concentration and periodically transmit them over a network
to the monitoring center. The development of this technology is allowing for the
potential use of environmental control in the future.

(b) *Visualization of crop growth*

Data collected using sensors based technology used by the monitoring center to
predict what effect changes in this environment will reflect on future crop growth
and harvests.

(c) *Visualization of production and sales*

This is concerned with factors such as the volume and timing of agricultural production, and also market prices.

4.3 Supporting Vegetable and Crop Production

M-Computing supports the production and management of agriculture by using historical data given by the farmers. The following examples describe the two main ways:

(a) In the case of a poor harvest, the reasons can be identified by analyzing historical data on the growing environment and this knowledge put to use in the cultivation of future crops.
(b) New entrants to the industry can obtain information about current conditions at the factory farm and make changes to the environment to bring it closer to optimal conditions.

5 M-Cloud Computing with Big Data

High speed, high volume, and high assortment data assets that require new forms of processing to empower upgraded decision making, insight discovery and process streamlining are called Big Data. It is clear that amount generated through agriculture is huge and need better system for processing system to uncover hidden information. Big data is expected to play significant roles in agriculture by utilizing M-cloud computing arm. It will provide weather forecasting, real-time optimization, real-time information resources for farmers, intelligent irrigation recommendations, price monitoring, logistic management, decision making on inventories and budgets.

Following are the characteristics of big data:

- *Volume*: The quantity of generated and stored data. The size of data determines the value, whether the value can be considered big data or not.
- *Variety*: The type and nature of data. This helps the people to analyze effectively use resulting insight.
- *Velocity*: The speed at which the data is generated and processed to meet demands and challenges that lie in the path of growth and development.
- *Variability*: Inconsistency of data set can hamper process to handle and manage it.
- *Veracity*: The quality of captured data can vary greatly, affecting analysis.

With particular yield history information on each field in the nation, agribusiness analysts would now be able to foresee their potential gathers better. Indeed, ranchers and agribusiness scientists could likewise profit by huge information to adapt trim

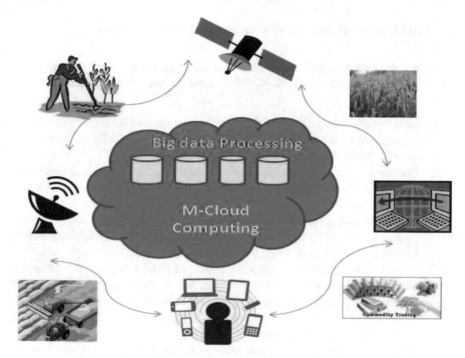

Fig. 3 An illustration of Big data processing using Cloud computing

yields. By conglomerating neighborhood estimating continuously and naturally fig-
uring transportation costs, ranchers can get the best costs for their items without a
middle person [9].

For instance, an IT services company has built up a framework that screens conditions, for example, temperature, stickiness, and oxygen levels of nourishment shipment compartments to screen and keep up the nature of horticultural deliver [9]
(Fig. 3).

There are many variables that add to famer's productivity such as Rainfall, Pes-
ticides, Ground Moisture, Market Data Water accessibility, Air dampness, Temperature, Price estimating, Government activities, and so on. From the mentioned
variables big data framework and cloud computing plays key role on in optimistic
decision making in the following:

- Optimum decisions in farming
- Weather reporting
- Commodity prices
- Crop recommendations
- Hybrids crop selection
- Pesticide recommendation
- Tools recommendations
- Farming practices
- Profitability Analysis

6   Advantages of M-Cloud Computing in Agriculture

There are various advantages of using cloud computing in agriculture management. Following are the main contribution of M-Cloud computing in agriculture sector [10–14]:

   (i) *Data management*: The data obtained from crops in different scenarios will be overseen by the specialist organization, a group of experts. That ensures a superior and sorted out administration of information.
  (ii) *Data readiness*: This gives information from the e-information bank databases to its whole partner whenever and at any area during the time on a $24 \times 7 \times 365$ premise.
 (iii) *Local and global Communication*: This makes the correspondence between various clients significantly speedier, less demanding and less expensive. Likewise, the correspondence will be secured.
  (iv) *Rural-urban migration*: A noteworthy issue of India is provincial urban relocation. It can be lessened as this gives its administrations everywhere throughout the state and may likewise everywhere throughout the nation whenever regardless of how remote the spots. This will likewise help in controlling joblessness issue in the nation.
   (v) *Technical advantages*: There are some technical advantages also such as No capital investment Servers, Software, No Maintenance, No Data backups, No Infrastructure, Easily Share his information with farmers planting the same crops to exchange expertise.
  (vi) *Motivation*: It will persuade ranchers and analysts to get included increasingly in horticulture as any correspondence will be result situated. It will bring about a general improvement of the division in the country.
 (vii) *Security*: It provides an enhanced security as the resources will be stored in the cloud and will be maintained centrally by the service providers. Thus, it is not a cause of concern for its users.
(viii) *Reduction of technical issues*: It cuts short the manpower, maintenance, and infrastructure requirement drastically, as it will be provided by the service providers.
  (ix) *Overall economy*: Execution of distributed computing in the farming part will help in inspiring the rural segment of the nation. That will support the general improvement of the economy. It is because of the mass contribution of various partners, as the framework will screen and convey advance report at whatever point and wherever required.

7   Conclusion

Indian agriculture has been traditionally been maintained by farmers and their families, it is important to pass and share the knowledge towards for better efficiency and productivity. Using M-Cloud Computing we should able to improve our agriculture practices in drastically and definitely is provide a pathway to out of poverty. M- Cloud computing saves costs by eliminating the need for high infrastructure expenses in the field of agriculture and also provides an easy to use, cost-efficient, flexible, dynamic and secure environment for farmers.

References

1. Vouk MA (2008) Cloud computing—issues, research, and implementations. J Comput Inf Technol 16(4):235–246
2. Mell P, Grance T (2011) The NIST definition of cloud computing (draft). National Institute of Standards and Technology Special Publication
3. Capobianco F (2010) Five reasons to care about mobile cloud computing. Int Free Open Source Softw Law Rev 1(2):139–142
4. Jain VK, Kumar S (2015) Big data analytic using cloud computing. In: Second international conference on advances in computing and communication engineering (ICACCE 2015), pp 667–672
5. Buyya R, Yeo CS, Venugopal S, Broberg J, Brandic I (2009) Cloud computing and emerging IT platforms: vision, hype, and reality for delivering computing as the 5th utility. Future Gener Comput Syst 25(6):599–616
6. Chen MJ, Wang J, Lu Z, Niu Y (2011) Research on agricultural information resources sharing system construction based on cloud computing. Agric Netw Inf 11(4)
7. Tian M, Xia Q, Yuan Hao (2013) Discussion on the application of cloud computing in agricultural information management. Res J Appl Sci Eng Technol 5(8):2538–2544
8. Onyegbula F, Dawson M, Stevens J (2011) Understanding the need and importance of the cloud computing environment within the National Institute of Food and Agriculture, an agency of the United States Department of Agriculture. JISTP 4(8):17–42
9. Big Data: Revolutionizing Agriculture. https://www.wipro.com/blogs/wipro-insights/bi
10. Kaloxylos A, Groumas A, Sarris V, Katsikas L, Magdalinos P, Antoniou E, Politopoulou Z, Wolfert S, Brewster C, Eigenmann R (2014) A cloud based farm management system: architecture and implementation. Comput Electron Agric 100
11. Prasad MSVKV, Kumar GJ, Naidu VVS, Nagaraju GJ (2013) Use of cloud computing in agriculture sector, a myth or reality. Int J Eng Res Technol (IJERT) 2(10)
12. Hori M, Kawashima E, Yamazaki T (2010) Application of cloud computing to agriculture and prospects in other fields. Fujitsu Sci Tech J 46(4)
13. Sahoo G, Mehfuz S, Rai R (2014) Applications of cloud computing for agriculture sector, Technical trends. www.csi-india.org
14. Patel R, Patel M (2013) Application of cloud computing in agricultural development of rural India. Int J Comput Sci Inf Technol 4(6)

Part III
IOT & Mobility

Detection and Analysis of Drowsiness in Human Beings Using Multimodal Signals

C. Anitha

Abstract The state of drowsiness can be characterised as an intermediate state of mind which occurs between the alert state and the sleep state. The alertness of the mind is reflected immediately through the sense organs and other body parts. Automatic detection and analysis of drowsy state of mind is essential in applications where the human's mental status is important. One such scenario is monitoring driver's alertness while he is driving. A multi-modal approach is analysed for detecting the drowsy state in humans. Two modalities are considered here, video information and bio signals, for analysis. Visual information conveys a lot about the human alertness. The precise indicators from the video information need to be identified and captured for analysis and detection. The bio signal that indicates human brain alertness is EEG signal. The physical and mental alertness are analysed for detecting drowsiness state of a human being. A framework is proposed for drowsiness detection of humans in real-time.

Keywords Drowsiness detection · EEG signal · Video signal · Framework
Real-time

1 Introduction

Driver's behaviour monitoring system finds its application in many real-time situations that include alertness detection and surveillance. In fact, detecting how alert a driver is very essential and crucial to avoid mishaps and save lives. The factors influencing mishaps during driving can be categorized under: Speed, emotional disturbance, and drowsiness. The first cause is apt whenever two or more automobiles are involved. The second and the third is purely an individual's state of mind. To determine the alertness of a person, different types of information can be considered. These include visual, audio, vehicular, bio/physiological indicators and

C. Anitha (✉)
Department of CSE, The National Institute of Engineering, Mysuru, India
e-mail: anithac.cse@nie.ac.in

© Springer International Publishing AG, part of Springer Nature 2019
S. Patnaik et al. (eds.), *Digital Business*, Lecture Notes on Data Engineering
and Communications Technologies 21, https://doi.org/10.1007/978-3-319-93940-7_7

so on. Visual information can be captured using cameras, audio information require microphones to capture them. Various sensors related to the movement of the vehicle can be captured using a set of associated vehicular sensors. While the bio-signals or the physiological signals require sensors that are intrusive to human body. The analysis first begins with capturing these signals using related sensors, followed by extraction of important features or parameters and finally analysing these features and categorizing accordingly. Many intermediate steps like; pre-processing, training, and decision making need to be followed in order to analyse and conclude the drowsiness state. In this chapter, the focus is on two indicators for drowsiness detection; the visual indicator and the physiological indicator. Using the visual information, yawning is detected. Yawning is an early indicator for drowsiness or sleepiness. Repeated yawning most of the time, leads to drowsiness. If a driver's drowsiness has to be detected, an analysis of his/her brain activity is important. The bio-signal that reflects the alertness of a human mind is EEG (Electro Encephalo Gram). In order to capture and measure the EEG signals, wearable gadgets have to be used. The alternative for analysis of EEG signals would the use of databases.

As per the statistics provided by the National Highway Traffic Safety Administration (NHTSA) [1], about 72,000 crashes, 800 fatalities and 44,000 injuries have occurred in 2013 due to drowsy driving. Hence, there is a need for automatically alerting the driver in advance. With the increasing rate of fatal accidents on road, efforts are made to avoid them. The identification of the driver's emotional disturbance or drowsiness can be a critical means of accident avoidance.

2 Related Work

The consideration of two modalities (video and bio-signals) is an effective combination. The related work discussed in this section considers the use of one of the indicators (information) mentioned in the previous section. Most of the research carried out use visual information alone for drowsiness detection through yawning. In [2], three different approaches for yawning detection are described. The colour segmentation technique, the Snake Contour method and the third method was usage of Viola-Jones theory [3] for face and mouth detection for detecting and locating the mouth region. The yawning detection was based on the openness of the mouth and the number of frames it was open. A comparative histogram was used between the closed mouth condition and the yawning condition. In [4], an SVM classifier was used to detect yawning condition. The classifier was given the width to height ratios of eyes and mouth. Both the eyes and the mouth features were considered for concluding the yawning condition. In [5], the location of the chin and nostrils were used to detect yawning. The distance between the chin and the nostrils was considered for yawning detection. The localization of the chin and the nostrils were done using directional integral projection method. The Viola-Jones technique [3] for face and mouth detection was used in paper [6]. A Support Vector Machine (SVM) classifier was trained with mouth and yawning images. The mouth region was detected from the face using a cascade of classifiers during fatigue. SVM was used to classify the mouth regions as yawning or alert. In [7], mouth corners were detected by grey projections and extracted using Gabor wavelets. Least Discriminant Analysis (LDA) was applied to classify the Gabor wavelets into yawning and not yawning condition. In [7], the authors detect yawning based on the mouth's height to width ratio. Most of the techniques discussed for yawning detection make use of classifiers. The use of classifiers adds to the complexity of the system. The proposed system does not use any classifiers for yawning detection; this reduces the computational time and complexity. Another feature of the proposed system is that the yawning detection is performed in two folds; this improves the accuracy.

The bio-signal is seldom considered for analysis in real-time environment as capturing the EEG bio-signal requires the use of wearable/intrusive gadgets. Most of the analysis considers data from an already existing dataset. A thorough analysis of multichannel signals in time series is described in [8]. The MIT-BIH Polysomnographic database [9], used for evaluation in the proposed system, is one of a kind as it provides a combination of seven physiological signals for study and evaluation. The usage of this database can be found in various bio-signal analyses including sleep apnea study [10], stroke related study [11], analysis of respiration and heart beat relationship [12]. The identification of sleep stages warrants the need of biomedical sensors which are not only expensive but also intrusive. But nowadays, several wearable gadgets are available which can be used on the driver's head to extract the EEG data. However, research contribution such as reported in [13] shows a selection of Action Units (AUs) from Facial Action Coding System (FACS) for drowsiness detection. There are evidences in [14, 15] showing the possible use of image processing techniques in order to detect drowsiness in drivers. A wide variety of applications are related with drowsiness detection. One of the major application areas includes the automotive field, especially the application to detect the alertness of a driver during an automobile drive. The statistics [16] provided by the National Highway Traffic Safety, about 100000 police-reported crashes are direct result of driver fatigue each year. This results in an estimated 1550 deaths, 71000 injuries and $12.5 billion in monetary losses. Thus it is all and all important to detect drowsiness and alarm the drivers before a mishap occurs. Even though other modals (such as image processing) are used for detection, EEG signal usage in drowsiness detection is one of the most precise ways of detection, as it directly reads the brain activity.

3 Proposed Work

The main objective of the proposed work is to combine two different type of modalities to detect drowsiness in human beings. Two reasons for choosing the video signal and bio-signal as the two types of signals; one was to come up with a framework that considers different types of signals to detect drowsiness in future. The second reason for this combination is novel.

A multimodal approach for detecting drowsiness in humans is proposed. Both of them are employed for the early detection of a drowsy mind. Defining the drowsy state is fuzzy. It can be termed as an intermediate state of mind between sleep state and wake state. The human mind expected to be in one of the states at any given time according to human sleep cycle: Wake, REM (Rapid Eye Movement) and NREM (Non-Rapid Eye Movement) state.

Drowsiness can be detected using visual signals at the earliest stage using effective indicators. One such indicator is yawning. When a person yawns continuously, he tends to get drowsy. Here, a two-fold yawning detection system is used [17]. Yawning is detected after two confirmations. Firstly, yawning detection is based on the skin tone detection and secondly based on blob dimensions and face containment.

3.1 Databases

The signals considered for analysis are from two datasets: one containing the video signals and the other, containing the physiological signals (EEG and others) with annotations. Both of these datasets are benchmark datasets used for research analysis and evaluation.

3.1.1 MIT-BIH Polysomnographic Database

The MIT-BIH Polysonmographic database [9, 11] is exclusive, as it provides a combination of 7 physiological signals, including EEG, for the study of sleep stages and sleep disorders. These signals include the ECG (Electro Cardio Gram), EEG (Electro Encephalo Gram), BP (Blood Pressure), respiration, SV (stroke volume), SO_2 (oxygen saturation), EOG (Electro Oculo Gram) and EMG (Electro Myo Gram). The data extraction from the MIT-BIH Polysomnographic database required separation of these signals. The extracted EEG signals from the database were used for detection of drowsiness. After the basic pre-processing, the frequency and amplitude features were extracted to form a feature set. This feature set was trained and tested using SVM classifiers with two kernel functions, the Sigmoidal and the Gaussian Radial Basis Function (RBF).

The extraction of these individual physiological signals becomes crucial for researchers in the field of signal and bio-signal processing. The signals are downloaded subject-wise along with the header file and the annotation file. The main problem arises when we have to experiment on individual signals. The downloaded file contains a combination of 4 to 7 signals depending on the subject's recording. The signal files that are downloaded require perfect segregation. The database is a collection of recordings of multiple physiologic signals during sleep. Subjects were monitored in Boston's Beth Israel Hospital Sleep Laboratory for evaluation of chronic obstructive sleep apnea syndrome, and to test the effects of constant positive airway pressure (CPAP), a standard therapeutic intervention that usually prevents

Table 1 Annotation symbols with meaning [9]

Aux	Meaning
W	Wake stage
1	NREM Sleep stage 1
2	NREM Sleep stage 2
3	NREM Sleep stage 3
4	NREM Sleep stage 4
R	REM sleep
H	Hypopnea
HA	Hypopnea with arousal
OA	Obstructive apnea
X	Obstructive apnea with arousal
CA	Central apnea
CAA	Central apnea with arousal
L	Leg movement
LA	Leg movement with arousal
A	Unspecified arousal
MT	Movement time

or substantially reduces airway obstruction in these subjects. The database contains over 80 hours worth of four-, six-, and seven-channel polysomnographic recordings, each with an ECG signal annotated beat-by-beat, and EEG and respiration signals annotated with respect to sleep stages and apnea.

Each record includes a header (.hea) file, a short text file that contains information about the types of signals, calibration constants, the length of the recording, and (in the last line of the file) the age, gender, and weight (in kg) of the subject. In this database, all 16 subjects were male, aged 32–56 (mean age 43), with weights ranging from 89 to 152 kg (mean weight 119 kg). Records slp01a and slp01b are segments of one subject's polysomnogram, separated by a gap of about one hour; records slp02a and slp02b are segments of another subject's polysomnogram, separated by a ten-minute gap. The remaining 14 records are all from different subjects.

All recordings include an ECG signal, an invasive blood pressure signal (measured using a catheter in the radial artery), an EEG signal, and a respiration signal (in most cases, from a nasal thermistor). The six- and seven-channel recordings also include a respiratory effort signal derived by inductance plethysmography; some include an EOG signal and an EMG signal (from the chin), and the remainder include a cardiac stroke volume signal and an earlobe oximeter signal. Each record includes two annotation files. The '.ecg' files contain beat annotations, and the '.st' files contain sleep stage and apnea annotations. Annotations in the '.st' files contain the sleep staging and apnea information in their aux fields. Each annotation in the '.st' files applies to the thirty seconds of the record that *follow* the annotation. The coding scheme is indicated in Table 1.

3.1.2   YAWDD—Yawning Detection Database

YawDD [18] is the first yawning database available for non-commercial research purpose. This database contains two datasets of drivers that can be used for testing algorithms for yawning detection. Illumination variations are present in the videos. The recording camera is fixed under the rear mirror for the first dataset. The subjects were asked to perform three actions during the recording; simply drive, talk or sing and yawn while driving.

For the second dataset the camera was fixed on the car's dashboard. As in the first dataset, the subjects here were also asked to perform the three actions. A total of 29 recordings comprise the second dataset. The subjects are both male and female from different authenticities. The database includes a table indicating the action performed by the subjects, the duration of the recording and the publication consent is provided for researchers' use. The videos are in 640×480 24-bit true colour (RGB) 30 frames per second AVI format without audio. The total data size is about 5 gigabytes. The complexity lies in the data size and number of frames. Hence, the video size was reduced to almost 50% so that the computation time was minimized [18].

3.2   Drowsiness Detection

For detecting drowsiness with EEG and video signals, some amount of pre-processing is needed. As the signals from the datasets need to be extracted in accordance with the analysis to be carried out, the EEG signal had to be separated from the other six signals' information. This step of data extraction can be skipped if the EEG signal is taken in real time.

3.2.1   Data Extraction—EEG Signal

The bio-signal of interest is only the EEG signal that needs to be extracted from the dataset. With slight modification done to the code contributed by Robert Tratnig from Mathworks exchange library [19], a successful attempt was made to extract all the bio-signals into separate vectors. The code execution requires a thorough understanding of the header file. A sample header file is given below.

1. slp66 7 250/0.033333333 3300000 0:23:30 4/1/1990
2. slp66.dat 212 500 12 0 66 52440 0 ECG
3. slp66.dat 212 7.71711(-963)/mmHg 12 0-9 57501 0 BP
4. slp66.dat 212 10021/mV 12 0-18 21720 0 EEG (C3-O1)
5. slp66.dat 212 1654/l 12 0 305 28677 0 Resp (nasal)
6. slp66.dat 212 1628/l 12 0 132 22306 0 Resp (abdomen)
7. slp66.dat 212 9.957/ml 12 0 147 20748 0 SV
8. slp66.dat 212 19.25(-1006)/% 12 0 755 32914 0 SO2
9. # 33 M 95 08-01-90

The first line, numbered 1, gives the subject id, number of signals, sampling rate, length of the integrated signal, time and date of recording. Line numbers 2 to 8 specify the details of the individual signals in the following order: data file name, data format type, gain, bit resolution, zero-value of the signal, first value of the signal, signal length, end delimiter and the signal name. The last line, numbered 9, gives details about the subject such as age, gender, weight and date of completion of record.

Algorithm 3.1: Segregation of signals from dataset
Input: MIT-BIH Polysomnographic data set containing a set of bio-signals for each subject.
Output: Segregated signals (4 to 7 vectors) subject-wise.

1. Open the header file (.hea file) and read the first line of the header file and extract the following information: number of signals, sampling rate, length of the signal.
2. Depending on the number of signals, extract the details of the individual signals into vectors containing: data format, gain, bit resolution, zero value of the signal, first value of the signal, signal length. Close the header file.
3. Verify the data format vector values for all the signals. Exit if format in not equal to 212. Else, Open the data file and read the contents into 3D array.
4. Perform basic logical operations to convert the data array into signed integer values. Save these values into a 2D array.
5. Depending on the number of signals, pre-allocate the signal vectors (4 or 6 or 7 vectors). If the number of signals is 4, copy the signal values from the 2D array into the individual signal vectors in the order specified in the header file.

If the number of signals is 6 or 7, convert the 2D array of signed signal values into a 1D vector to avoid inconsistency. Copy the signed 1D vector values into individual signal vectors in the order specified in the header file.

6. Verify the first value of each signal vector with the first value vector extracted from the header file. Stop if there is a mismatch.
7. Convert the vector values into original signals values using the gain and zero value vectors obtained from the header file.
8. The computed values in Step 7 are the signals extracted and are ready for viewing and analysis.

3.2.2 Data Extraction—Video Signal

The most important assumptions made for yawning detection:

- Driver's face is expected to be frontal to the camera (located on the dashboard of the car).
- Head rotation maximum up to 45° permitted, not beyond that (practically not feasible to drive with head turned beyond 45°)

In case of extracting the video signal for yawning detection, the video sequences from dataset I of the YAWDD dataset [18] is considered. The region of interest is

Table 2 Frequency range for the sleep stages

Stage	Parameter
W	Alpha rhythm 8–13 Hz; Beta range > 13 Hz; Gamma waves < 40 Hz
N1	Theta range, 4–7 Hz
N2	Sleep spindles with K complex—a sharp negative edge followed by a positive edge spike
N3	Delta range, 0.1–3 Hz
R	Transient muscle activity

the lower part of the face which needs to be detected and tracked in each video frame to avoid loss of continuity. The first step involves detecting the face in the video sequence. The Joint Viola-Jones and Mean-Shift algorithm [20] is used for face detection and tracking. The non-skin regions on a human face are the eyes and mouth region which defines the blobs (black region) in the binary image being tracked and detected.

Algorithm 3.2: Segmenting lower part of the face
Input: Detected and tracked face from pre-processing.
Output: Segmented lower part of the face containing the mouth region.

1. Detect and track the face at the centre of the image frame: high speed face detection and tracking algorithm is used [20]. The algorithm works faster as face detection algorithm is judiciously combined with the tracking algorithm.
2. Check the existence of blob: Applying the skin-tone algorithm, the human skin is detected as the white pixels in the binary image and the non-skin regions appear as a black spot or blob. The non-skin regions inside the face are eye and mouth region.
3. Verify blob position: Only those image frames need to be considered that contain the blob at the lower part of the face.
4. Segment face: If blob exists in lower part of the face, segment the face region and retain only the lower part of the face for further processing.

3.3 EEG Signal Classification

The EEG signals from the MIT-BIH Polysomnographic database required basic pre-processing for elimination of DC noise. Hence a simple filter was used to filter out all high frequency components. Using algorithm 3.1 the required bio-signal, EEG, was extracted subject-wise, in different sleep stages. With the help of the annotation file, provided by the database corresponding to each data file, the EEG data was separated stage-wise. The study and identification of drowsiness detection restricts us to the Wake stage and the first stage of NREM sleep. The sleep stages and the corresponding brain activity with frequency range are given in Table 2.

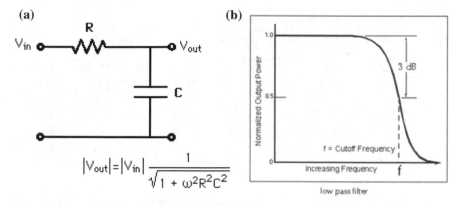

Fig. 1 **a** Low-pass filter design, **b** Frequency response

The sleep cycle involves the following stages: W, N1 N2, N3 and R according to AASM's manual on scoring of Sleep and associated events [21]. When a person is awake (W stage), he is conscious, relaxed but not drowsy. When a person is in the first NREM stage (N1), he is drowsy or his mind is dazed in imagination. So, the objective is to detect N1 and W stages. The frequency range for N1 stage is between 4 and 7 Hz while the frequency range for stage W is 8 to 13 Hz. The frequencies above 13 Hz (up to 40 Hz) are also in the W stage, typically associated with high alertness and high level processing (brain) stage. So, if the frequency in 4 to 7 Hz range and the above 8 Hz is differentiated, the two stages N1 and W are identified.

The EEG signal is first passed through a low pass filter with a cut-off frequency of 50 Hz to eliminate the high frequency noise components. The low pass filter also acts as a smoother which can produce slow changes in the input thereby allowing us to discover important patterns in it. The Fig. 1 shows a typical low pass filter.

The EEG signal is now transformed using the Fourier transformation.

$$X(k) = \sum_{j=1}^{N} x(j)\omega_N^{(j-1)(k-1)} \tag{3.1}$$

where, $\omega_N = e^{(-2\pi i)N}$ is the Nth root of unity.

The top four frequency components and their absolute amplitude values are stored in the respective vectors. Since the MIT-BIH Polysomnographic database has annotation file with every data signal file, the verification of the sleep stage is done according to the time period in the annotation file. This file includes sleep stage identification after every 30 s. With the input sampling frequency of 250 Hz, the number of samples for every 30 s will be 7500. So, a minimum of 7500 samples have to be considered in order to cross validate the sleep stages with the annotation file. The extraction of the frequency and its absolute magnitude is done for every second, i.e. for every 250 samples. This combination of maximum 4 frequency components and their respective magnitude components are the features considered for sleep stage recognition.

Table 3 Typical feature set values for the SVM classifier

Class	Feature 1	Feature 2	Feature 3	Feature 4	Feature 5	Feature 6	Feature 7	Feature 8
N1	2.344	1.953	2.734	1.563	73.25	72.02	59.88	51.68
N1	1.953	2.344	1.563	2.735	72.33	61.10	60.69	41.09
N1	1.953	1.563	2.344	1.172	96.86	81.35	76.31	44.19
W	45.31	13.28	44.92	13.67	46.41	41.00	38.43	38.03
W	46.09	36.33	36.72	35.94	65.37	60.63	58.93	58.73
W	3.75	44.14	12.89	13.28	111.7	103.9	94.94	94.46

The database contains EEG signal samples of 16 subjects out of which 5 subjects were considered for analysis. The reason for this was the remaining 11 subjects did not have the both W and the N1 stages in their recordings. A total of approximately 18,000 samples were considered for evaluation. About 70% of them were used for training and the remaining 30% were used for testing. The 8 features extracted from the transformation were given to an SVM classifier. Typical feature values extracted are as shown in Table 3.

The features 1 to 4 are the frequency values while 5 to 8 represent the corresponding absolute magnitudes.

3.4   SVM Classifier

SVM classifier is well known classifier used in the field of pattern recognition and regression problems. Given a training set of instance-label pairs (x_i, y_i), $i = 1, \ldots, l$ where $x_i \in R^n$ and $y_i \in \{1, -1\}^l$, support vector machines [89] requires the solution of the following (primal) optimization problem:

$$\min_{\omega, b, \xi} \frac{1}{2} \omega^T \omega + C \sum_{i=1}^{l} \xi_i \tag{3.2}$$

Subject to $y_i (\omega^T z_i + b) \geq 1 - \xi_i$, $\xi_i \geq 0$, $i = 1, \ldots, l$
Equation (3.2) is solved by solving the following dual problem:

$$\min_{\alpha} F(\alpha) = \frac{1}{2} \alpha^T Q \alpha - e^T \alpha \tag{3.3}$$

Subject to $0 \leq \alpha_i \leq C$, $i = 1, \ldots, l$, $y^T \alpha = 0$ where $e =$ vector of all ones,
$Q = [l \times l]$ semi definite matrix
The $(i, j)^{th}$ element of Q is given by,

$$Q_{ij} = y_i y_j K(x_i, x_j) \tag{3.4}$$

where $K(x_i, x_j) \equiv \varphi(x_i)^T \varphi(x_j)$ is called the kernel function

Then, $\omega = \sum_{i=1}^{l} \alpha_i y_i \varphi(x_i)$

And $sgn(\omega^T \phi(x) + b) = sgn\left(\sum_{i=1}^{l} \alpha_i y_i K(x_i, x) + b\right)$ is called the decision function

Two kernel functions were used for evaluation of the feature set:

Gaussian RBF kernel with the kernel function,

$$K(x_i, x_j) = e^{\frac{-1}{2}\left[(x_i - x_j)^T \Sigma^{-1}(x_i - x_j)\right]} \tag{3.5}$$

$$K(x_i, x_j) = exp\left(-\frac{\|x_i - x_j^2\|}{2\sigma^2}\right) \tag{3.6}$$

where $\|x_i - x_j\|^2 = squared\ euclidean\ distance\ between\ x_i\ and\ x_j$ with $= -\frac{1}{2\sigma^2}$,

$$K(x_i, x_j) = exp\left(\gamma \|x_i - x_j\|^2\right) \tag{3.7}$$

Equation 3.3 is the Gaussian RBF kernel function.

For Sigmoidal kernel or MLP kernel, the kernel function used is,

$$K(x_i, x_j) = \tanh\left(\gamma x_i^T x_j + c\right) \tag{3.8}$$

where $\gamma = \frac{1}{number\ of\ features}$ and $c = constant$

The training process was performed using different cross-validation values and the best value was chosen to achieve the maximum accuracy.

3.5 Yawning Detection

Using algorithm 3.2, the lower part of the face is segmented and considered for yawning detection. The method adopted is simple computationally, yet efficient and accurate. Simple as there is no use of classifiers for yawning detection. The proposed system is a two-agent expert system to detect yawning. The first agent detects yawning based on skin tone detection. The second agent detects yawning based on the blob dimensions and containment within the face region.

From the segmented lower face region, the presence of blobs may indicate yawning. But the position and the dimension of the blob has to be verified for positive detection of yawning. The first step involves bounding the blobs. The blobs confirmed to be inside the face are bounded in rectangles as per their dimensions. The histograms from the vertical projection of the bounded blobs are measured. The next step is to verify yawning through histogram. The histogram of blobs is obtained through vertical projection of the segmented face. Histograms are considered for blobs present only at the centre and near-centre regions of the segmented face. This is in accordance with the assumptions stated earlier. The histogram's length and sum values are verified with the threshold values. If the values satisfy the thresholds specified, yawning is confirmed.

3.5.1 Dealing with Occlusions

The occlusions on driver's face may be many, such as mouth covered by hand while yawning; wearing sunglasses while driving, hair covering the face while driving and so on. Among these, the mouth covered by hand while yawning is one which cannot be handled. Frames with the hand covering the mouth, even though yawning, would indicate not yawning. This can be overcome by monitoring the frames' output before and after the occluded frames.

Wearing of sunglasses or any other type of glasses does not bother the yawning detection as face is segmented, and only the lower part of the face is considered for yawning detection.

Covering of face with hair also causes occlusion to some extent as the skin region covered cannot be detected. But this is not as serious as it does not affect the yawning decision made finally.

4 Results and Conclusion

4.1 Using EEG Signals

The feature set extracted for classification were the highest 4 frequencies and their corresponding magnitudes. Figure 2 shows a sample feature set corresponding to stage W and N1.

The classification of the two stages was performed subject-wise. For the Gaussian RBF kernel with the following parameters were considered: $\gamma = 1/8$, $c = 2^0 = 1$ (cross-validation) and degree $= 3$. Using the sigmoidal kernel, with the parameters: $\gamma = 1/8$, $c = 0$ (intercept constant) and degree $= 3$. The accuracy obtained for subject-wise classification using Gaussian RBF kernel and sigmoidal kernel is as below (Table 4).

The results obtained indicate that driver's drowsiness can be effectively recognized with the help of a fast frequency transformation and the use of SVM classifier.

4.2 Using Video Signals

The methodology adopted for yawning detection was tested on the YAWDD data set [18]. Different cases were considered for evaluation. These include image sequences

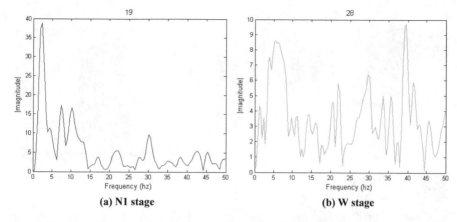

Fig. 2 Graph corresponding to the feature set used for classification

Table 4 Subject-wise accuracy obtained

Sl no.	Subject name	Accuracy (%)	
		RBF	Sigmoidal
1	M41	86.7	87.2
2	M48	64.6	61.5
3	M59	80.5	78.8
4	M66	82.0	83.3
5	M67	84.2	85.4

with frontal faces, sequences with sunglasses, sequences with prescribed specta-
cles, sequences with mouth occlusion, and sequences with non-frontal faces. Three
categories were considered for evaluation:

- Non-occlusive, frontal faces with/without glasses
- Occlusion with hand, frontal with/without glasses
- Non-frontal faces with/without glasses.

4.2.1 Non-occlusive, Frontal Faces With/Without Glasses

The detection is accurate for non-occlusive, frontal faces. The driver's sunglasses do
not affect the response of the yawn detector, since only the lower part of the face is
considered (Fig. 3 illustrates some true positives). The histogram values indicated
in the rightmost column of Fig. 3 indicates the blob measurements in the lower
segmented region of the face. The presence of multiple bins indicates the presence
of other blob regions. But mouth blob is identified based on the location and the
containment inside the face.

(a) **(b)** **(c)** **(d)**

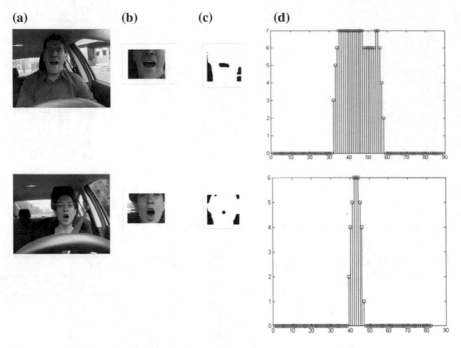

Fig. 3 True Positives; **a** input image, **b** Segmented region, **c** yawn detector's output, **d** histogram

4.2.2 Occlusion with Hand

The detection in this case was not possible in the frames where the hand completely covered the mouth region. But, due to the detection of yawning in the previous and continuing frames, this occlusion could be easily handled. Figure 4 shows the input and output resulting in this case.

When the driver occludes the mouth region during yawning, the blob in the mouth region is covered and the vertical projection does not reflect any value. Hence the histogram does not reflect any variations. During these frames, the yawning condition goes undetected. As our system increments a count value for every frame when yawning is detected based on the histogram values. The yawning is detected the moment the hand is removed away from the mouth region. The proposed system could not overcome this occlusion as the feature used for yawning detection was the based on the mouth region.

4.2.3 Non—frontal Faces

The frames with subject completely frontal to the camera are the actual and typical way a driver drives a car. But some cases where the face was not completely in front of the camera during yawning were considered. The success rate was good for faces

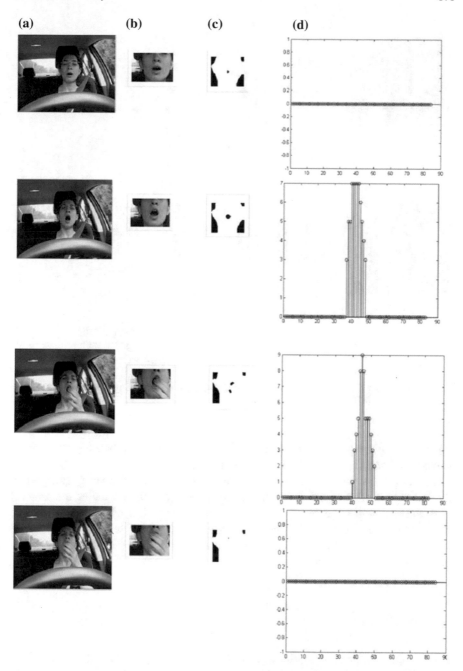

Fig. 4 Occlusion with hand, Yawn Undetected **a** Input image, **b** Segmented region, **c** Yawn detector's output, **d** histogram

(a)   **(b)**   **(c)**   **(d)**

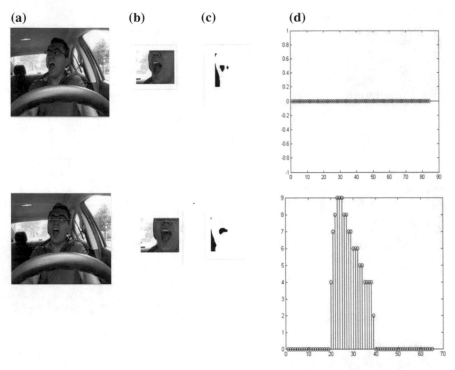

Fig. 5 Non—frontal frames, less than 45° detectable **a** Input image, **b** Segmented region, **c** Yawn detector's output, **d** Histogram

up to 45° inclination, but went bad for faces inclined beyond that. Figure 5 shows some of the non-frontal frames from the database.

As per the assumptions made earlier, the proposed system detects yawning when the driver's face is completely parallel to the camera (completely facing the camera) or slightly rotated. The extent to which yawning is detected with head rotation is 45°. Beyond this the blob's containment in the face becomes false, leading to undetected yawning condition. Moreover, the driver cannot afford to keep his head away from the windshield all the time.

5 Conclusion

An attempt was made to consider two different types of signals to detect the alertness of the driver. The signals used were; one, brain activity related bio-signal EEG and two, video signal. This combination enabled double—check on driver's alertness detection. Real—time implementation of such multi-modality driver's alertness detection system can be made possible only if the EEG bio-signal measuring equipment is non-intrusive to the maximum extent. Video signal analysis can further be enhanced in the real—time implementation by introducing multiple cameras to capture the driver's face image sequence. The limitation of only frontal or near—frontal image restriction can be completely eliminated. Considering a typical travel time, no driver can afford to look away from front (road in front) for a much longer time. The proposed system is only an attempt to avoid mishaps caused due to non—alert drivers. The proposed work can be the initial framework for a real-time multimodal based drowsiness detection system to which other modalities such as audio signals, vehicular signals etc. can be considered to make the system completely robust.

References

1. NHTSA—National Highway Traffic Safety Administration, Washington DC, Online: http://www.nhtsa.gov/Driving+Safety/Drowsy+Driving
2. Abtahi S, Hariri B, Shirmohammadi S (2011) Driver drowsiness monitoring based on yawning detection. In: 2011 IEEE international instrumentation and measurement technology conference, Binjiang, pp 1–4
3. Viola P, Jones M (2001) Rapid object detection using a boosted cascade of simple features. In: Proceedings of the IEEE computer society conference on computer vision and pattern recognition CVPR, pp I-511–I-518
4. Azim T, Jaffar M, Mirza A (2009) Automatic fatigue detection of drivers through pupil detection and yawning analysis. In: Fourth International Conference on Innovative Computing, Information and Control, pp 441–445
5. Yufeng U, Zengcai W (2007) Detecting driver yawning in successive images. In: International Conference on Bioinformatics and Biomedical Engineering, pp 581–583
6. Saradadevi M, Bajaj P (2006) Driver fatigue detection using mouth and yawning analysis. IJCSNS Int J Comput Sci Netw Secur 8(6):183–188
7. Fan X, Yin B, Sun Y (2007) Yawning detection for monitoring fatigue. In: Sixth international conference on machine learning and cybernetics, Hong Kong, pp 664–668
8. Rigney DR, Goldberger AL, Ocasio WC, Ichimaru Y, Moody GB, Mark RG (1993) Multi-channel physiological data: description and analysis. In: Weigend AS, Gershenfeld NA (eds) Time series prediction: forecasting the future and understanding the past. Addison-Wesley, Reading, MA, pp 105–129
9. Ichimaru Y, Moody GB (1999) Development of the polysomnographic database on CD-ROM. Psychiatry Clin Neurosci 53:175–177
10. Lee J-M, Kim D-J, Kim I-Y, Park K-S, Kim SI (2002) Detrended fluctuation analysis of EEG in sleep apnea using MIT/BIH polysomnography data. Comput Biol Med 32(1):37–47. ISSN 0010-4825
11. Garpestad E, Katayama H, Parker JA, Ringler J, Lilly J, Ya-suda T, Moore RH, Strauss HW, Weiss JW (1992) Stroke volume and cardiac output decrease at termination of obstructive ap-neas. J Appl Physiol 73(5):1743–1748
12. Ichimaru Y, Clark KP, Ringler J, Weiss WJ (1990) Effect of sleep stage on the relationship between respiration and heart rate variability. Comput Cardiol 17:657–660
13. Vural E, Cetin M, Ercil A, Littlewort G, Bartlett M, Movellan J (2007) Drowsy driver detection through facial movement analysis. In proceedings of the IEEE international conference on Human-computer interaction (HCI). Springer, Berlin, Heidelberg, pp 6–18
14. Ji Qiang, Yang Xiaojie (2002) Real-time eye, gaze, and face pose tracking for monitoring driver vigilance. Real-Time Imaging 8(5):357–377

15. Grace R, Steward S (2001) Drowsy driver monitor and warning system. In: First international driving symposium on human factors in driver assessment, training and vehicle design, Aug 2001, pp 64–69
16. Drowsydriving.org/about/facts-and-stats/ by the National Sleep Foundation
17. Anitha C, Venkatesha MK, Suryanarayana Adiga B (2016) A two fold expert system for yawning detection. In: Second International Conference on Intelligent Computing, Communication and Convergence (ICCC-2016). Published by Elsevier Procedia of Computer Science, Bhubaneswar, India, Jan 2016, pp 63–71
18. Abtahi S, Omidyeganeh M, Shirmohammadi S, Hariri B (2014) YawDD: a yawning detection dataset. In: Proceedings of ACM multimedia systems, Singapore, 19–21 Mar 2014, pp 24–28
19. https://in.mathworks.com/
20. Anitha C, Venkatesha MK, Suryanarayana Adiga B (2015) High speed face detection and tracking. Int J Soft Comput Artif Intell 3(2):84–90
21. Iber C, Ancoli-Israel S, Chesson A, Quan SF (2007) For the American Academy of sleep medicine. In: The AASM manual for the scoring of sleep and associated events: rules, terminology and technical specifications, 1st edn, American Academy of Sleep Medicine, Illinois, Westchester

Enhancing Security and Privacy in Enterprises Network by Using Biometrics Technologies

Abdallah Meraoumia, Hakim Bendjenna, Yahia Dris and Mohamed Amroune

Abstract Of course, in all world countries, commercial and industrial enterprises play an important role in the development of their economics. Recently, the change in enterprise operation mode, such as their activities speed in terms of production and transport, requires integration of these activities into a private and/or public networks structure in order to achieve many benefits such as the time and the cost. So, for an effective development, these enterprises must be connected with them in order to exchange necessary information. Thus, the resultant of this leads to emergence of special structure, called Enterprises Network (EN), which allows such enterprise to more effectively interact with others enterprises inside and outside their activities. The challenges that oppose this structure are the concerns of transmitted data security which are exchanged between them as well as the privacy issues, which ensure the secrecy of enterprise information. In contrast with other approaches presented in literature, a complete biometric cryptosystem is presented. Based on the obtained results, it can be inferred that the proposed authentication system is highly secure, effective and cheap, which are vital for any authentication system to gain enterprise confidence for implementation in real time secure access control systems.

A. Meraoumia · H. Bendjenna (✉) · M. Amroune
LAboratory of Mathematics, Informatics and Systems (LAMIS),
University of Tebessa, Tébessa, Algeria
e-mail: hbendjenna@gmail.com

A. Meraoumia
e-mail: ameraoumia@gmail.com

M. Amroune
e-mail: medamroune@gmail.com

Y. Dris
Management Sciences Department, University of Tebessa, Tébessa, Algeria
e-mail: y.ddris@gmail.com

© Springer International Publishing AG, part of Springer Nature 2019   175
S. Patnaik et al. (eds.), *Digital Business*, Lecture Notes on Data Engineering
and Communications Technologies 21, https://doi.org/10.1007/978-3-319-93940-7_8

1   Introduction

Among several basic points which can be used as indicators on the positive development of the countries economics, the size as well as the number of companies remains the most important indicator of these economics. Indeed, the successful of these companies depends necessarily on the interaction between various parameters such as information, materials, manpower and money [1]. Unfortunately, the majority of companies require many technical and operational resources which are, generally, difficult to exist within a single company. Thus, to overcome these limitations; companies must coordinate with others companies in or out their activity or/and country. Basically, the Business Network (BN) can speedy and effectively develops innovations, business strategy and deliver superior value to end customers [2]. Additionally, through the BN, enterprises can also achieve profitable growth and a new source of competitive advantage at the international market level [3].

As previously mentioned, information exchange is the utmost important and essential thing to do between these enterprises where nothing can be managed without it. Now there are several techniques being used for accomplishing such tasks and these involve tools such as Intranet and internet [4]. These tools both are a kind of connection between several systems. However, Intranet is a connection within a circle of very shorter radius (in the case of EN, It is essential that the connected enterprises should be belongs to this circle), whereas, internet covers the whole world. Moreover, another tool, called Extranets, allows to link a given Intranet of set of enterprises over the internet by providing access to Intranet of others set of enterprises. It combines the privacy and security of Intranet with the global reach of the Internet. Generally speaking, enterprises are still restricted to use Internet as a means of information exchange due to their dissatisfaction and their distrust to this means. Thus, in order to ensure the security and confidentiality in enterprises network, such a service should provide more credible and stronger security measures that can withstand the different risks of the possible attacks for a reasonable period in practice. For that, several methods have been proposed in the last few years in order to, firstly, the guarantee of the enterprise confidential information and secondly, authenticate effectively the enterprise requesting the service. The first point is solved by encrypting in safc way any information via the networks, whereas, the second point is solved by authenticating the service applicant (enterprise delegate). So, combination of cryptography and identity identification are considered among the safest means which are recommended as the most interesting solution to ensure the safety of enterprise data and their authentication. In the proposed chapter, we propose a new security mechanism which uses the biometrics in the encryption process of the exchanged data by some enterprises. The idea is to use the biometric trait of some employees to identify the enterprise and for securing the encryption key.

2 Inter-entreprises Cooperation

To cope with several business conditions varying like changes in customer demands, competition increasing and communications performance, enterprises must migrate to inter-organizational relationships [5, 6] in order to adapt to their new environment, to achieve competitive advantage and to increase their efficiency. Thus, the enterprises, which want to integrate into this new configuration, should define the new objectives as precisely as possible and then inventory all necessary resources and specific competencies to reach their objectives.

In this configuration, enterprises skills and resources are united under cooperation to respond to new opportunities. Basically, cooperation is the action or process of working together to the same end [7]. Indeed, several challenges can be faced today in the enterprise, so, one of the possible solutions, if she doesn't have all the sufficient, necessary skills and resources (human, technological and financial) to deal with these challenges, is the cooperation with another's enterprises. This process allows enterprises to act together to achieve an otherwise unattainable business goals. So, the main objective of this process is to design and produce faster and better innovative products and services to their customers which can produce several benefits for each enterprise [8]. The advantages of cooperation are:

1. Search for economies of scale and the profitability of synergies.
2. Modification and mitigation of the competition rules.
3. Fight against uncertainty and the sharing of risks inherent in market transactions.
4. Sharing of costs related to major development projects between the contracting parties, with a view to their accumulated competitiveness.
5. Search for access to new markets at the international level previously closed.
6. Transfer of know-how and technologies without abandoning the associated property rights.

3 Interest of Business Networks

The high bargaining power of consumers as well as the need to reach new markets, influenced by globalization and the policy of open international markets is majors concerns of our modern institutions [9]. Hence, rethink the value chain in its traditional form by institutions is one means to overcome potential risks. As result, the rigid value chain is transforming to a dynamic business network, from customers, partners and suppliers [10]. Indeed, business networks are defined as the sets of connected exchange relationships where one relationship affects another [11].

Thus, for the enterprises involved in business network, several advantages can be expected. Firstly, enterprises, which are employed the power of the business network, will be able to develop innovations faster, develop business strategy, and deliver superior value to the customers. Secondly, through the business network, enterprises can achieve profitable growth and new source of competitive advantage at the international market level.

3.1 Business Networks and Internet Worldwide

In fact, due to the great need of business networks, various ways which based on the available technology, are developed. Thus, linking different enterprises across a network is the most powerful solution. Furthermore, as mentioned later, Intranet possesses limitation which is the short distances between the concerned enterprises. Due to the greater development in technologies, business networks using Internet and Extranet [12] are veritable alternative solution to overcome the limitations produced by Intranet. This solution provide several characteristics:

1. Low cost implementation, especially for the geographically distant enterprises across business network (Internet is an available and public network).
2. The rapid transfer and exchange of data, which is very important in business.
3. Ability to transfer and exchange large and various amount of data and information (Text, image, audio, video, etc.) between enterprises connected in network.

However, the obsession with security and confidentiality remains a worrying aspect. So, due to rapid development in information technology, business networks are facing new challenges regarding electronic fraud through internet.

3.2 Sensitive Business Information

In business networks a set of information are exchanged between a finite set of entreprises. In general, the exchange of Information between these entreprises is covered by an agreements previously concluded between them. Hence, each entreprise can request new information, additional information, confirmation or cancelation information and/or corrective information. Consequently, much of these information require protection, so that others can not access them, especially, the competitors, who want to know and benefit from these information. Generally, the information to be protected in business networks are:

1. **Confidential information**: The disclosure of this information affects negatively the enterprises position in the market. Especially the information related to the volume of activity and business or customer list.
2. **Financial information**: In business, this information is very sensitive and it must be sure that is complete and accurate. So, an exchange protocol is needed to verify the integrity of this information.
3. **Business and technical information**: Generally, each enterprise has a high degree of transparency in the business, but enterprise would not like to leak of some very important business information such as their business plans.
4. **Human resources information**: This information is related to the enterprise members who need special attention for their protection. Thus, employee information and personal data, including salaries, and insurance data, and health status and performance reports must be secured during transmission and storage.

5. **Customer Information**: Companies always need to keep detailed information about their customers and the nature of their work. If some of this information is sensitive, enterprises must maintain it with confidentiality. Furthermore, other information, such as market data, is also confidential.

6. **Security Information**: This information is related to the protection mechanism the company's data. Basically, this information is the most important and sensitive one among these cited above.

4 Information Security and Privacy

Security is an essential part of any transmission operation of important information and data that takes place over the internet. Enterprise will lose her faith in *e*-business if its security is compromised. To meet this purpose, enterprises are embracing digital technology, such as *e*-security, to help secures their business during transmission (for example via Internet). Indeed, electronic information security or e-security means protect an important data and information from any fraud attempt during transmission via Internet [13]. This process must guarantee the essential requirements for safe transmission such as confidentiality and authenticity. Hence, the first requirement must guarantee the not-accessible information to an unauthorized enterprise (not be intercepted during the transmission which is the role of *cryptography*), whereas, the second requirement must authenticate an enterprise before giving her access to the required information (identity identification of the enterprise delegate which is the role of *biometrics*).

4.1 Cryptography Process

Cryptography is a set of mathematical methods and techniques ensuring the messages security [14]. For the time being, cryptography is the very effective and practical way to safeguard the data being transmitted over the network. The principle of cryptography is to encode the message, using a key, in such a way as to make it secret (encryption operation). The decryption operation requires a key to obtain the original message. In fact, without the knowledge of the encryption key, the decryption operation is impossible. There are two cryptography methods [15]: Symmetric encryption methods in which the same key is used for encryption as well as for decryption, and asymmetric encryption methods, in this case the encryption key is different from the decryption key. In both cases, it must avoid any kind of the encryption key sharing during the communications made between enterprises, which is an important criterion for effectiveness of any encryption method.

4.2 Biometrics Process

Logically, each enterprise can be recognized by some employees, such as its director as well as there assistants (enterprise delegate). So, it is sufficient to recognize the identity of these employees in order to identify the required enterprise. In fact, due to the great need for such identity recognition, researchers have developed several ways that are related to the information that a person has provided (*What you have*, e.g., a card, or a key), or knows (*What you know*, e.g., a password or a personal identification number (PIN)). Unfortunately, passwords can be guessed by an intruder; cards can be stolen or lost. However, to overcome the limitations associated with such means, other means of security has been developed that allow obtaining the specific or intrinsic information of the person [16]. It is about the biometrics-based recognition (*What you are/do*, e.g., fingerprint, iris or voice). A biometric technology offers a natural and reliable solution to the problem of identity determination by recognizing personals based on their physical or behavioral characteristics [17]. Compared with traditional security means, the biometric-based security offers more properties and several advantages due to biometric characteristics of an individual are not transferable, unique and are not lost, stolen or broken [18].

4.3 Crypto-Biometric Security System

Generally, in the cryptographic process, the shared key (random key) must be exchanged between the different enterprises in order to decrypt the transmitted information (symmetric cryptography). For that, one of efficient solution is to use biometric traits of the user (enterprise delegate) to secure this key during transmission in order to attain a higher security against cryptographic attacks [19]. These systems, called biometric cryptosystems or crypto-biometric systems, can benefit of the advantages of both fields (cryptography and biometrics). In such systems, while cryptography endows with high and modifiable security levels, biometrics ensured that enterprise requesting the information is legitimate. Furthermore, biometrics can secure effectively the cryptographic key. Among several scenarios, the cryptographic commitment [20] scheme draws greater attention from researchers. In this scenario, the cryptographic key is embedded in a feature vector without disclosing it (hiding the cryptographic key). Thus, the fuzzy commitment [21] is one of the used commonly method of this scenario.

5 Crypto-Biometric Based EN Security

The main objective of this study is to design and develop an e-security system for enterprises information exchange. The proposed system uses symmetric cryptographic and multi-factor authentication methods. In our system, users (enterprises delegates) authentications are based on a card combined with a PIN code and a biometric trait. The proposed e-security system based on multi-factor authentication is shown in Fig. 1. In any biometric system, there are two essential phase: enrolment and identification. Thus, all the cooperated enterprises must be previously enroled (registered) in each others. For each enterprise, during the enrollment process, the feature vectors (or template) for their delegate is generated from their biometric modalities (e.g., Palmprint modality) and stored in all databases of the cooperated enterprises for later use (identification process) as well as in the delegate ID card. In addition, a PIN code is randomly generated and stored in this card. Hence, the PIN code is used for security purposes and to authenticate the delegate in the electronic transmission device (logical access control).

In the enterprise requesting the information (terminal system), the delegate sent over the network an encrypted data format. In this system, two steps can be found: As the first step of the sending process, verification and feature vector generation are the foundation on which the ID card PIN code is verified, if it is true, the feature extraction technique is applied on the delegate palmprint modality for generate the feature vector (V_F) and then compared with the stored feature in ID card. On the other hand, if the card PIN code or the extracted feature does not match with these in the ID card, the delegate access is denied. After the preliminary verification in which the feature vector is extracted, a fuzzy commitment scheme (where an encryption key (c) is protected using the extracted vector) is applied. In this case, a combination function is used in order to associate the encryption key with a person and to compute an offset (V_{FC}) which is a combination between the key and the feature vector. The fuzzy commitment is then represented by the pair ($V_{FC}, h(c)$), where $h(c)$ is a one way hash function. It is worth to notice that neither the biometric feature (V_F), nor the key (c) are publicly transmitted. Finally, the enterprise sent: *i)* the offset V_{FC}, *ii)* the hash function $h(c)$ and *iii)* the encryption message ($\widehat{m}$) to the central server of the destination (other enterprise).

In the destination enterprise, the authentication process is correctly performed if a fresh feature vector reading V'_F allows the computation of a binary string sufficiently close to c and the comparison between their hash values succeeds. In this case, the central system can decrypt the delegate data ($\widehat{m}$). Thus, in the key retrieval process, the feature vector of the stored delegate (V'_F) in the system database is used with the binary string (V_{FC}). Now, the key can be recovered. To achieve that, bits of result binary vector are used to retrieving the key. After that, the extracted key is also used to obtain the feature vector V_F. Then, the obtained feature vector V_F is matched with feature vector V'_F, if the obtained distance is below a predefined security threshold the key must be verified. Thus, to check whether the retrieve key (c') is identical to the original key, the system checks to see whether $h(c) = h(c')$.

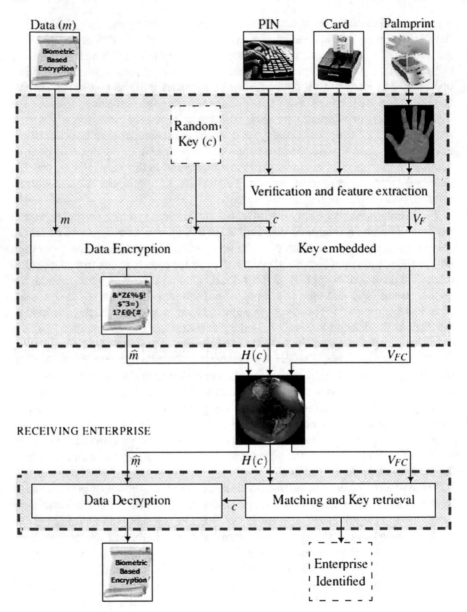

Fig. 1 Block diagram of the palmprint based biometric cryptosystem for EN security

6 Proposed Security Solution for Enterprises Information

Traditionally, personal identity recognition plays an important role in ensuring security of logical and physical access due to their effectiveness as well as their robustness. Recently, biometrics has attracted an increasing amount of attention due to their effectiveness as one of the most powerful personal identity recognition methods up to date. So, for several years, and among many biometric modalities, these extracted from the human hand have been consistently used to help make identification easier [22]. Commonly used modalities to identify person include the palmprint [23] which is relatively stable and can be extracted from a low resolution image. Therefore, over the last decades, palmprint has received a great amount of attention and it has proven suitable to be used as a unique biometric identifier. Furthermore, the vast majority of studies for improving the performance of palmprint identification system uses images captured under the visible light, which means that several and important others features are not used. However, during the past few years, some researchers have used more features from the palm, such as the veins of the palm, to improve the effect of these systems. The veins of the palm mainly refer to the inner vessel structures beneath the skin and the palm-vein images can be collected using near-infrared light. Obviously, the palm-veins [24] are a low risk of falsification, difficulty of duplicated and stability because they lie under the skin. Moreover, the availability of a device that can acquire palmprint and palm-vein images simultaneously has promoted research to constrict a multimodal system based on the fusion of these modalities in order to overcome some of the limitations imposed by unimodal systems like as insufficient accuracy caused by noisy data acquisition in certain environments [25]. For that, in our proposed system, we use, as biometrics modalities, the palmprint (PLP) and palm-vein (PLV) of the enterprise delegate in charge of communication. We give hereafter three sub-sections which described in detail the principal tasks in the server of each enterprise (receiver, transmitter).

6.1 Preliminary

Feature extraction is an important task in the biometrics applications due to the large amount of different existing features in signal (especially in image) [26]. Due to this necessity, a well considerable effort has been made by the researchers in this direction. Indeed, all issues related to the final conception of a biometrics system are generally related to the feature extraction task. In this sub-section, our feature extraction method and the corresponding feature matching are discussed.

6.1.1 Feature Vector Extraction

In our biometric system, the feature vectors are generated from the ROI sub-images by filtering it with 2D Gabor filter [27]. In pattern recognition fields, this technique is widely used and it proves their efficiency vis a vis the simplicity of implantation and the system accuracies. Specifically, a 2D Gabor filter $G(x, y; \theta, \mu, \sigma)$ can be formulated as follows:

$$G(x, y; \theta, \mu, \sigma) = \frac{1}{2\pi\sigma^2} e^{-\frac{1}{2}(\frac{(x^2+y^2)}{\sigma^2})} e^{2\pi j \mu (x cos\theta + y sin\theta)} \tag{1}$$

where $j = \sqrt{-1}$, μ is the frequency of the sinusoidal signal, θ controls the orientation of the function, and σ is the standard deviation of the Gaussian envelope. Thus, the response of a Gabor filter $(G(\theta, \mu, \sigma))$ to an image (I) is obtained by a 2D convolution operation.

$$I_f(\theta, \mu, \sigma) = I * G(\theta, \mu, \sigma) \tag{2}$$

In our work, the $N \times N$ Gabor filter, $N = 16$, at an orientation, $\theta = \frac{\pi}{4}$, will convolute with the ROI sub-images. The results of a pair of a real and an imaginary filtered image are combined into a module response $\mathscr{A}$ as follows:

$$\mathscr{A} = \sqrt{\{Re(I_f(\theta, \mu, \sigma))\}^2 + \{Im(I_f(\theta, \mu, \sigma))\}^2} \tag{3}$$

The Gabor module response is encoded as "0" or "1" based on the binarized threshold (T_{th}). Therefore, the binary vector, $V_F(i, j)$, is represented by the following inequalities:

$$V_F(i, j) = \begin{cases} 1, & \text{if} \quad \mathscr{A}(i, j) \geq T_{th} \\ 0, & \text{if} \quad \mathscr{A}(i, j) < T_{th} \end{cases} \tag{4}$$

The binarized threshold value is given as follow:

$$T_{th} = k \cdot \rho \tag{5}$$

where ρ denotes the mean value of the module response $(\mathscr{A})$ and the $k \in [0.25 : 0.25 : 3.00]$. Finally, it is important to note that, in our series of experiments, the 2D Gabor filter parameters: μ and σ are set as 0.010 and 1.200 and the T_{th} is chosen empirically (by varying the k from 0.25 to 3.00).

6.1.2 Feature Vector Matching

Each delegate palmprint is represented by a unique feature vector. The task of major concern was to appoint an unknown palmprint to one of the possible classes. We will assume that a set of reference palmprints are available to us and we have to decide which one of these reference patterns the unknown one (the test pattern) matches best. A reasonable first step to approaching such a task is to define a score or distance between the (known) reference features and the (unknown) test feature, in order to perform the matching operation known as feature matching. Systems then make decisions based on this score and its relationship a predetermined threshold. If the similarity is below the threshold, a match is declared and a no match is declared otherwise. By changing the security threshold, system errors can be adjusted. Because our feature vectors is binaries, the suitable matching task is based on a normalized Hamming distance [28]. It is defined as the number of places where two vectors differ. We can define the Hamming distance d_h as:

$$d_h = \frac{1}{H \times W} \sum_{i=1}^{H} \sum_{j=1}^{W} V_F^T(i, j) \oplus V_F^R(i, j) \qquad (6)$$

where V_F^T and V_F^R denote the test and reference feature vector and $H \times W$ represents the feature vector size. However, it is noted that Hamming distance d_h is between 1 and 0. For perfect matching, the obtained matching score is zero.

6.2 Encryption and Key Embedded Process

The following algorithm is executed at the enterprise requesting information level (at the transmitter level).

1. Perform Region Of Interest (ROI) extraction method on the original PLP and PLV images.
2. Based on 2D Gabor filter, generate, from the ROI sub-image, a 2D binary feature vector which is denoted V_F.
3. Compare the ID card PIN with this enter by user (enterprise delegate); if not equal go to end (denial access).
4. Perform feature matching between the feature vector V_F and this stored in ID card (V_{F_s}); if the score is below a predefined threshold go to end (denial access), else take out V_F (see Fig. 2) which is used as input for the key embedded task.
5. Generate a random encryption key; we denote it by c and their length by ℓ_c.
6. Encrypt message (m) by the key c, resulting an encrypted message denoted by $\widehat{m}$.
7. Embed the key c in the feature vector V_F using the fuzzy commitment scheme, the result is denoted by V_{FC}. For that, transform V_F into a 1D binary vector (denoted by $\overline{V}_F$), concatenate the key c in order to obtain a vector C with same size of V_F (size of $n \times \ell_c$ and n denotes the number of c within C) and then *XORed* C with $\overline{V}_F$ to obtain the commitment $V_{FC} = \overline{V}_F \oplus C$.
8. Generate the Hash function of c, denoted by $h(c)$.
9. Transmitting $\widehat{m}$, $h(c)$ and V_{FC}.

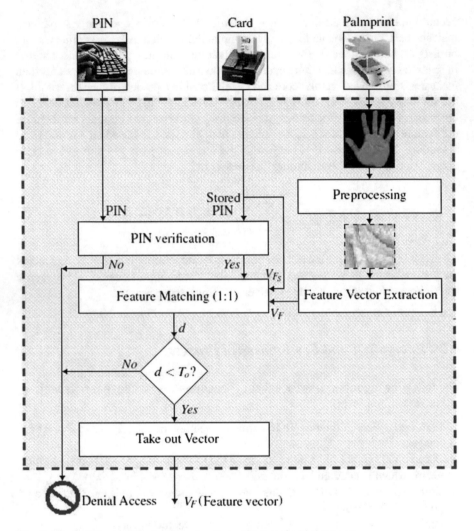

Fig. 2 Block diagram of the feature extraction and key embedded process

6.3 *Decryption and Key Retrieval Process*

At the reception (at the receiver level), the enterprise must execute the following algorithm in order to retrieve the key and authenticate the transmitted enterprise.

1. Perform a *XORed* function between a stored feature vector (V_F^i) and the received commitment V_{FC} to obtain a vector $\tilde{C}$ ($\tilde{C} = V_F^i \oplus V_{FC}$), see Fig. 3.
2. Reshape the obtained vector $\tilde{C}$ into matrix M_c with size $n \times \ell_c$.
3. Take the majority voting among M_c, as result we obtain $\tilde{c}$.

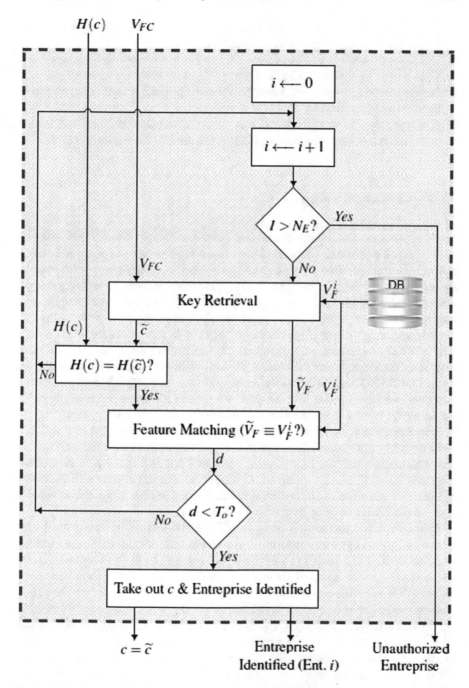

Fig. 3 Block diagram of the key retrieval and enterprise identification process

4. Compare the Hach functions $h(c)$ and $h(\tilde{c})$. If not equal go to another person ($i = i + 1, i <$ Number of stored enterprises, N_E)
5. If they are the same keys, concatenate the key $\tilde{c}$ in order to obtain a vector $\tilde{C}'$ with same size of V_{FC}, and then *XORed* it with V_{FC} to obtain $\tilde{V}_F$ ($\tilde{V}_F = V_{FC} \oplus \tilde{C}'$).
6. Perform feature matching between the $\tilde{V}_F$ and V_F^i; if the score is above a predefined threshold go to another person ($i = i + 1$).
7. Finally, decrypt $\widehat{m}$ using the key $\tilde{c}$. It is important to note that i denote the enterprise which requested the information (represented by the feature vector V_F^i).

7 Experimental Setup

In our experiments, we use palmprint and palm-vein images dataset [29] with size of 300 class (or persons), which is similar to the number of enterprises in small to medium sized cooperative. In this dataset, each delegate (person) has twelve samples for each modality. Thus, we randomly select three samples for each modality in order to construct the system database (enrolment phase). The remaining nine samples were used to test the system performance. However, by comparing nine test samples with the corresponding class in the database, we can obtain the client experiments. So, a total of 2700 scores were made. Similarly, by comparing these samples with each class in the database (except their class), we can obtain the impostor experiments. So, a total of 403650 scores were made. In our work, the set of experiments are divided into three sub-parts. In the first sub-part, we present a comparison study between the unimodal biometric system (without encryption key) by varying everytime the binarized threshold for Gabor template in order to choose the best one which is yield the best performance. The second sub-part focusing on the biometric system performance, in which the encryption key is embedded. In this sub-part, several key lengths (32 to 512 bits by a step of 32 bits) were tested. Note that, in these two sub-parts we use unimodal biometric systems. Finally, the last sub-part is devoted to evaluate the performance of the multimodal systems.

Generally, in pattern recognition application such as biometric system, the feature extraction task has a greater impact on the recognition rate of the classification system. Because our feature extraction method depends not only on the Gabor filter size, but also on the binarizing threshold, for that a series of experiments were carried out using PLP and PLV modalities in order to select the suitable threshold of binarization. Thus, after the feature, $\mathscr{A}$ is formed, each value of $\mathscr{A}$ ($\mathscr{A}(i, j)$) is compared to a threshold, T_{th}, to quantize it to '1' or '0'. Hence, this threshold depends on the mean value of the extracted feature (ρ) and a parameter k. The change of k makes it possible to give several feature vectors, as result, we can empirically choose a k which can enhance effectively the precision of the feature vector. An example of a binarized feature for palmprint image obtained in this manner is shown in Fig. 4. From this Figure, it is clear that a good choice of k allows extracting only the discriminative characteristics of the image. For example, if $k = 0.5$, the resulting feature vector contains several non-discriminative characteristics which cause a greater correlation

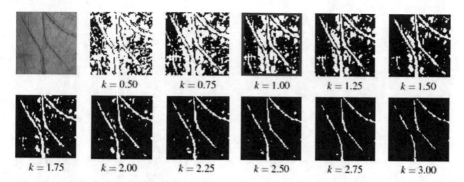

Fig. 4 An example of binary features of palmprint image under different value of k

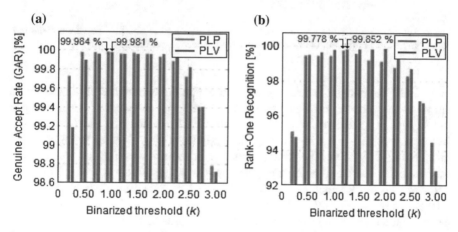

Fig. 5 Relationship between the identification rate and the values of k. **a** *open-set* identification system, **b** *closed-set* identification system

between the inter-class. Whereas, if $k = 1.50$, the non-discriminative characteristics decreased and then correlation between the inter-class also decreased. In our e-security scheme, before using the proposed biometric system in a crypto-biometric protocol, it must, firstly, choose the best value of k in order to obtain an efficient biometric system. In our study, we test twelve values of k (0.25 to 3.00 by step 0.25) using the PLP and PLV modalities. The series of tests are performed by comparing all the given identification rate and finding the value that gives the best rate. The problem we address is as follows: we want chosen the k such that both of Genuine Acceptance Rate (GAR), for open-set identification, and Rank-One Recognition (ROR), for closed set identification, are maximized.

In Fig. 5, we plot the system performance as a function of k for PLP and PLV modalities. The reason Fig. 5 was generated to show how k might have an effect on the performance of our system. We observe that the identification accuracy (GAR and ROR) becomes very high at certain values of k, where it actually exceeds 99% and slight decreases in identification accuracy as we go to higher values of k. Thus, a value

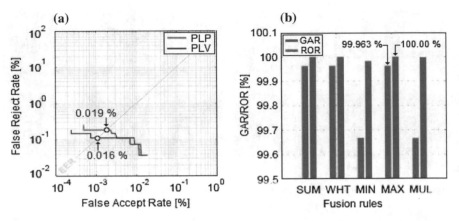

Fig. 6 Biometric systems test results. **a** comparison between PLM and PLV based unimodal biometric system, **b** Multimodal open-set/closed-set identification system

of 1.00 of k is enough to achieve good accuracy in the two modes of identification. However, the open-set identification system (see Fig. 5a) can work with a GAR equal to 99.984% at a security threshold T_o equal to 0.2267 in the case of PLP modality. In the case of PLV modality, this system operates with a GAR of 99.981% and a threshold of 0.2714. In other hand, when the closed-set identification (see Fig. 5b) was used, the PLP based biometric system works with a ROR equal to 99.778% and a Rank of Perfect Recognition (RPR) equal to 7. This rate is equal to 99.852% with a RPR = 41, for the PLV based biometric system. It is concluded that the system performance under $k = 1$ is a very efficient and yields better results than the others values of k. Thus, the threshold of binarization is equal to the mean value of the amplitude of filtered image (ρ).

In Fig. 6a, we have compared the performance of PLP and PLV based unimodal biometric systems in the case of open-set identification mode and the results show that PLV modality has a better performance than PLP modality. Thus, in the PLV modality an improvement about 16% is remarked compared with the PLP modality in the unimodal system. In this case, the system can give an Error Equal Rate (EER) equal to 0.016% and 0.019% for respectively PLP and PLV modalities. Furthermore, biometric system performance in the case of unimodal scenario produces some errors. For that, in this sub-part, we try to improve their performance by using the data fusion principal. However, in the proposed multimodal system, each modality, PLP & PLV, are operated independently and their results are combined using score level fusion scheme. In the two cases, open-set and closed-set identification modes, to find the best fusion rules (in our scheme, we examined five fusion rules which are: SUM, WHT (weighted sum), MUL (multiplication), MAX and MIN) an experimental result at the GAR point as well as ROR point are shown in Fig. 6b. Thus, this Figure relates to a comparison of the identification rates by varying the fusion rules, in the open/closed-set identification system. From this Figure, it is clear that our open-set identification system achieves the best GAR in all fusion rules with expected the

MIN rule. However, it can operate with a perfect recognition (GAR = 100.00%) in SUM, WHT, MAX and MUL rules at a threshold T_o, respectively, equal to 0.0772, 0.0794, 0.0661 and 0.0583.

Always, from this Figure, it can be seen also by combining PLP and PLV, the closed-set identification performance is in general improved. These results show that SUM, WHT and MAX fusion rules perform better than the MIN and MUL fusion rules and improves the original performance by a ROR equal to 99.963% and RPR equal to 4 for SUM and WHT rules and 2 for MAX rule. Finally, through a serious analysis of all obtained results, it can be concluded that in general the performance of the unimodal biometric system is significantly improved by using the proposed feature extraction method. In addition, these results are also demonstrated that multimodal fusion (using MAX fusion rule) performs better than systems accuracies in the two modes of identification (open-set and closed-set identification).

8 Security Analysis

Generally, the typical practice in a biometric cryptosystem is that a key must randomly generate and then binds in the feature vector for the user (delegate). Thus, this method always makes the system more secure. For that, several tests are presented in this sub-section in order to evaluate the key retrieval task as well as the biometric system at the receiver end. Basically, at the reception of the transmitted data, the system must be identified concerns this enterprise is an authorized enterprise (genuine enterprise) or not. For that, the system must execute firstly an open-set identification task. If the enterprise is accepted, then a closed-set identification is performed in order to authenticate exactly this enterprise (see Fig. 7).

8.1 Unimodal Biometric Cryptosystem

In this sub-part, we examine the performance of the biometrics cryptosystem when single modality is used. Indeed, the biometric system performance can be affected by the variation of the length of the key, because this key is combined with the biometric template at the terminal users level and retrieved at the receiver end level

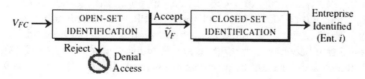

Fig. 7 Flowchart of the proposed enterprise authentication

Fig. 8 Variation of the GAR as function with the key lengths

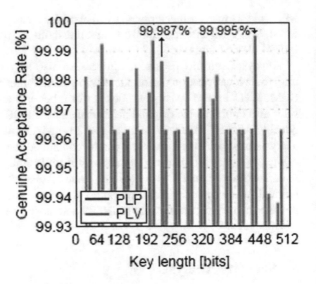

and then used again to generate the biometric template. For that, a series of tests are performed in order to evaluate the biometric system performance when the key length is varied. The open-set identification system performance, for PLP and PLV modalities, as function with the key lengths is plotted in Fig. 8. The experimental results in this figure indicate that, firstly, the integration of the key in feature vector affects considerably the accuracy of the system. Secondly, with the best choice of the key lengths, the system works with a performance near of this given without key. So, in the PLP modality, a GAR equal to 99.987% at a threshold $T_o = 0.3019$ is obtained for a length of 224 bits. This performance is almost obtained without key (GAR = 99.984%). When the PLV is used, GAR was 99.995% instead 99.981% for a length of 448 bits. Regarding these results, we cannot judge the system without seeing their performance of the point of view of closed-set identification performance as well as the key retrieval rate. For that, we test the key retrieval task for various key lengths (32, 64, 96, 128, 160, 192, 224, 256, 288, 320, 352, 384, 416, 448, 480, 512 bits). Thus, we compute the key retrieval rate as well as the ROR for the PLP and PLV modalities and the results are illustrated respectively in Fig. 9a and Fig. 9b. From Fig. 9a, we observe that the maximum key retrieval rate is equal to 99.778% for key lengths of 160 bits, but in this case a poor ROR, equal to 72.852% and a RPR equal to 21, is obtained. Thus, when a key length of 224 bits is used, the system retrieves the key with a rate of 99.667%. At this point, the closed-set identification system operates with a ROR = 99.704% and a RPR = 21. In other hand, when PLV modality is used (see Fig. 9b), the first observation which can be made is that the key retrieval rate becomes very high at key lengths of 480 bits (99.852 %). In this scenario, the closed-set identification system works with a ROR equal to 99.704% and RPR = 21.

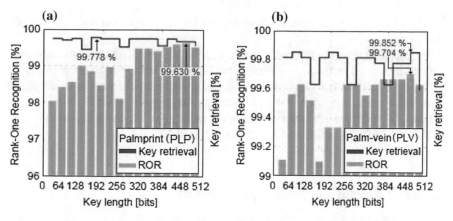

Fig. 9 Key retrieval rate. **a** PLP based system performance, **b** PLV based system performance

Virtually, a perfect biometric identification system must not accept any impostors (unauthorized enterprise), with a very few rejection of clients (authorized enterprise). For that, our scheme was tested at the False Accept Rate (FAR) equal to 0.000% and the results are illustrated in Fig. 10. So, for the best case of PLP modality (224 bits), the system rejects a 0.259%, at a threshold $T_o = 0.190$, of authorized enterprises but it can work with False Reject Rate (FRR) equal to 0.185% at a threshold T_o equal to 0.0071 (key of 192 bits). Furthermore, for the best case of PLV modality, the system rejects a 0.148% of impostors but it can work with FRR equal to 0.111% at a threshold T_o equal to 0.0510 (key of 64 bits). Note that, in the two best cases for

Fig. 10 Crypto-biometric system performance at FAR $= 0$

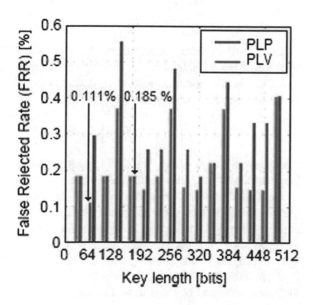

PLP and PLV, the errors rates are very reasonable for a database size equal to 300 enterprises.

Generally, biometric systems that use a single modality for identification are often affected by several practical problems like noisy sensor data, non-universality and/or lack of distinctiveness of the biometric trait and unacceptable error rates. Multimodal biometric systems overcome some of these problems by consolidating the evidence obtained from different sources. For that, the objective of the next sub-section is to test our system when multimodality principal is used.

8.2 Multimodal Biometric Cryptosystem

Generally, biometric system performance when using single biometric trait produces some errors. For that, in this study, we try to improve their performance by using the data fusion principal. However, in the proposed multimodal system, the modalities used, PLM & PLV, are operated independently and their results are combined using score level fusion scheme [30]. Thus, as our main goal is to improve the e-security system, it is imperative that we evaluate the performance of the identification system based on several key lengths. For that, several key binding schemes are proposed using the best fusion rule. Thus, two schemes are evaluated. In the first one, the same key is bound to the two templates produced by the PLP and PLV modalities (duplicated key). Thus, in this case, the entire encryption key is given by:

$$c = \begin{cases} \widetilde{c}_{PLP}, & \text{if} \quad h(c) = h(\widetilde{c}_{PLP}) \\ \widetilde{c}_{PLV}, & \text{if} \quad h(c) = h(\widetilde{c}_{PLV}) \\ \widetilde{c}_{PLP} \text{ or } \widetilde{c}_{PLV}, & \text{if} \quad h(c) = h(\widetilde{c}_{PLP}) = h(\widetilde{c}_{PLV}) \end{cases} \quad (7)$$

where $\widetilde{c}_{PLP}$ is the retrieved key from PLP and $\widetilde{c}_{PLV}$ is the retrieved key from PLV. Whereas, in the second scheme, each template (for each modality) contains a different key. Thus, in this case, the entire encryption key is given by the concatenated of the two extracted keys, as follow:

$$c = [\widetilde{c}_{PLP}, \widetilde{c}_{PLV}] \quad (8)$$

It is important to note that, the first scheme is dedicated for the small to medium length key, whereas, the second scheme is devoted for the greater length key.

Until now matching score level is the efficient scheme of fusion due to their simplicity, ease implementation and practical aspects [31]. In this scheme, the individual matching scores from the two subsystems are combined to generate a single scalar score, which is then used to make the final decision. Thus, the aim of this sub-part is to investigate whether the system performance could be improved by using the fusion of information from each modality. Generally, rule-based technique, for fusing the scores produced by the different unimodal identification systems, is used. Thus, in our multimodal system, we limited the test to only the maximum rule (MAX). The first test is performed when the key is duplicated in the two modalities and the experimental result for the open-set identification system at the EER points indicates that our system can effectively work with a very minimum EER equal to $2 \times 10^{-4}\%$ at a threshold 0.0360. So, for this case, the system can achieve a higher open-set identification rate (ROR $= 99.852\%$ with RPR $= 4$). Furthermore, a key retrieval rate, equal to 99.963%, is obtained. Also, at FAR $= 0.000\%$ point, the system can reach a FRR equal to 0.148% and a $T_o = 0.0061$. Finally, when the key is shared in the two modalities, the EER, ROR and FRR (FAR $= 0.000\%$) remain unchanged but the key retrieval rate becomes equal to 99.556%.

It is important to note that, regarding the PLP images, it is obvious to find several persons presenting almost the same feature vector due to the high inter-class correlation. Thus, if an attacker possesses a feature vector similar to that of the enterprise delegate, then the attacker could retrieve the embedded key. As a result, it can decrypt the transmitted data in the network. Thus, a very little probability that two enterprise delegates have a very similar PLP and PLV, for that, the use of multimodality scheme with sharing the key in the tow modalities allows decreasing the probability that an attacker retrieves the encryption key.

9 Conclusion and Further Works

Today, biometrics has been vigorously promoted around the world as a means to strengthen network security. The objective of this study is to enhance the security and privacy of enterprises data by developing a biometric cryptosystem. This system implements the concepts of fuzzy-commitment scheme combined with palmprint/palm-vein biometrics. In this work, we use feature extraction method based on Gabor filter with thresholding to enhance the discriminating capability of the palmprint feature vector. Subsequently, the enterprise data is encrypted using a random key (AES encryption), then this key is binding in palmprint (or palm-vein) modality of the enterprise delegate charged of transmission using fuzzy commitment scheme. After that, in the enterprise destination, a new scheme for the key retrieval is implemented in order to determine the cryptographic key which will be used to decrypt the message. The results of experimental research using a database of 300 enterprises delegates show the advantage of our proposed method. Future perspectives of this study are numerous. This work can be continued to study and test the properties and efficiencies of the proposed approach and also extend the study to other biometrics, e.g., evaluate the performance of this proposed methods on face, iris, etc. One critical area of future research is the implementation and the testing of our scheme on a commercial scale.

Acknowledgements The authors are grateful to the anonymous referees for their valuable and helpful comments. This research has been carried out within the CNEPRU project (Grant: B*02920-140028) of the Department of mathematics and computer sciences, University of Laarbi Tebessi.

The authors thank *Miss. Wissal Bouakal* for her valuable help as well as staff of LAMIS laboratory for helpful comments and suggestions.

References

1. Forrester JW (1958) Industrial dynamics: a major breakthrough for decision makers. Harv Bus Rev 36(4):37–66
2. Bendjenna H, Zarour N, Charrel PJ (2010) Eliciting requirements for an inter-company cooperative information system. J Syst Inf Technol 12(4):303–335
3. Ford D, Mouzas S (2013) The theory and practice of business networking. Ind Mark Manag 42(3):433–442
4. Benade SJ, van Waveren CC (2012) Technology management for emerging technologies. In: Technology management for emerging technologies proceedings of PICMET '12, pp 2395–2404
5. Amroune M, Inglebert JM, Zarour N, Charrel PJ (2011) AspeCis: an aspect-oriented approach to develop a cooperative information system. In: Bellatreche L, Mota Pinto F (eds) Model and data engineering. MEDI 2011. Lecture notes in computer science, vol 6918. Springer, Berlin, Heidelberg
6. Andersen KV, Debenham JK, Wagner R (eds) (2005) How to design a loose inter-organizational workflow: an illustrative case study. In: Lecture notes in computer science, vol 3588. Springer
7. Grefen PWPJ, Ludwig H, Dan A, Angelov S (2006) An analysis of web services support for dynamic business process outsourcing. Inf Softw Technol 48(11):1115–1134
8. Grefen PWPJ, Mehandjiev N, Kouvas G, Weichhart G, Eshuis R (2009) Dynamic business network process management in instant virtual enterprises. Comput Ind 60(2):86–103
9. Jackson L, Young L (2016) When business networks "kill" social networks: a case study in Bangladesh. Ind Mark Manag 58:148–161
10. Matinheikki J, Pesonen T, Artto K, Peltokorpi A (2017) New value creation in business networks: the role of collective action in constructing system-level goals. Ind Mark Manag
11. Törnroos JÅ, Halinen A, Medlin CJ (2017) Dimensions of space in business network research. Ind Mark Manag 61:10–19
12. Park KH, Favrel J (1999) Virtual enterprise-Information system and networking solution. Comput Ind Eng 37(1–2):441–444
13. Godbole N, Belapure S (2011) Cyber security: understanding cyber crimes, computer forensic and legal perspective, 1st edn. Wiley India Pvt Ltd
14. Matthews R (1989) On the derivation of a "chaotic" encryption algorithm. Cryptologia 13(1):29–42
15. Schneier B (1996) Applied cryptography: protocols, algorithms and source code, C, 2nd ed. Wiley
16. Meraoumia A, Kadri F, Chitrob S, Bouridane A (2014) Improving biometric identification performance using PCANet deep learning & multispectral palmprint. In: Book of biometric security and privacy-opportunities & challenges in the big data Era. Published by Springer, pp 51–69
17. Faundez-Zanuy M, Elizondo DA, Ferrer-Ballester MA, Travieso-González CM (2007) Authentication of individuals using hand geometry biometrics: a neural network approach. Neural Process Lett 26(3):201–216
18. Khan MK, Zhang J, Tian L (2007) Chaotic secure content-based hidden transmission of biometrics templates. Chaos Solitons Fractals 32:1749–1759
19. Uludag U, Pankanti S, Prabhakar S (2004) Biometrics cryptosystems: issues and challenges. Proc IEEE 92(6):948–960
20. Failla P, Sutcu Y, Barni M (2010) Esketch: a privacy-preserving fuzzy commitment scheme for authentication using encrypted biometrics. In: Proceedings of the 12th ACM workshop on multimedia and security, pp 241–246

21. Hoang T, Choi D, Nguyen T (2015) Gait authentication on mobile phone using biometric cryptosystem and fuzzy commitment scheme. Int J Inf Secur 14(6):1–12
22. Charfi N, Trichili H, Alimi A, Solaiman B (2015) Personal recognition system using hand modality based on local features. In: 11th international conference on information assurance and security (IAS2015), pp 13–17
23. Kumar A, Wong DCM, Shen HC, Jain AK (2003) Personal verification using palmprint and hand geometry biometric. In: Proceedings of the 4th international conference on audio-and video-based biometric person authentication, pp 668–678
24. Ladoux P, Rosenberger C, Dorizzi B (2009) Palm vein verification system based on sift matching. In: 3rd international conference ICB, pp 1290–1298
25. Wang WC, Chen WS, Shih SW (2009) Biometric recognition by fusing palmprint and hand-geometry based on morphology. In: Proceedings of the IEEE international conference on acoustics, speech and signal processing, pp 893–896
26. Choras M (2007) Image feature extraction methods for Ear biometrics. In: University of technology & life sciences, Bydgoszcz. Computer information systems and industrial management applications. IEEE Explore, pp 28–30
27. Wang X, Lei L, Wang M (2012) Palmprint verification based on 2D-Gabor wavelet and pulse-coupled neural network. Knowl-Based Syst 27:451–455
28. Yasuda M (2017) Secure Hamming distance computation for biometrics using ideal-lattice and ring-LWE homomorphic encryption. J Inf Secur Glob Perspect 26(2):85–103
29. PolyU. database. http://www4.comp.polyu.edu.hk/~biometrics/
30. Thepade SD, Bhondave RK (2015) Bimodal biometric identification with palmprint and Iris Traits using Fractiona coefficients of Walsh, Haar and Kekre transforms. In: International conference on communication, information and computing technology (ICCICT), Mumbai, India
31. Meraoumia A, Chitroub S, Bendjenna H, Bouridane A (2016) An automated finger-knuckle-print identification system using jointly RBF & RFT classifiers. In: 15th international conference on ubiquitous computing and communications and international symposium on cyberspace and security (IUCC-CSS), Granada, Spain, pp 17–22

Lightweight Context-Based Web-Service Composition Model for Mobile Devices

Roshan Fernandes and G. L. Rio D'Souza

Abstract With the widespread use of mobile devices having powerful processing capabilities, there is an ever-increasing demand to localize essential mobile services on these mobile devices to increase convenience. Most of the current computation concentrates on collecting service information from mobile devices and processing it at the server side. This is because semantic analysis of the service description in the mobile device is a resource intensive process. One of the main processes involved in the semantic analysis is the Parts of Speech (POS) Tagging. Currently, POS tools are not available for mobile devices. POS tagging however is a resource intensive application which is a challenge in the context of mobile devices due to limited availability of resources such as power, memory and processing capability. This chapter discusses a new, lightweight, context based web service composition model for mobile devices. The main idea is to build a lightweight POS tagger in the mobile device itself. The POS tagger finds its application in the context of identifying services requested by users in the form of natural language queries. Once the service names are identified in the mobile device, the request is sent to the web-service providers for their response and these responses are composed in the mobile device.

Keywords Mobile web-service · Parts of speech tagger (POS)
Markov chain probabilistic model · JSON · XML · Trie

R. Fernandes (✉)
Department of Computer Science and Engineering, NMAM Institute of Technology,
Udupi District, Nitte 574110, Karnataka, India
e-mail: roshan_nmamit@nitte.edu.in

G. L. Rio D'Souza
Department of Computer Science and Engineering, St. Joseph Engineering College,
Vamanjoor, Mangalore 575028, Karnataka, India
e-mail: riod@sjec.ac.in

© Springer International Publishing AG, part of Springer Nature 2019
S. Patnaik et al. (eds.), *Digital Business*, Lecture Notes on Data Engineering
and Communications Technologies 21, https://doi.org/10.1007/978-3-319-93940-7_9

1 Introduction

In the recent past, mobile devices are becoming more and more advanced in terms of processing power, battery and memory. The mobile users expect multiple services to be accessible easily, efficiently and quickly. The users will request for multiple services by entering through natural language in a mobile device. Most of the existing works in literature, the entire query from the user is transferred to the server and the semantic analysis of this sentence is performed at the server. This is because of the limitation of processing power of the mobile devices. Parts of Speech (POS) Tagging is a very important process involved in the semantic analysis of the sentence. This process is resource consuming and hence most of the development on POS Tagger is implemented on the servers. Currently, no POS Tagger is available for mobile devices. In this chapter, algorithm for a lightweight POS Tagger for mobile devices is discussed by using a limited corpus. The tagging is performed by using a probabilistic model, namely Markov Chain. Such a system also uses Rule-Based Hidden Markov Model to identify various categories of words in the Parts of Speech. Once the services are identified in the mobile device, these service requests are sent to the web-service providers. The responses are combined in the mobile device and presented to the user. To conclude, this chapter discusses a new approach to perform semantic analysis in the mobile device. This will enable an individual to save network bandwidth to send data to the server and also saves processing time in the server. The main objectives of this chapter include:

- To design a lightweight context–based POS tagger on mobile devices.
- The POS tagger must be able to accurately identify service requests from users' input in Natural Language by designing the probabilistic models. It must effectively train the corpus with certain rules and implement a Markov-chain probabilistic model to optimize POS tagging and achieve faster results.
- To identify the service and send requests to the web service providers and compose the responses accordingly.
- To achieve better space efficiency as well as have more processing power to achieve better time efficiency.

Most of the text processing problems need the Parts of Speech (POS) tagging as the first step. The POS tagger will identify each words in a sentence as a parts of speech and will tag them appropriately. There are various tools used to perform this task. The popular tool used for this is Stanford POS tagger. The Stanford POS tagger has a huge corpus associated with it and also uses Hidden Markov Model (HMM) to implement the parts of speech tagging process. Unfortunately this tool cannot be used in the mobile devices because of the resource constraints. Hence the semantic analysis of natural language sentences on the mobile devices become a challenge. This chapter proposes a technique to design a light weight POS tagger for the mobile devices. This work uses a light weight corpus database to store the words against their parts of speech. In this chapter, the words stored in the corpus are of context based. The context used in this chapter is related to travel or tourism. The Markov chain probabilistic model is used to further predict the parts of speech of a word which is not present in the corpus. Once the POS tagging part is completed, the sentence is semantically analysed and identifies the service name in an intelligent way. The service name will be present in the sentence provided by the user. The sentence may have multiple service names. In such cases all the service names must be identified. In this chapter, the Google map API service is used for returning places details and Zomato API service is used for returning restaurant details. In case of multiple service requests, the response is combined and presented to the user. The proposed work is lightweight in terms of the resource consumption.

Section 2 contains the related work, Sect. 4 describes the methodology and implementation, Results and Discussion is described in Sects. 5 and 6 concludes the chapter.

2 Related Work

The most common research task in Natural Language Processing is Part-of-Speech (POS) Tagging. Every word has an associated part of speech. POS tagging can be defined as a technique of assigning each term in the text to a part-of-speech. Commonly used ones are adjectives, nouns, verbs, pronouns, adverbs, prepositions and conjunctions [1, 2]. It is possible for a word to be associated with multiple parts of speech depending on the context. Wind for example is a noun meaning breeze and is a verb meaning breathlessness or gasping for breath [3]. English is one such language which is prone to ambiguity. Resolving such ambiguity is the important aspect of POS tagging. Table 1 gives a brief literature survey on the POS taggers.

Since there hasn't been a light weight mobile application for processing Natural Language queries, this work mainly focuses on localizing the process of POS tagging on a mobile device and also developing a user friendly, light weight mobile application to achieve the same.

3 Overview

This section of the chapter gives an overview of Parts of Speech (POS) tagging and Markov Chain probabilistic model.

3.1 Parts of Speech (POS) Tagging

Parts-of-speech (POS) tagging is the way of labeling a word in a text content as comparing to a specific parts of speech, in view of its definition and its unique circumstance i.e., its association with contiguous and related words in an expression, sentence or passage. Generally parts of speech include nouns, verbs, adverbs, pro-

Table 1 Literature survey related to POS tagging

Author	Techniques used	Accuracy achieved	Advantages	Drawbacks
Xiao et al. [4]	Maximum entropy Markov model (MEMM) and non stationary MEMM	Greater than that of HMM model	The conditional probability of the data in the sequence is increased	Space complexity problem of MEMM results in curse of dimensionality
Tursun et al. [5]	Semi-supervised method is proposed for the Uyghur language	94%	Accuracy surpassed the visible Markovian model	Since Uyghur is a resource poor language, its linguistic characteristics posed many challenges which made the task a complex one
Elahimanesh et al. [6]	Associative classifier in combination with HMM algorithm	98%	Effectively identifies and tags unknown words or the out of vocabulary words	Complexity is more
Lv et al. [7]	GEP (genetic expression programming) model for POS tagging	97.40%	The accuracy rate of GEP is proven to be greater than all other taggers like GA, HMM, and neural networks	Only limitation is its processing speed
Rattenbury et al. [8]	Baseline methods like Naive scan, Spatial scan and TagMaps TF-IDF	Hybrid method yielded greater accuracy	User contributed Flickr tags where used to extract the place semantics effectively	Ambiguity on the forming of place tags
Hamzah et al. [9]	HMM based POS tagger	95%	The proposed method effectively determines if a sentence is just an opinion or a simple statement and also finds the target of the sentence	Rule sets needs to be modified according to the type of corpus used since it affects the performance

(continued)

Table 1 (continued)

Author	Techniques used	Accuracy achieved	Advantages	Drawbacks
Yi [10]	Maximum entropy model	94%	Combines the features of Statistical methods and Rule based methods to achieve high accuracy	Data preprocessing stage can be further improved for basic noun phrase recognition
Piao et al. [11]	Historical Thesaurus Semantic Tagger for semantic annotation	77.12% to 91.08%	Effectively annotates historical English data and also reflects their meanings	Better algorithms can be used for word sense disambiguation
Sun et al. [12]	Pre-classification hidden Markov model	72.59%	Pre-classification done with Naive Bayes model increases the overall performance	Yields better performance only when huge data is considered
Liu et al. [13]	Latent semantic analysis using SVM classifier	Highly accurate	SVM yields greater accuracy	Fluency problem exists in the summary
Rathod et al. [14]	Rule based, Stochastic and hybrid methods of tagging	Rule based methods-88.3% Stochastic methods- 93.82% Hybrid methods-87.33%	Rule based methods make use of a small set of simple rules. Stochastic methods are highly accurate and Hybrid methods yields higher accuracy than the individual models	Rule based methods are less accurate. Stochastic methods are slightly complex and Hybrid methods doesn't assign right tags to the words unknown
Kadim et al. [15]	Bidirectional HMM based POS tagger	Reverse tagger is more accurate compared to the direct tagger	Since both the direct and reverse taggers used same resources and had same operations, the overall cost was reduced	Using a huge tag set increased the complexity of calculations

(continued)

Table 1 (continued)

Author	Techniques used	Accuracy achieved	Advantages	Drawbacks
Brill [16]	Rule based POS tagger	95%-99%	The tagger is portable, simple and yields accuracy which is almost equivalent to that of the stochastic taggers	Rule based systems are usually difficult to construct
Fonseca et al. [17]	Neural network based POS tagger	93.5%	Accurately tagged the Out of Vocabulary (OOV) words	Training the neural network model is time consuming
Xu et al. [18]	Sensing based model for mining social media data	Accurately mined user opinions on geographic locations	Mining data from the geographic locations would help the government present intended assistance in cases of emergency	As the data increases over a period of time, managing the data becomes challenging
Du et al. [19]	Vehicle GNSS and mobile RFID technology	The positioning accuracy was greater than ± 5 m with n distance of 160 m	Useful in outdoor positioning	Error rate is high in case of vehicle moving away from line-of-sight
Zhao et al. 20]	Combine Rule based methods and stochastic methods	95.2%	Combined approach of Rule based approach and stochastic approach reduces the ambiguity while tagging	The only limitation to be rectified is the speed of the system
Seyyed et al. [21]	Probabilistic approach with maximum likelihood and TnT approaches	98%	Prepares the text for tagging so that higher tagging accuracy can be achieved	Compound words with more than 3 sections are ignored
Ratnaparkhi [22]	Maximum entropy model	96.5%	It is a non-context based tagger that will work for any context	Not suitable for the implementation on a mobile device given the training set and limited resources

Table 2 Part of speech tag set

Tag	Description	Example
NN	Noun, singular or mass	Cat, ship, hero, baby
NNS	Noun, plural	Cats, ships, heroes, babies
VB	Verb, base form	Play, go, think, see
VBD	Verb, past tense	Saw, ran, came,
JJ	Adjective	Good, best, bad, dark, half
RB	Adverb	Angrily, deeply, clearly
PRP$	Possessive pronoun	My, his, hers, you

nouns, conjunctions and many more. Parts of speech tagging is an essential problem in the field of natural language processing. That means if a word is the essential unit of a language then part of speech is the most vital element of vocabulary. It plays an important role in natural language analysis and understanding. For instance errors in POS tagging may prompt wrong comprehension of sentence in Machine translation. The natural language processing tasks like data extraction, data retrieval and classification depend on part of speech tagging to eventually accomplish the desired outcomes. For example in a sentence "I fish a fish", the word fish will be marked with same tag without POS tagging. The sentence after POS tagging is "I/PRP fish/VBP a/DT fish/NN". Some of the parts of speech tag set is shown in Table 2.

There are two types of POS taggers: Rule based tagger and Stochastic taggers. Rule based POS tagging makes use of hand-written rules for tagging. Stochastic tagger makes use of gigantic amount of data to set up the language of every situation and it doesn't need any information about the rules of the language. The POS tagger steps consists of Tokenization, Ambiguity lookup, Ambiguity resolution. This is shown in Fig. 1. In Tokenization process, the given text or content is partitioned into tokens which can be utilized for further analysis. Here token may refer to words, punctuation marks or articulation limits. Ambiguity lookup uses guesser and lexicon. The lexicon is used to provide words list and their possible parts of speech, guesser is used to analyze the tokens. Ambiguity resolution is known as Disambiguation which depends on the information about word, for example, the likelihood of the word. For instance, in most of the cases the word power is used as noun rather than verb.

3.1.1 Rule Based Tagger

Rule based methodology uses predefined language rules to increase the precision of tagging. It obtains the possible tags for each word from a dictionary or lexicon. Hand-written rules are useful in finding a right tag when a word has multiple tags. Disambiguation is accomplished by examining the topographies of the word, its first word, its following word and different perspectives. For instance, if the previous word is adjective then the word being referred to must be noun. This information is

Fig. 1 Steps involved in
POS tagging

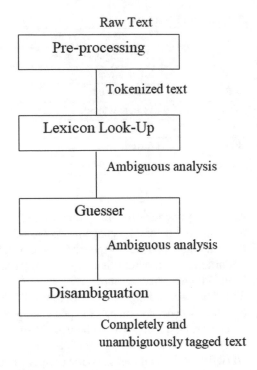

Raw Text

Pre-processing

Tokenized text

Lexicon Look-Up

Ambiguous analysis

Guesser

Ambiguous analysis

Disambiguation

Completely and
unambiguously tagged text

coded as rules. Another example can be if a possible verb does not agree in number with the preceding NP, then eliminate the verb tag.

Basically Rule based tagger needs supervised training. Recently there has been a lot of interest for automatic generation of rules. One method to deal with automatic rule generation is to apply POS tagging to the raw text, then checking the result for accuracy. Then manually correcting the erroneously tagged words. This correctly tagged content is submitted to the tagger. At this point the tagger learns correction rules. Sometimes it may require several iterations.

3.1.2 Stochastic Tagger

Stochastic tagger uses various ways to deal with the problem of POS tagging. The model incorporates frequency or probability to appropriately label the words. The disadvantage of stochastic tagger is tagging a sentence which is incorrect according to the grammatical rules. The tag which appeared more number of times in training set is allocated to an ambiguous instance of that word. This is known as word frequency measurement. It may also yield the forbidden tag sequence which is considered as an issue. Another option to the word frequency is to predict the probability of tag sequence. This is called n-gram approach.

Another level of complexity that can be brought into a stochastic tagger joins the past two methodologies. This is called as Hidden Markov Model. It is used to find the probability of specific words and to find the probability of remaining words in the segment. Consider an example where there is an article such as 'an' or 'the' then the possibility of next word being noun or adjective is 40%, and possibility of being a number is 20%. With this logic we can say that the word 'Park' in 'the park' has the probability of being noun than verb. A similar strategy can be utilized for tagging in a given context.

In this chapter, the light weight POS tagger is designed for the mobile devices using the Markov Chain model concept. The next section gives an overview of Markov Chain model.

3.2 Markov Chain Probabilistic Model

Markov Chain process was named after Andrey Markov, a Russian Mathematician. It satisfies *memorylessness* property most popularly termed Markov property. Markov property states that given the present, the future is independent of the past. Markov Chain is the simplest Markov Model. The Markov chain process can be explained with a simple example of weather prediction. Suppose we have access to the weather data of 10 years. Initially if we note that Day 1 was sunny, Day 2 was sunny, Day 3 was cloudy, Day 4 was rainy and so on. With this data in hand we can compute the probability of next day's weather based on current day's weather information. For example if today is a sunny day there are 50% chances that tomorrow would be sunny, 40% chances that tomorrow would be cloudy and 10% chances that tomorrow would be rainy. We can have an entire system of probabilities that can be used to predict the weather in future. Markov chain models can be used in agricultural sector in order to predict how much to plant depending on the state of the soil and weather conditions. It can also be used in finance to decide how much to invest in stocks. Another example would be word predictions in messaging applications smartphones. In Google keyboards, an option called Share snippets is available that asks what we type and how we type in order to improve the keyboard. These words are analyzed first and then used in the application's Markov chain probabilities. This is how words are predicted based on what we type. Other real world applications of Markov Chain model would be, Google's page rank algorithm, random walk example, gene prediction in biological sequences, stocks, business analytics and many more.

Markov chain can be specified as a set of states say S= $\{s_1, s_2 \ldots s_n\}$. It starts with a starting state s_1 and moves from one state to another. If the chain is in the current state of s_j and moves to a state s_i, the probability can be represented by P_{ji}. As per the Markov property this probability doesn't depend on the previous states. P_{ji} is also termed as transition probability i.e. the probability of moving from state s_j to s_i. Initial probability also needs to be specified. This is usually defined on some starting state S.

Markov process is an integral part of probabilistic theory. Markov chain models are most popularly used in the queuing systems to analyze the long term behavior of

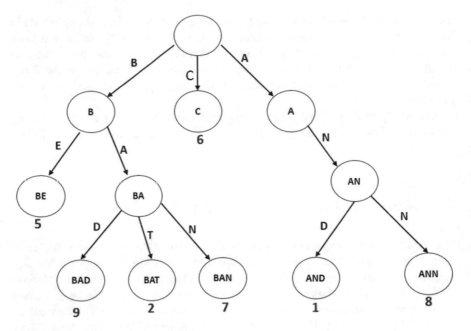

Fig. 2 A *Trie* for sample keys with values

the system and also to analyze the sequential and temporal data like that of biological DNA sequence.

3.3 Trie Data Structure

Trie is a tree data structure used to store a collection of strings. This data structure is basically used to store the key-value pairs. Unlike BST's (Binary Search Tree), the key associated with a node is defined by the position of the node. If two strings have a common prefix then they will have a same ancestor in this tree. If we have thousands of strings, trie data structure can be used to store all the strings, which makes it easy to search if a string exists or not. Trie can be an ideal data structure for storing dictionaries. Another alternative would be hash table, but it occupies more space compared to trie, hence trie is usually preferred. Figure 2 shows a typical Trie data structure to represent the words "BE" with a value associated 5, "BAD" with a value associated 9, "BAT" with a value associated 2, "BAN" with a value associated 7, "AND" with a value associated 1, "C" with a value associated 6 and "ANN" with a value associated 8. For trie data structure the values to the keys are randomly assigned.

Let m be the length of a string. Then trie data structure will search the string faster in worst case which is $O(m)$, as compared to the hash table. When a hash function of different keys map to the same position, key collision is said to occur. The search time taken in the hash table in the worst case is O(N) time (where N specifies total key length), yet significantly more ordinarily is O(1), with O(m) time spent in assessing the hash. Unlike hash table. The trie data structure avoids the key collision. Hash table buckets are the ones that are used to store the key collisions. Trie data structure requires buckets whenever a key is associated with multiple values. As the number of keys are added on to the trie structure there is no need to make changes to the structure as in hash tables. The trie also provides the alphabetically sorted order of the entries with respect to the key.

In some cases where the data is directly accessed from the secondary storage or hard drive, trie can be slower when compared to hash tables. Floating point numbers that form the keys can result in lengthier chains. However, a trie data structure with bitwise operation can be used in such situations. In some cases, trie data structure can occupy more memory when compared to the hash tables, as memory is allocated to each character in the string, instead of an entire chunk for the whole string.

In this chapter, Trie data structure is used to store the context based corpus in a mobile device.

4 Methodology and Implementation

This work proposes an approach to model a context based POS tagger for mobile devices. This system is built on a light weighted corpus. Space efficiency is achieved using Markov chain probabilistic model. The main aim of implementing this system is to make the process of identifying and tagging each word with the right Parts-of-Speech and also identifying the relevant context behind the search query in very less time which would bar the problem of limited availability of memory, processing and power which is why it can be used by any android device efficiently. The proposed architecture is shown in Fig. 3.

The proposed system contains 3 main functional components: POS tagging, location tracking and web services. POS tagging includes splitting the user query into tokens, tagging them with appropriate parts-of-speech, comparing with the rules and passing the relevant words to the web service. To optimize the application in terms of processing, probability is calculated and parts of speech are assigned to a sentence that doesn't match any of the rules. Location tracking includes detecting the current location of the mobile device, retrieving the details of the current location and identifying the list of services available. The system tries to identify the location of the user with the help of the service/network provider. If this fails, then the GPS is used. The user query after being processed by the POS tagger is passed to a particular web service which would then process the request and pass it on to the user to view. API's used in web service are Google maps API and Zomato API. Implementation of the system involves following modules.

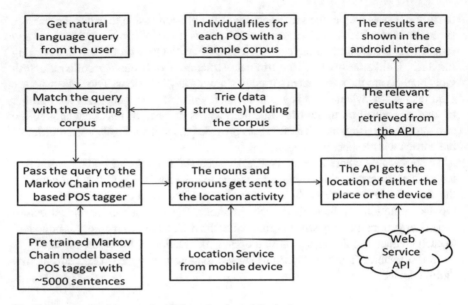

Fig. 3 Proposed architecture for POS tagging in mobile devices

4.1 Get Natural Language Query from the User

The user has to enter the query in the form of a natural language query. The language considered in this chapter is English. The user will be entering the query in the user interface provided in the final Application. Typical sentences the user may enter would be:

- *Show restaurants near Nitte*
- *Show ATMs near Nitte*
- *Show restaurants, ATMs and Theatres near Mangalore*
- *Theatres near me*

and many more such sentences. The sentences are collected from the various users at NMAM Institute of technology, Nitte campus. Most of the users are using the above mentioned patterns of sentences to find the services in the context of tourism. In most of the recent works these sentences are being submitted to the server and the linguistic analysis of these sentences are performed at the server. The analysis requires Parts of Speech (POS) tagging as the first step. The POS tagging process requires more memory to store the corpus of language words. Also the processing of the POS tagging step needs more resources. Because of these reasons so far no POS tagger in the mobile device is available. In this chapter, the user query is processed at the client side (mobile) itself there by reducing the burden at the server. The service names like *restaurants, ATMs, theatres* and many such words are recognized in the mobile device and the details of these are sent to the web service providers. The web service providers used in this chapter include *Google Maps* and *Zomato*.

Table 3 Details of the POS tag set

Tags Used	Abbreviation
DT	Determiner
IN	Preposition or conjunction
JJ	Adjective
NN	Noun/common nouns
NNP	Proper noun
VB	Verb
WP	Pronoun

The response from these web service providers will be in JSON (java Script Object Notation) format. The details are combined at the mobile side to give the complete response to the user. The input from the user to this system is a natural language query requesting service details and the response is the composed details of these services.

4.2 Preparing the Corpus

Corpus refers to a collection of words likely to be used by the POS tagger. Initially, all the words relevant to the context of tourism are manually found and written. Once the words are found, they are grouped into their respective parts-of-speech and each of these groups are saved in respective text files. Details of the POS tag set used in the context of tourism is specified in Table 3.

For instance, if we consider noun, the relevant words would be ATM's, Hotels, Restaurants, Theatres and many more.

The *Trie* data structure is used to implement the corpus which is discussed in detail in Sect. 3.3.

4.3 Framing Sample Sentences

Once the corpus is ready, all the possible sentences to be used in the application as user query are written. These sentences are collected from various users in the NMAM Institute of Technology, Nitte campus. Around 5000 sentences are collected from various users through an online portal system. These sentences are used in designing the probabilistic model.

Example sentences: *Show restaurants and ATM's near Mangalore, I want theatres near Mangalore.*

Table 4 Transition Probability calculations for class C1 sentences

Base state	From state	To state	Transition probability
Noun	Noun	Preposition	582/5000 = 0.12
	Preposition	Proper noun	102/5000 = 0.02
	Proper noun	Noun	40/5000 = 0.008
	Noun	Conjunction	216/5000 = 0.043
	Conjunction	Noun	865/5000 = 0.173
	Conjunction	Determiner	180/5000 = 0.036
	Preposition	Pronoun	263/5000 = 0.053
	Pronoun	Proper noun	112/5000 = 0.022

4.4 Using Markov Chain Probabilistic Model

Each word of the sentence, entered by the user, is mapped into the corpus. If there is a match with the adjectives, pronouns, prepositions and nouns (standard nouns for the context of tourism), then the corresponding word is tagged with the Parts-of-Speech tag. However, for words which were not matched with the corpus, Markov-chain probabilistic model is used to perform the same. It uses probabilistic approach and assigns the parts-of-speech based on probability. It predicts the probability of next word's parts-of-speech based on the current word's parts-of-speech. In order to predict the next word's probability, the initial step is to build a probability based finite automata. The finite automaton specifies the arrival of different words after a specific state. Suppose the very first word in the sentence doesn't match the words in the corpus, then frequency based approach is used to tag such words. The tag which is most frequently used in the corpus is assigned to such words. For instance, if the query is *Eateries in Mangalore* and the very first word *Eateries* does not exist in the corpus and suppose the most frequently used tag in the corpus is Noun/Common Noun. Then the tag NN is assigned to the word *Eateries* and the other tokens in the sentence are tagged using the Markov chain probabilistic model.

In the proposed implementation, two classes of sentences have been identified. The first class of sentence starts with a Noun and the second class of sentence starts with a Verb. Let the class of sentence starting with a Noun be C1 and let the class of sentence starts with a Verb be C2. The transition probability calculations for class C1 are shown in Table 4.

The finite automata for the class C1 sentences with the transition probabilities is shown in Fig. 4.

Given the current state, the transition probability of the next state is calculated by analyzing the existing sentences. In the finite automata shown in Fig. 4, noun forms the initial state. Given the current state is noun, the probability of the next state being preposition is 0.12. Therefore 0.12 is the transition probability. This transition probability is computed by analyzing the existing sentences. In this experiment, there are 582 sentences in which the *Noun* is followed by a preposition. The transition

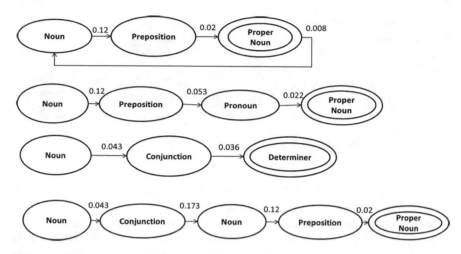

Fig. 4 Finite automata for class C1

Table 5 Transition Probability calculations for class C2 sentences

Base state	From state	To state	Transition probability
Verb	Verb	Noun	622/5000 = 0.124
	Verb	Pronoun	342/5000 = 0.068
	Noun	Preposition	537/5000 = 0.107
	Pronoun	Noun	126/5000 = 0.025
	Noun	Noun	71/5000 = 0.014
	Preposition	Pronoun	223/5000 = 0.045
	Preposition	Proper noun	116/5000 = 0.023
	Proper noun	Preposition	554/5000 = 0.111

probability from the state Noun to Preposition is computed as 582/5000 = 0.12 where the number of total sentences considered in this experiment is 5000. The base state (start state) of this class is *Noun*. The sample sentences which belong to class C1 include *theatres near me, restaurants and ATM's near me* and many more.

The transition probability calculations for class C2 is shown in Table 5.

The finite automata for the class C2 sentences with the transition probabilities is shown in Fig. 5.

In the finite automata shown in Fig. 5, *Verb* forms the initial state. Given the current state is Verb, the probability of the next state being *Noun* is 0.124. Therefore 0.124 is the transition probability. In this experiment, there are 622 sentences in which the *Verb* is followed by a *Noun*. The transition probability from the state *Verb* to *Noun* is computed as 622/5000 = 0.124 where the number of total sentences considered in this experiment is 5000. The base state (start state) of this class is *Verb*. The sample sentences that belong to the class C2 include *show restaurants near me, show me ATM's and restaurants in Mangalore*. In the above example, verb forms the initial

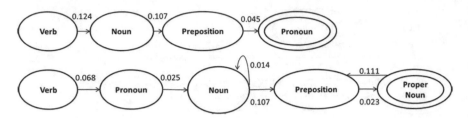

Fig. 5 Finite automata for class C2

state. Given the current state is verb, the probability of the next state being noun is 0.13. Therefore 0.13 is the transition probability.

In order to identify the word to be of a particular Parts of Speech, the token with highest probability is considered as the final Parts of Speech tag. Once the Parts of Speech tagging is done, then next challenge is to identify the service names from the input sentence. This step is discussed in the next section.

4.5 Identifying Service Names from the Sentence

Once the light weight POS tagging process is completed in the above mentioned step, the words tagged with noun, proper noun and pronoun tags are extracted from the sentence and are passed to a Web-Service which will validate the words and return relevant results in JSON format. Common Nouns are used to generalize items rather than being specific. In the context considered, the *Restaurants, ATM's, theatres* represent Common nouns. Pronouns are the words that refer to a Noun. Hence pronouns like *me, us* are also extracted. Proper nouns are used for a specific place or organization. Hence, in this context, place names such as *Mangalore, Nitte* represents Proper Nouns. In case of Pronouns, location co-ordinates are extracted to locate the exact location of the user in order to render required information. Once the results from the Web-Service is received in the mobile device, the application program in the device will parse the JSON data, combines the result if multiple JSON objects are received and then it renders to the user.

4.6 API's and Libraries Used in Web Service

At the server side two APIs are used to access the Web Service. In this chapter the context used is tourism hence the Web services are used to get services namely location, restaurant, ATM and boarding. The two APIs used are Google Map API and Zomato API.

4.6.1 Google Maps API

Google Map API is used to access the location information by passing the parameters related to a location using latitude and longitude.

This API returns details of organizations, interesting information about a specific place, establishments and various geographic settings.

The place requests in Google map API used in this chapter are:

- *Place Searches*—gives a set of data related to place information based on a user's location or explicit place query request.
- *Place Details*—gives a detailed data of a particular place including the user's feedback.

Both the above mentioned services are accessed using HTTP request and they in turn return the response in JSON or XML format. Also the requests must use the https://protocol, and include an API key.

Place Search Service

The Google Places API permits users to gather various information like geographic settings, details of a place, interesting information about a place, feedback from users for a specific location or place, and much more data about a place or organization. The data can be automatically gathered by taking user's location or by explicitly providing the place information in the request query.

Nearby Search (Proximity) Requests:
A Nearby search allows us to find places within a particular location. The search can be more fine-tuned by providing keywords or giving information of the type of place. The Nearby search HTTP request has the following form:

https://maps.googleapis.com/maps/api/place/nearbysearch/output?parameters
where the output may either be in JSON (Java Script Object Notation) or XML.

The required parameters are discussed below. All the parameters are separated using ampersand (&) character in the URL.

API key—The Application's API key.

Location—The latitude/longitude around whose place information is to be retrieved. This must be specified as *latitude* followed by *longitude*. After the location parameter, the radius information must be specified.

Radius—Indicates the radius value within which the place information must be returned. The maximum value of radius is 50000 meters.

The optional parameters are discussed below:

Keyword—A term used by Google to match the content using indexing technique.

Language—Refers to the language code, in which the response must be returned.

Minprice and Maxprice—Specifies the most affordable or most expensive information about a place. The value 0 (zero) indicates most affordable place and 4 (Four) represents the most expensive place information. The exact amount which is indicated by a specific value will differ from one region to another.

Opennow—Gives location details which are open for business during the query time.

Rankby—Lays down the order of listing the results. Possible values comprise of:

- *Prominence (default)*. This option value results in displaying the responses based on their weightage. It depends on global popularity and other factors including importance of Google indexing.
- *Distance*. This option value ensures that the results from the query are in ascending order of the distance with reference to the current specified location.
- *Type*. This option value indicates the specific type of search.

Pagetoken—The *Pagetoken* value specifies the next 20 results from the previous search query.

The following example demonstrates a search request for place with latitude -54.3407844 and longitude 243.2463018 of type 'restaurant' containing the word 'nonveg'.

https://maps.googleapis.com/maps/api/place/nearbysearch/json?location=-54.3407844,243.2463018&radius=440&type=restaurant&keyword=nonveg&key=API_KEY

To get the API key we need to follow the steps given below:

Step 1: Move to the **Google API Console**.

Step 2: Create or select a project.

Step 3: Click **Continue** to enable the API.

Step 4: On the **Credentials** page, get an **API key**. Set the necessary API key restrictions.

Step 5: Do not use this key outside of the server code.

Text (String) Search Requests:

The Google Places API for Text (String) Search requests returns the location information based on text—for example "Pizza in Mangalore" or "shoe stores near Udupi". The response will be a set of location names matching the query text. We can request for more details about the location received in the response.

The format of the text search HTTP request is as follows:

https://maps.googleapis.com/maps/api/place/textsearch/output?parameters

where the result may be in JSON or XML format. The parameters used are discussed below. The parameters are separated using ampersand (&).

Query—Specifies the key text for search operation. For example: "restaurant". The Google Places service will return the matches found.

Key—The Application's API key.

The optional parameters are discussed below:

Region—Specifies the two character region code which is defined as a *ccTLD* (country code top-level domain. Most of these codes are similar to ISO 3166-1 codes.

Location—The latitude/longitude around which place the data is to be retrieved. This must be specified as *latitude* followed by *longitude*. When we specify a location parameter, we also need to notify a radius parameter.

Radius—Indicates the radius value within which the place information must be returned. The maximum value of radius is 50000 meters.

Language—Refers to the language code, in which the response must be returned.

Minprice and Maxprice—Specifies the most affordable or most expensive information about a place. The value 0 (zero) indicates most affordable place and 4 (Four) represents the most expensive place information. The exact amount which is indicated by a specific value will differ from one region to another.

Opennow—Gives location details which are open for business during the query time.

Pagetoken—Displays the next 20 results from an earlier run search.

Type—This option value indicates the specific type of search.

The ensuing example demonstrates a text search service for *hotels near Udupi*.

https://maps.googleapis.com/maps/api/place/textsearch/xml?query=hotels+in+Udupi&key=API_KEY

Place Details Service

Once we receive place_id from the location search, we can request for more details on this place by sending place_id in the place details query. The additional details we can get using place_id parameter include full address of the place, telephone numbers of the organizations, and the ratings from various users including their feedback.

A Place Details query is an HTTP URL having the following form:

https://maps.googleapis.com/maps/api/place/details/output?parameters

The output may be either in JSON or XML format. The required parameters are discussed below. The parameters are separated using ampersand (&).

Key—The Application's API key.

Placeid—A string that identifies a location very uniquely. This information is returned from the *Place Search* service.

The optional parameters are discussed below:

Region—Specifies the two character region code which is defined as a *ccTLD* (country code top-level domain. Most of these codes are similar to ISO 3166-1 codes.

Language—Indicates the language code for the language in which the results should be returned.

The following example demonstrates a request for the specifics of a location by *placeid*:

https://maps.googleapis.com/maps/api/place/details/json?placeid=PlaceId1&key=API_KEY

The JSON response has three elements:

- *Status*—contains metadata of the request
- *Result*—contains comprehensive data about the location requested
- *html_attributions*—contains a set of attributions. These attributions must be specified to the one who use this service.

The *status* field may comprise of one of the following values:

OK - Specifies that the place name was successfully identified and there was no error detected in the process.

UNKNOWN_ERROR - Specifies that there is a problem in the server side. It also specifies that trying to connect the server after some time may result in the successful response.

ZERO_RESULTS - Shows that there is no valid result found. This refers to no business with the specified name.

OVER_QUERY_LIMIT - Indicates that the quota assigned to a user has reached the maximum limit.

REQUEST_DENIED - It specifies that the request was denied. This may result due to the wrong key value specification.

INVALID_REQUEST - This indicates that the query is not present.

NOT_FOUND - This value specifies that the site name mentioned in the query was not available in the location or pace database.

4.6.2   Zomato API

This API provides access to the Zomato library which allows us to use their database which consists of details of restaurants. Zomato API gives access to the information related to more than 1.5 million restaurants across 10,000 cities globally.

Zomato APIs are used to:

- Search for hotels by their name or place.
- Get the rating information along with place details
- Choose best areas in a city to have food using Zomato Foodie Index.

In order to use the Zomato services, first an API key must be requested. This API key is required to use the various services provided by Zomato. The various services provides by Zomato include:

Common services—Get list of Categories, Get city details, Get Zomato collections in a city, Get list of all cuisines in a city, Get list of restaurant types in a city, Get location details based on coordinates.

Location services—Get Zomato location details, Search for locations.

Restaurant services—Get daily menu of a restaurant, Get restaurant details, Get restaurant reviews, Search for restaurants.

In this chapter the common and restaurant services are used to get information within a specified city.

Most of the services provided by Zomato requires the following parameters.

- *User-key*—Zomato API key
- *City_id*—id of the city
- *Latitude* of any point within a city
- *Longitude* of any point within a city

The response will be in the form of JSON.
The next section discuss the results.

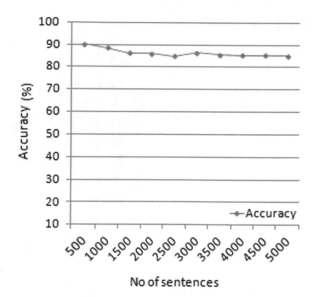

Fig. 6 Accuracy analysis

5 Results and Discussion

In this work, an attempt is made to design a light weight POS tagger for mobile devices. As discussed in the methodology section, the POS tagger is designed using a light corpus and also by using Markov-chain probabilistic model. This gives better efficiency in terms of both time and space consumption. These two factors are important for the mobile device. The processing part is completely offloaded from the server and is performed in the mobile device itself. For the performance analysis purpose, 5000 sentences are collected from the users of Nitte University apart from the sentences collected for the analysis using Markov-chain probabilistic model. The performance analysis is conducted on the Android platform mobile device. Figure 6 depicts a line graph which shows the number of sentences with the accuracy percentage. Accuracy is calculated based on the number of statements that were identified correctly based on the tourism context. As seen in the graph, the accuracy is almost a linear straight line. However, there are minor changes with the accuracy; as the number of sentences increases, the accuracy goes down by a small value. The results were nearly 90% accurate.

Accuracy is also analyzed with respect to the number of tokens considered as well. Figure 7 and Table 6 show the details of accuracy with respect to the number of tokens.

The above graph shows that the accuracy of the POS tagger increases with the increase in the number of tokens. In this work, the context based POS tagger is tested on various tokens. For about 1200 tokens, the performance achieved was almost 90%. With a greater training set in hand and larger tokens, the performance of the tagger could be further increased. Using already existing standard corpus' such as

Fig. 7 Performance analysis

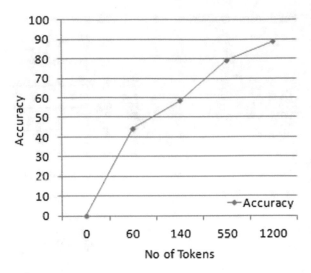

Table 6 Performance of POS tagger

Number of tokens	Accuracy (%)
0	0
60	44.35
140	58.6
550	79.2
1200	88.92

Brown corpus, *PennTree Bank* would further increase the accuracy, but the space occupied would be more. Hence the corpora prepared in the current work according to the context, achieves satisfactory accuracy of about 90% and the space occupied is also less.

6   Conclusion and Future Work

This work has achieved its purpose of creating a light weighted context based POS tagger for mobile devices. The application can efficiently process natural language queries and identify the context and give appropriate results quickly. This is achieved by implementing Markov-chain probabilistic model which ensures faster processing. The benchmarks of having a space efficient and time efficient application that doesn't consume much of memory and processes service request much faster is achieved.

Future work includes developing a system for POS tagging of multiple languages, few of which are highly ambiguous and complex on mobile devices and also making the system compatible on various platforms such as IOS. Attaining greater accuracy and efficiency would be the main focus.

References

1. Hasan FM, UzZaman N, Khan M (2007) Comparison of different POS tagging techniques (n-gram, HMM and Brill's tagger) for bangla. In: Elleithy K (ed) Advances and innovations in systems, computing sciences and software engineering. Springer, Dordrecht
2. Alva P, Hegde V (2016) Hidden Markov model for POS tagging in Word Sense Disambiguation. In: International conference on computational systems and information systems for sustainable solutions
3. Diesner J, Part of speech tagging for english text data. School of Computer Science Carneige Mellon University Pittsburgh, PA 15213
4. Xiao J, Wang X, Liu B (2007) The study of a non-stationary maximum entropy Markov model and its application on the POS tagging task. ACM Trans Asian Lang Inf Process 6:2 (Article No. 7)
5. Tursun E, Ganguly D, Osman T, Yang Y-T, Abdukerim G, Zhou J-L, Liu Q (2016) A semi-supervised tag-transition-bsed Markovian model for Uyghur Morphology analysis. ACM Trans Asian Low-Resour Lang Inf Process 16:2 (Article No. 8)
6. Elahimanesh MH, Minaei-Bidgoli B, Kermani F (2014) ACUT: an associative classifier approach to unknown word POS tagging. In: Movaghar A, Jamzad M, Asadi H (eds) Artificial intelligence and signal processing. AISP 2013. Communications in Computer and Information Science, vol 427. Springer, Cham
7. Lv C, Liu H, Dong Y (2010) An efficient corpus based part-of-speech tagging with GEP. In: Sixth international conference on semantics, knowledge and grids
8. Rattenbury T, Naaman M (2009) Methods for extracting place semantics from flickr tags. ACM Trans Web 3:1 (Article 1)
9. Hamzah A, Widyastuti N (2015) Document subjectivity and target detection in opinion mining using HMM-POS-tagger. In: International conference on information, communication technology and system (ICTS)
10. Yi C (2015) An english pos tagging approach based on maximum entropy. In: International conference on intelligent transportation, big data & smart city (ICITBS)
11. Piao S, Dallachy F, Baron A, Demmen J et al (2017) A time-sensitive historical thesaurus-based semantic tagger for deep semantic annotation. Comput Speech Lang 46:113–135
12. Sun S, Liu H, Lin H (2012) Twitter part-of-speech tagging using pre-classification hidden markov model. In: IEEE international conference on systems, man, and cybernetics, 14–17 Oct 2012
13. Liu, C.-L., Hsaio W.-H, Lee C.-H, Lu G.-C, Jou E (2012) Movie rating and review summarization in mobile environment. IEEE Trans Syst Man Cybern Part C (Appl Rev) 42(3)
14. Rathod S, Govilkar S (2015) Survey of various POS tagging techniques for Indian regional languages. Int J Comput Sci Inf Technol 6(3):2525–2529
15. Kadim A, Lazrek A (2016) Int J Speech Technol 19:303. https://doi.org/10.1007/s10772-015-9303-7
16. Brill E (1992) A simple rule-based part of speech tagger. In: Proceedings of the third conference on Applied natural language processing
17. Fonseca ER, Rosa JLG, Aluisio SM (2015) Evaluating word embeddings and a revised corpus for part-of-speech tagging in Portuguese. J Braz Comput Soc 21(2). https://doi.org/10.1186/s13173-014-0020-x
18. Xu Z, Zhang H, Sugumaran V, Raymond Choo K-K, Mei L, Zhu Y (2016) Participatory sensing-based semantic and spatial analysis of urban emergency events using mobile social media. EURASIP J Wirel Commun Netw
19. Du M, Jing C, Du M (2016) Tag location method integrating GNSS and RFID technology. J Glob Position Syst
20. Zhao L, Men J, Zhang C, Liu Q, Jiang W, Wu J, Chang Q (2010) A combination of statistical and rule-based approach for mongolian lexical analysis. In: 2010 international conference on asian language processing

21. Seyyed SR, Fakhrahmad M, Sadredini MH (2015) PTokenizer: POS tagger tokenizer. In: 2015
    2nd international conference on knowledge-based engineering and innovation (KBEI)
22. Ratnaparki A (1996) A maximum entropy model for part-of-speech tagging. In: Proceedings
    of the conference on empirical methods in natural language processing, vol 1
23. https://developers.google.com
24. https://developers.zomato.com
25. en.m.wikipedia.org

On Weighted Extended Cumulative Residual Entropy of k-th Upper Record

Rajesh Moharana and Suchandan Kayal

Abstract Generalized cumulative residual entropy has been shown an alternative uncertainty measure to cumulative residual entropy in the literature. We find its applications in the theory of communication, actuarial and computer science. In this chapter, we consider a shift-dependent version of the generalized cumulative residual entropy for the k-th upper record and its dynamic version. The advantage of the proposed measure over the existing measure is discussed. Various results including some useful properties, effect of affine transformations, bounds, stochastic ordering and aging properties are obtained. In addition, inequalities based on the proportional hazard rate model are derived. Further, we obtain some characterization results for various probability models based on the proposed information measure. A nonparametric estimator of the proposed measure is provided and then, its asymptotic normality is established.

Keywords Weighted extended cumulative residual entropy · k-th upper record value · Affine transformation · Characterization · Empirical function

1 Introduction

To overcome various drawbacks of the differential entropy (see [24]), Rao et al. [22] introduced an information measure. This is known as the cumulative residual entropy (CRE). Consider a nonnegative random variable T which represents survival time

R. Moharana
Department of Mathematics, National Institute of Technology Rourkela,
Rourkela, India
e-mail: rajeshmoharana31@gmail.com

S. Kayal (✉)
Faculty of Department of Mathematics, National Institute of Technology Rourkela,
Rourkela, India
e-mail: suchandan.kayal@gmail.com; kayals@nitrkl.ac.in

© Springer International Publishing AG, part of Springer Nature 2019
S. Patnaik et al. (eds.), *Digital Business*, Lecture Notes on Data Engineering
and Communications Technologies 21, https://doi.org/10.1007/978-3-319-93940-7_10

of a working component. We respectively denote F_T, $\bar{F}_T$ and f_T for the cumulative distribution function (CDF), survival function (SF) and probability density function (PDF) of T. The CRE of T is given by

$$\mathcal{E}(T) = \int_0^\infty \left[-\ln(\bar{F}_T(u))\right] \bar{F}_T(u) du. \tag{1}$$

One important property of CRE is that it is defined for the random variables having no well-defined PDF. Eq. (1) is consistent. It is always nonnegative. Several other properties of $\mathcal{E}(T)$ are studied by Rao et al. [22], Rao [21], Asadi and Zohrevand [1] and Navarro et al. [15]. Asadi and Zohrevand [1] introduced a dynamic version of (1), which is defined as $\mathcal{E}(T;t) = \mathcal{E}(T_t)$, where $T_t = (T - t|T > t)$ is known as the residual life. [1, 15] obtained characterizations, orderings and properties of aging-based classes for $\mathcal{E}(T_t)$. Recently, [8] proposed relations between $\mathcal{E}(T)$ and T_t on using the relevation transform.

Psarrakos and Navarro [18] proposed an extension of (1). Let a sequence of independent and identically distributed (iid) random variables is available. The corresponding records are also known. The authors relate their proposed concept to the notion of relevation transform and the average time between records. The generalized CRE of T is given by

$$\mathcal{E}_n(T) = \frac{1}{n!} \int_0^\infty \left[-\ln \bar{F}_T(u)\right]^n \bar{F}_T(u) du, \tag{2}$$

where $n = 1, 2, \ldots$. These authors have also considered its dynamic version for residual lifetime T_t as $\mathcal{E}_n(T;t) = \mathcal{E}_n(T_t)$. For the generalized CRE given by (2) and its dynamic version, authors obtained some characterizations, stochastic ordering and aging classes properties. Further, they derived various relationships with some reliability functions. Very recently, Navarro and Psarrakos [16] derived several characterizations based on $\mathcal{E}_n(T)$. In particular, they obtained characterization of some specific probability distributions based on the generalized CRE given by (2). Psarrakos and Toomaj [19] further explored new features of (2). It is showed that (2) can be applied in measuring risk in actuarial science. Psarrakos and Economou [17] introduced a sequence of weighted random variables upon using generalized CRE. For generalized cumulative entropy and its properties, one may see the manuscript by [9].

Very recently, Tahmasebi et al. [25] proposed further extension of the generalized CRE $\mathcal{E}_n(T)$. They relate this concept with the k-th upper record values. Next, we briefly describe the k-th upper record value. Consider a sequence of iid random variables $\{T_n; n \geq 1\}$ with SF $\bar{F}_T$ and PDF f_T. An observation T_j is said to be an upper record if $T_j > T_i$ for all $i < j$. For an integer $k \geq 1$, let us define a sequence $\{U_k(n); n \geq 1\}$ of k-th upper record times of $\{T_n; n \geq 1\}$ as

$$U_k(1) = 1, \quad U_k(n+1) = \min\{j > U_k(n); T_{j:j+k-1} > T_{U_k(n):U_k(n)+k-1}\}, \ n \geq 1.$$

Then, $\{T_{n:k}, n \geq 1\}$, where $T_{n:k} = T_{U_k(n):U_k(n)+k-1}, n \geq 1$ is known as the sequence of the k-th upper record values of $\{T_n; n \geq 1\}$. It is noted that $T_{1:k} = \min\{T_1, \ldots, T_k\}$ and $T_{n:1} = T_{U(n)}, \ n \geq 1$ are upper record values. The SF of the k-th upper record $T_{n:k}$ is given by

$$\bar{F}_{T_{n:k}}(u) = [\bar{F}_T(u)]^k \sum_{i=0}^{n-1} \frac{[-k \ln \bar{F}_T(u)]^i}{i!}. \tag{3}$$

For relevant details on k-th upper record value, we refer to [4]. Further, if $m_{n:k}(T) = \int_0^\infty \bar{F}_{T_{n:k}}(u)du$ denotes the mean value of $F_{T_{n:k}}$, then

$$k[m_{n+1:k}(T) - m_{n:k}(T)] = \frac{k^{n+1}}{n!} \int_0^\infty [-\ln(\bar{F}_T(u))]^n [\bar{F}_T(u)]^k du.$$

Motivated by this representation and the concept of generalized CRE given by (2), [25] defined an information measure as

$$\mathcal{E}_n^k(T) = \frac{k^{n+1}}{n!} \int_0^\infty \left[-\ln(\bar{F}_T(u))\right]^n \left[\bar{F}_T(u)\right]^k du, \tag{4}$$

where $n = 1, 2, \ldots$ and k is a fixed positive integer. The authors call this measure as extended cumulative residual entropy (ECRE). They have also considered its dynamic version for residual lifetime T_t. Based on these concepts, [25] obtained several properties on stochastic ordering and aging classes which are analogous to the generalized CRE.

Note that the measure in (4) is shift-independent, i.e., $\mathcal{E}_n^k(T+b) = \mathcal{E}_n^k(T)$, where $b \geq 0$. However, there are some situations (see [3]), where a shift-dependent measure is useful. Thus, similar to [3, 10–14], in this chapter we study a shift-dependent form of $\mathcal{E}_n^k(T)$. For T, the shift-dependent ECRE is expressed as

$$\mathcal{E}_{n,w}^k(T) = \frac{k^{n+1}}{n!} \int_0^\infty u \left[-\ln(\bar{F}_T(u))\right]^n \left[\bar{F}_T(u)\right]^k du, \tag{5}$$

where $n \in \{1, 2, \ldots\}$ and $k \geq 1$. Henceforth, it is dubbed as the weighted extended CRE of order n (WECRE$_n$). When $k = 1$, WECRE$_n$ given by (5) reduces to the weighted generalized CRE of order n considered by Kayal [10]. Further, when $k = 1$ and $n = 1$, (5) becomes weighted CRE proposed by Kayal and Moharana [11]. Also, for $k = 1$ and $n = 0$, $\mathcal{E}_{0,w}^1(T) = \int_0^\infty u \bar{F}_T(u)du = E(T^2)/2$. Below, we consider an example to illustrate the advantage of $\mathcal{E}_{n,w}^k(T)$ upon $\mathcal{E}_n^k(T)$. In particular, the example shows that even though the measures given by (2) for two probability distributions are equal, the WECRE$_n$s are not same. Thus, one must include qualitative characteristic (see (5)) in order to distinguish them.

Example 1 Consider two random variables T_1 and T_2 with SFs $\bar{F}_{T_1}(u) = 2 - u$, $1 < u < 2$ and $\bar{F}_{T_2}(u) = 3 - u$, $2 < u < 3$, respectively. Then, $\mathcal{E}_n^k(T_1) = \frac{k^{n+1}}{(k+1)^{n+1}} = \mathcal{E}_n^k(T_2)$. Further, for $n = 1, 2, \ldots$ and $k \geq 1$, we obtain $\mathcal{E}_{n,w}^k(T_1) = k^{n+1}[2(k + 1)^{-n-1} - (k + 2)^{-n-1}]$ and $\mathcal{E}_{n,w}^k(T_2) = k^{n+1}[3(k + 1)^{-n-1} - (k + 2)^{-n-1}]$, which gives $\mathcal{E}_{n,w}^k(T_1) \leq \mathcal{E}_{n,w}^k(T_2)$.

We consider dynamic form of the WECRE$_n$. For a random variable T, the dynamic WECRE$_n$ is

$$\mathcal{E}_{n,w}^k(T; t) = \frac{k^{n+1}}{n!} \int_t^\infty u \left[- \ln \left(\frac{\bar{F}_T(u)}{\bar{F}_T(t)} \right) \right]^n \left[\frac{\bar{F}_T(u)}{\bar{F}_T(t)} \right]^k du, \quad (6)$$

for $n = 1, 2, \ldots$ and $k \geq 1$. When $k = 1$, (6) reduces to the dynamic version of weighted generalized CRE of order n proposed by Kayal [10]. Further, when $k = 1$ and $n = 1$, the function given by (6) reduces to the dynamic weighted CRE introduced by Kayal and Moharana [11]. Also, $\mathcal{E}_{n,w}^k(T) = \mathcal{E}_{n,w}^k(T; 0)$.

In this part of the chapter, we recall definitions of stochastic orderings (see [23]), useful in the subsequent sections.

Definition 1 Let T_1 and T_2 have respective CDFs F_{T_1} and F_{T_2}, SFs $\bar{F}_{T_1}$ and $\bar{F}_{T_2}$, PDFs f_{T_1} and f_{T_2}, and failure rates $h_{T_1} = f_{T_1}/\bar{F}_{T_1}$ and $h_{T_2} = f_{T_2}/\bar{F}_{T_2}$. The random variable T_1 is smaller than T_2 in the

(i) likelihood ratio ordering, abbreviated by $T_1 \leq_{lr} T_2$, if $f_{T_2}(u)/f_{T_1}(u)$ is nondecreasing in u;
(ii) usual stochastic ordering, abbreviated by $T_1 \leq_{st} T_2$, if $P(T_1 > u) \leq P(T_2 > u)$ for all u;
(iii) failure rate ordering, abbreviated by $T_1 \leq_{fr} T_2$, if $h_{T_1}(u) \geq h_{T_2}(u)$ for all u;
(iv) increasing convex ordering, abbreviated by $T_1 \leq_{icx} T_2$, if $E(g(T_1)) \leq E(g(T_2))$ for all nondecreasing convex functions g.

Note that for all nondecreasing functions ϕ, we have $T_1 \leq_{st} T_2 \Leftrightarrow E(\phi(T_1)) \leq E(\phi(T_2))$, provided the expectations exist.

The chapter is arranged into 5 sections. In Sect. 2, we explore various merits of WECRE$_n$ as well as its residual form, which include the effect of affine transformations, bounds, stochastic orderings and aging properties. Section 3 presents some characterization results of various probability distributions. In Sect. 4, we propose an estimator of WECRE$_n$ via empirical approach and discuss its properties. Some concluding remarks and discussion on the future work are added in Sect. 5.

The random variables considered throughout the text are nonnegative and have absolute continuous distribution function. We use the terms increasing and decreasing in wide sense. We further assume that integrations and differentiations exist when they are used.

2 WECRE$_n$ and Its Dynamic Form

Here, we obtain some merits of WECRE$_n$ and its residual form. First, we derive expressions of WECRE$_n$ for some lifetime models.

Example 2 (i) Let T have uniform distribution in the interval $(0, \gamma)$. Then, $\mathcal{E}^k_{n,w}(T) = \gamma^2 k^{n+1}[(k+1)^{-n-1} - (k+2)^{-n-1}]$.

(ii) Assume that T is a Pareto distributed random variable associated with the CDF $F_T(u) = 1 - \left(\frac{a}{u}\right)^b$, $0 < a < u$, $b > 0$. Then, $\mathcal{E}^k_{n,w}(T) = \frac{k^{n+1}a^2b^n}{(bk-2)^{n+1}}$.

(iii) Let T follow Weibull distribution with CDF $F_T(u) = 1 - e^{-u^\alpha}$, $u > 0$, $\alpha > 0$. Then, $\mathcal{E}^k_{n,w}(T) = (k^{\frac{2\alpha n + \alpha - 2}{\alpha}}\Gamma(\frac{2+\alpha n}{\alpha}))/(n!\alpha)$.

(iv) If T follows Rayleigh distribution with CDF $F_T(u) = 1 - e^{-\frac{u^2}{2\sigma^2}}$, $u > 0$, $\sigma > 0$, then $\mathcal{E}^k_{n,w}(T) = \sigma^2$.

Let $T_{n:k}$ be the k-th upper record value. Then, the WECRE$_n$ given by (5) can be expressed as

$$\mathcal{E}^k_{n,w}(T) = E\left(\frac{T_{n+1:k}}{h_T(T_{n+1:k})}\right), \tag{7}$$

where $n = 1, 2, \ldots$ and $k \geq 1$. The following consecutive propositions show that $\mathcal{E}^k_{n,w}(T)$ is monotone with respect to n and k under conditions on T. A random variable T is said to have decreasing failure rate (DFR) property if $h_T(u)$ is decreasing with respect to u.

Proposition 1 *Let T have DFR. Then, for $n = 1, 2, \ldots$ and $k \geq 1$, we have $\mathcal{E}^k_{n,w}(T) \leq \mathcal{E}^k_{n+1,w}(T)$.*

Proof Note that $f_{T_{n+1:k}}(u)/f_{T_{n:k}}(u)$ is increasing in u. Thus, $T_{n:k} \leq_{lr} T_{n+1:k}$, which implies $T_{n:k} \leq_{st} T_{n+1:k}$. Further, under the assumption made, $\phi(u) = u/h_T(u)$ is increasing. Hence, from the Eq. (7), we obtain

$$E\left(\frac{T_{n:k}}{h_T(T_{n:k})}\right) \leq E\left(\frac{T_{n+1:k}}{h_T(T_{n+1:k})}\right), \tag{8}$$

which completes the proof. $\square$

Proposition 2 *If T is DFR, then $\mathcal{E}^k_{n,w}(T) \geq \mathcal{E}^{k+1}_{n,w}(T)$ for $n = 1, 2, \ldots$ and $k \geq 1$.*

Proof Here, $f_{T_{n:k}}(u)/f_{T_{n:k+1}}(u)$ is increasing in u. Now, the desired result follows analogous to Proposition 1. $\square$

To show the validity of Propositions 1 and 2, we consider an example presented below.

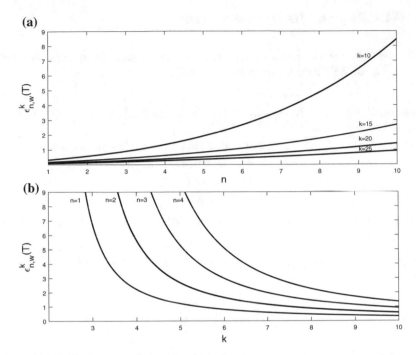

Fig. 1 **a** Plot of $\mathcal{E}^k_{n,w}(T)$ given by (9) for $k = 10, 15, 20$ and 25. **b** Plot of $\mathcal{E}^k_{n,w}(T)$ given by (9) for $n = 1, 2, 3$ and 4

Example 3 Let T be a nonnegative and continuous random variable with density function $f_T(u) = (1 + u)^{-2}$, $u > 0$. It easy to show that T has DFR and

$$\mathcal{E}^k_{n,w}(T) = k^{n+1} \left[(k - 2)^{-n-1} - (k - 1)^{-n-1} \right], \tag{9}$$

where $n \in \{1, 2, \ldots\}$ and $k > 2$. Here, the monotone behavior of $\mathcal{E}^k_{n,w}(T)$ with respect to n (for different values of k) and with respect to k (for different values of n) is shown graphically in Fig. 1a, b, respectively.

In various areas of probability and statistics, stochastic orderings and associated inequalities play an important role. It is used in portfolio selection. One may see the volume by [23] for many examples where stochastic orderings are treated as key mathematical tools. Recently, Kayal [10] established a new stochastic ordering based on weighted generalized CRE of order n. Similar to this, here, we introduce a new stochastic ordering using (5).

Definition 2 Consider two random variables T_1 and T_2 with SFs $\bar{F}_{T_1}$ and $\bar{F}_{T_2}$, respectively. Then, T_1 is smaller than T_2 in WECRE$_n$, denoted by $T_1 \leq_{WECRE_n} T_2$ if and only if $\mathcal{E}^k_{n,w}(T_1) \leq \mathcal{E}^k_{n,w}(T_2)$ for $n = 1, 2, \ldots$ and $k \geq 1$.

It is worth notifying that $\mathcal{E}^k_{n,w}(T_1) \leq \mathcal{E}^k_{n,w}(T_2)$ means that one's ability in predicting future outcome of T_1 is larger compared to that of T_2 in the sense of WECRE$_n$. Below, we consider an example to show that $\leq_{st}$ does not imply WECRE$_n$ order.

Example 4 Let T_1 and T_2 have density functions $f(u) = 2(1 - u)$, $0 < u < 1$ and $g(u) = 2u$, $0 < u < 1$, respectively. It can be easily shown that $\bar{F}_{T_1}(u) \leq \bar{F}_{T_2}(u)$ for all $u > 0$. Thus, $T_1 \leq_{st} T_2$. Further, $\mathcal{E}^2_{5,w}(T_1) - \mathcal{E}^2_{5,w}(T_2) = 0.043281$, $\mathcal{E}^6_{5,w}(T_1) - \mathcal{E}^6_{5,w}(T_2) = -0.087257$; and $\mathcal{E}^3_{2,w}(T_1) - \mathcal{E}^3_{2,w}(T_2) = -0.107006$, $\mathcal{E}^3_{5,w}(T_1) - \mathcal{E}^3_{5,w}(T_2) = 0.020306$. This observation implies that $T_1 \nleq_{WECRE_n} T_2$.

Now, question will arise. Is there any condition under which usual stochastic order implies WECRE$_n$ order. The next proposition gives answer in this direction. We omit the proof since it follows from Theorem 2.1 of [10].

Proposition 3 *Assume* $\mathcal{E}^k_{n,w}(T_1) < \infty$ *and* $\mathcal{E}^k_{n,w}(T_2) < \infty$. *If* $T_1 \leq_{fr} T_2$ *and* T_1 *has DFR, then* $\mathcal{E}^k_{n,w}(T_1) \leq \mathcal{E}^k_{n,w}(T_2)$.

Next, the effect of affine transformation on WECRE$_n$ given by (5) is discussed. It is worth pointing that in various computer applications, the affine transformations are used as the most fundamental and useful geometrical operations.

Proposition 4 *Let* T_1 *have the CDF* F_{T_1}. *If* $T_2 = aT_1 + b$, *where* $a > 0$ *and* $b \geq 0$, *then*

$$\mathcal{E}^k_{n,w}(T_2) = a^2 \mathcal{E}^k_{n,w}(T_1) + ab\, \mathcal{E}^k_n(T_1) \;\; for\; n \in \{1, 2, \ldots\}\; and\; k \geq 1. \tag{10}$$

Proof Proof is straightforward, and hence omitted. □

Next, consider the random variables T and T_τ having SFs $\bar{F}_T$ and $\bar{F}_{T_\tau}$, respectively such that the following relation holds

$$\bar{F}_{T_\tau}(u) = [\bar{F}_T(u)]^\tau, \;\; u > 0, \; \tau > 0. \tag{11}$$

Now, using (10) and (11), we have the following result. The proof is simple, and thus skipped for brevity.

Proposition 5 *Let* T *and* T_τ *be such that their SFs satisfy (11). Then, for* $n \in \{1, 2, \ldots\}$ *and* $k \geq 1$, *we have*

(i) $\mathcal{E}^k_{n,w}(T_\tau) \leq \mathcal{E}^k_{n,w}(\tau^{\frac{n}{2}} T)$ $(\geq)$ *if* $\tau \geq 1$ $(0 < \tau < 1)$,
(ii) $\mathcal{E}^k_{n,w}(T_\tau) \leq \tau^n \mathcal{E}^k_{n,w}(T)$ $(\geq)$ *if* $\tau \geq 1$ $(0 < \tau < 1)$.

Remark 1 As an application of the above proposition, we consider a series system comprising m components. Let $T_1, \ldots, T_m$; $(m \geq 1)$ be the lifetimes of m components of this system which are iid with a SF $\bar{F}_T$. Thus, the lifetime of this system is $T^* = \min\{T_1, \ldots, T_m\}$. The SF of T^* is $\bar{F}_{T^*}(u) = [\bar{F}_T(u)]^m$. Hence, from the Proposition 5, we obtain $\mathcal{E}^k_{n,w}(T^*) \leq \mathcal{E}^k_{n,w}(m^{\frac{n}{2}} T)$ and $\mathcal{E}^k_{n,w}(T^*) \leq m^n \mathcal{E}^k_{n,w}(T)$, where $n = 1, 2, \ldots$ and $k \geq 1$.

Table 1 Numerical values of the bounds of $\mathcal{E}_{n,w}^k(T)$ as in Proposition 6

n	k	$\mathcal{E}_{n,w}^k(T)$	Right hand side expression			
			(i)	(ii)	(iii)	(iv)
1	1	0.138885	0.742096	0.083333	0.332334	0.077649
1	2	0.194444	2.968380	0.133333	0.333331	0.115170
1	3	0.202500	6.678860	0.150000	0.300000	0.096088
2	5	0.214272	105.4060	0.124008	0.446429	0.051812
3	5	0.221945	531.7700	0.082672	0.992063	0.048768
4	5	0.215943	2460.130	0.046973	3.100130	0.034427
5	10	0.229576	671829.0	0.061213	1.503130	0.008722

Next result provides some lower as well as upper bounds of the WECRE$_n$.

Proposition 6 *Consider a random variable T with SF $\bar{F}_T$. Then, for $n = 1, 2, \ldots$ and $k \geq 1$, we have*

(i) $\mathcal{E}_{n,w}^k(T) \leq \frac{k^{n+1}}{n!} \int_0^\infty u[-\ln \bar{F}_T(u)]^n du$,

(ii) $\mathcal{E}_{n,w}^k(T) \geq \sum_{i=0}^n (-1)^i \frac{k^{n+1}}{i!(n-i)!} \int_0^\infty u \exp\{(k+i)\ln \bar{F}_T(u)\} du$,

(iii) $\mathcal{E}_{n,w}^k(T) \leq \sum_{i=0}^n (-1)^i \frac{k^{n+1}}{i!(n-i)!} \int_0^\infty u \exp\{(k+i-n)\ln \bar{F}_T(u)\} du$, *provided* $k \geq n - i$,

(iv) $\mathcal{E}_{n,w}^k(T) \geq \frac{k^{n+1}}{n!} D \exp\{S(T)\}$, *where* $D = \exp\{\int_0^1 \ln(z^k \bar{F}_T^{-1}(z)(-\ln z)^n) dz\}$ *and* $S(T) = -E(\ln f_T(T))$.

Proof The first part can be obtained using the fact that $\bar{F}_T(u) \leq 1$ for all $u > 0$. The second part follows from $\ln u \leq u - 1$ for $u > 0$. Third bound can be obtained upon using $u - 1 \leq u \ln u$ for $u > 0$. To prove the fourth result, we need to apply log-sum inequality. This completes the proof.   □

In Table 1, we derive numerical values of the bounds of $\mathcal{E}_{n,w}^k(T)$ as mentioned in Proposition 6 when T follows uniform distribution as in Example 2(i) with parameter $\gamma = 1$.

In this part of the chapter, we present various merits of the residual form of WECRE$_n$ (DWECRE$_n$) given by (6). It can be observed that $\mathcal{E}_{n,w}^k(T;t)$ takes non-negative values for $t > 0$. Further, as t tends to 0^+, $\mathcal{E}_{n,w}^k(T;t)$ equals to $\mathcal{E}_{n,w}^k(T)$ and when $t \to \infty$, $\mathcal{E}_{n,w}^k(T;t)$ reduces to zero. First, we discuss the effect of affine transformations on DWECRE$_n$, which is presented in the next proposition. Its proof is simple, thus omitted.

Proposition 7 *Let $T_2 = aT_1 + b$, $a > 0$, $b \geq 0$. For $t > b$, we get $\mathcal{E}_{n,w}^k(T_2; t) = a^2 \mathcal{E}_{n,w}^k\left(T_1; \frac{t-b}{a}\right) + ab\mathcal{E}_n^k\left(T_1; \frac{t-b}{a}\right)$, where $n \in \{1, 2, \ldots\}$ and $k \geq 1$.*

In the following example, we obtain closed-form expressions of DWECRE$_n$ for some probability distributions.

Example 5 (i) Let T follow uniform distribution as in Example 2(i). In this case, $\mathcal{E}_{n,w}^k(T;t) = k^{n+1}(\gamma - t)\left[\gamma(k+1)^{-n-1} - (\gamma - t)(k+2)^{-n-1}\right]$.

(ii) Consider a Pareto distribution with CDF given by Example 2(ii). Then, $\mathcal{E}_{n,w}^k$ $(T;t) = k^{n+1}t^2 b^n/(bk-2)^{n+1}$.

(iii) Let T follow Weibull distribution with distribution function as in Example 2(iii). Then, $\mathcal{E}_{n,w}^k(T;t) = (k^{\frac{2\alpha n+\alpha-2}{\alpha}}\Gamma(\frac{2+\alpha n}{\alpha}, kt^\alpha))/(n!\alpha)$, where $\Gamma(m,s) = \int_s^\infty u^{m-1}e^{-u}du$, $m, s > 0$, known as incomplete gamma function.

Next, we provide a probabilistic meaning of the dynamic weighted extended CRE of order n as

$$\mathcal{E}_{n,w}^k(T;t) = E\left[\frac{(T_t)_{n+1:k}}{h_T\left((T_t)_{n+1:k}\right)}\right] \tag{12}$$

The failure rate of T_t is $h_T(u+t)$, $u > 0$. Further, T_t is DFR when T is DFR. In the following results, we show the monotonicity result of DWECRE$_n$. We omit the proofs since these readily follow from that of Propositions 1 and 2.

Proposition 8 *Let T have DFR. Then, for $n \in \{1, 2, \ldots\}$ and $k \geq 1$,*

$$\mathcal{E}_{n,w}^k(T;t) \leq \mathcal{E}_{n+1,w}^k(T;t), \quad t > 0. \tag{13}$$

Proposition 9 *Let T be as in Proposition 8. For $n \in \{1, 2, \ldots\}$ and $k \geq 1$,*

$$\mathcal{E}_{n,w}^{k+1}(T;t) \leq \mathcal{E}_{n,w}^k(T;t), \quad t > 0. \tag{14}$$

The following proposition is analogous to Proposition 5.

Proposition 10 *Consider T and T_τ as described in Proposition 5. For $t > 0$, $k \geq 1$ and $n \in \{1, 2, \ldots\}$, we obtain $\mathcal{E}_{n,w}^k(T_\tau; t) \leq \tau^n \mathcal{E}_{n,w}^k(T;t)(\geq)$ if $\tau \geq 1(0 < \tau < 1)$.*

We end this section with the discussion on nonparametric classes of the lifetime distributions. These classes are based on the monotone nature of $\mathcal{E}_{n,w}^k(T;t)$. In the literature, various aging-based classes of lifetime distributions are introduced (see [5]). Below, we prove a result, which is essential to obtain subsequent results. The following representation of $\mathcal{E}_{n,w}^k(T;t)$ is useful in this direction. Using binomial expansion in (6),

$$\mathcal{E}_{n,w}^k(T;t) = \frac{k^{n+1}}{\left[\bar{F}_T(t)\right]^k} \sum_{i=0}^{n} \frac{(-1)^{n-i}}{i!(n-i)!}\left[-\ln \bar{F}_T(t)\right]^{n-i}$$

$$\times \int_t^\infty u\left[-\ln \bar{F}_T(u)\right]^i \left[\bar{F}_T(u)\right]^k du, \tag{15}$$

where $n \in \{1, 2, \ldots\}$ and $k \geq 1$.

Proposition 11 *For $n \in \{1, 2, \ldots\}$ and $k \geq 1$, we obtain $\frac{d}{dt}\left[\mathcal{E}_{n,w}^k(T;t)\right] = k h_T(t)$* $\left[\mathcal{E}_{n,w}^k(T;t) - \mathcal{E}_{n-1,w}^k(T;t)\right]$.

Proof From (15) we have

$$\mathcal{E}_{n,w}^k(T;t)\left[\bar{F}_T(t)\right]^k = k^{n+1} \sum_{i=0}^{n} \frac{(-1)^{n-i}}{i!(n-i)!}[-\ln\bar{F}_T(t)]^{n-i}$$

$$\times \int_{t}^{\infty} u[\bar{F}_T(u)]^k[-\ln\bar{F}_T(u)]^i\,du. \qquad (16)$$

By convention, consider $\sum_{i=0}^{n} \frac{(-1)^{n-i}}{i!(n-i)!} = 0$. On differentiating (16) with respect to t, and then simplifying further we obtain

$$\frac{d}{dt}\left[\mathcal{E}_{n,w}^k(T;t)\right][\bar{F}_T(t)]^k - k[\bar{F}_T(t)]^{k-1}f_T(t)\mathcal{E}_{n,w}^k(T;t) = -k[\bar{F}_T(t)]^k h_T(t)$$

$$\times \mathcal{E}_{n-1,w}^k(T;t).$$

Hence, the desired result follows.   □

Below, we construct DWECRE$_n$-based aging classes.

Definition 3 A random variable T is increasing (decreasing) DWECRE$_n$, denoted by IDWECRE$_n$ (DDWECRE$_n$) if for $k \geq 1$ and $n \in \{1, 2, \ldots\}$, $\mathcal{E}_{n,w}^k(T;t)$ is increasing (decreasing) with respect to $t > 0$.

It is easy to see that the above constructed (DWECRE$_n$-based) aging classes are not empty. Pareto distribution falls into IDWECRE$_n$ class and uniform distribution belongs to DDWECRE$_n$ class, which are clear from Figs. 2 and 3.

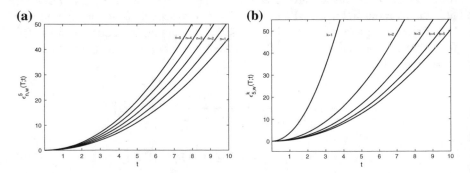

Fig. 2 **a** Graph of $\mathcal{E}_{n,w}^5(T;t)$, where T follows Pareto distribution as in Example 5(ii) with parameter $b = 3$ for $n = 1, 2, 3, 4, 5$. **b** Graph of $\mathcal{E}_{5,w}^k(T;t)$, where T follows Pareto distribution with parameter $b = 3$ given by Example 5(ii) for $k = 1, 2, 3, 4, 5$

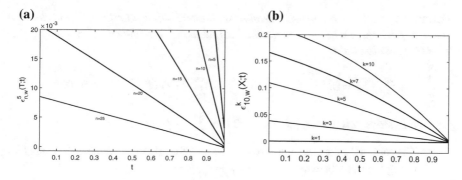

Fig. 3 **a** Graph of $\mathcal{E}^5_{n,w}(T;t)$, where T follows $U(0,1)$ for $n = 5, 10, 15, 20, 25$. **b** Graph of $\mathcal{E}^k_{10,w}(T;t)$, where T follows $U(0,1)$ for $k = 1, 3, 5, 7, 10$

The following proposition provides necessary and sufficient conditions such that $\mathcal{E}^k_{n,w}(T;t)$ is monotone.

Proposition 12 *T is said to be IDWECRE$_n$ (DDWECRE$_n$) if and only if $\mathcal{E}^k_{n,w}(T;t) \geq (\leq)\mathcal{E}^k_{n-1,w}(T;t)$ for all $n = 1, 2, \ldots$ and $k \geq 1$.*

Proof Proof follows from Proposition 11. Thus, it is omitted.   □

3 Characterization Results

Here, we provide characterizations based on $WECRE_n$. To derive some characterizations based on the proposed measure, the following results are useful. Let $a, b \in \mathbb{R}$ and $n = 1, 2, \ldots$. Then, a sequence of functions f_n is said to be complete on $L(a, b)$, if for all functions $g \in (a, b)$, $\int_a^b g(u) f_n(u) du = 0$ for any $n = 1, 2, \ldots$ implies $g(u) = 0$ a.e. on (a, b). Now, we state the following lemmas (see [6, 7]), which are required to prove characterization results in this section.

Lemma 1 *For a sequence of positive integers $\{n_s, s \geq 1\}$, the sequence of polynomials $\{u^{n_s}\}$ is complete on $L(0, 1)$ if and only if*

$$\sum_{s=1}^{\infty} \frac{1}{n_s} = \infty. \qquad (17)$$

Lemma 2 *Let $f(u)$ be an absolutely continuous function defined in (a, b) such that $f(a)f(b) \geq 0$. Further, let $f'(u) \neq 0$ a.e. on (a, b). Then, under the condition given in (17), the sequence $\{f^{n_s}, s \geq 1\}$ is complete on $L(a, b)$ if and only if the function $f(u)$ is monotone on (a, b).*

First, we derive a characterization result for the Weibull distributions.

Theorem 1 *Let T be a random variable with distribution function F_T and density function f_T. Then, F_T belongs to the Weibull family of distributions if and only if the following relation holds:*

$$n! \, \mathcal{E}_{n,w}^k(T) = ck^{n+1} E\left[T^2\left(-\ln \bar{F}_T(T)\right)^{n-1}\left(\bar{F}_T(T)\right)^{k-1}\right], \quad c > 0, \qquad (18)$$

for fixed k and for all $n = n_s$, $s \geq 1$ such that $\sum_{s=1}^{\infty} \frac{1}{n_s} = \infty$.

Proof The 'only if' part is easy to prove. To show the 'if' part, we assume that (18) holds. Using the transformation $1 - v = F_T(u)$, we get

$$\mathcal{E}_{n,w}^k(T) = \frac{k^{n+1}}{n!} \int_0^1 \frac{F_T^{-1}(1-v)v^k(-\ln v)^n}{f_T\left(F_T^{-1}(1-v)\right)} dv \qquad (19)$$

and

$$E\left[T^2\left(-\ln \bar{F}_T(T)\right)^{n-1}\left(\bar{F}_T(T)\right)^{k-1}\right] = \int_0^1 \left(F_T^{-1}(1-v)\right)^2 v^{k-1}(-\ln v)^{n-1} dv. \qquad (20)$$

Thus, from the Eqs. (18), (19) and (20), and after some simplification we obtain

$$\int_0^1 v^k(-\ln v)^{n-1}\left(\frac{F_T^{-1}(1-v)}{f_T(F_T^{-1}(1-v))} + \frac{c\left(F_T^{-1}(1-v)\right)^2}{v\ln v}\right) dv = 0. \qquad (21)$$

Note that under the assumption made, (21) holds for $n = n_s$, $s \geq 1$ such that $\sum_{s=1}^{\infty} \frac{1}{n_s} = \infty$. Thus, from the Lemma 2 and then using $\frac{d}{dp}F_T^{-1}(p) = \frac{1}{f_T(F_T^{-1}(p))}$, we obtain the ordinary differential equation as

$$\frac{d}{dp}F_T^{-1}(p) + \frac{cF_T^{-1}(p)}{(1-p)\ln(1-p)} = 0, \quad p \in (0,1). \qquad (22)$$

Solving (22) we get $F_T^{-1}(p) = b(-\ln(1-p))^c$, where $b > 0$ is an integrating constant. Thus, we have $F_T(u) = 1 - e^{-(u/b)^{1/c}}$, $u > 0$. □

Remark 2 In particular, if we consider $c = 1$ in (18), we get a characterization result for exponential distribution from Theorem 1.

Theorem 2 *Consider a nonnegative random variable T with distribution function F_T and density function f_T. Then, F_T follows Pareto distribution if and only if*

$$n! \, \mathcal{E}_{n,w}^k(T) = \left(\frac{k^{n+1}}{c}\right) E\left[T^2\left(-\ln \bar{F}_T(T)\right)^n\left(\bar{F}_T(T)\right)^{k-1}\right], \quad c > 0, \qquad (23)$$

for fixed k and for all $n = n_s$, $s \geq 1$ such that $\sum_{s=1}^{\infty} \frac{1}{n_s} = \infty$.

Proof The necessary part follows directly from Example 1. To prove the sufficient part, we assume that the relation (23) holds. Now proceeding along the similar arguments of the proof of Theorem 1, we obtain the following differential equation

$$\frac{dF_T^{-1}(p)}{dp} - \frac{F_T^{-1}(p)}{c(1-p)} = 0, \quad p \in (0, 1). \tag{24}$$

Solving this we get $F_T^{-1}(p) = b(1-p)^{-1/c}$, where $b > 0$. Hence the desired result follows. $\qquad\qquad\qquad\qquad\qquad\qquad\qquad\qquad\qquad\qquad\qquad\qquad\qquad\qquad\qquad\square$

Baratpour [2] obtained some characterization results based on CRE for first order statistic. In the following, we provide characterizations of the family of Weibull and Pareto distributions based on the $WECRE_n$ for first order statistic of independent and identically distributed random observations.

Theorem 3 *Let $T_1, \ldots, T_m$ be a random sample of size m drawn from a population with distribution function F_T and density function f_T. Let $c > 0$. Then, F_T belongs to the family of Weibull distributions if and only if*

$$n! \, \mathcal{E}_{n,w}^k(T_{1:m}) = ck^{n+1} E\left[T_{1:m}^2 \left(-\ln \bar{F}_{T_{1:m}}(T_{1:m}) \right)^{n-1} \left(\bar{F}_{T_{1:m}}(T_{1:m}) \right)^{k-1} \right], \tag{25}$$

for fixed n, k and for all $m = m_s$, $s \geq 1$ such that $\sum_{s=1}^{\infty} \frac{1}{m_s} = \infty$, where $T_{1:m}$ denotes the first order statistic.

Proof The "only if" part is easy to obtain. In proving "if" part, we first assume that (25) holds. Note that under the given assumption, the SF of the first order statistic is $\bar{F}_{T_{1:m}}(u) = [\bar{F}_T(u)]^m$. Using the probability integral transformation $1 - v = F_T(u)$ we obtain

$$\mathcal{E}_{n,w}^k(T_{1:m}) = \frac{m^n k^{n+1}}{n!} \int_0^1 \frac{F_T^{-1}(1-v) v^{mk}(-\ln v)^n}{f_T(F_T^{-1}(1-v))} dv \tag{26}$$

and

$$E[T_{1:m}^2(-\ln \bar{F}_{T_{1:m}}(T_{1:m}))^{n-1}(\bar{F}_{T_{1:m}}(T_{1:m}))^{k-1}] = m^n \int_0^1 [F_T^{-1}(1-v)]^2 v^{mk-1}$$

$$\times (-\ln v)^{n-1} dv. \tag{27}$$

Substituting (26) and (27) in (28) and then further simplification leads to

$$\int_0^1 v^{mk}(-\ln v)^n \left(\frac{F_T^{-1}(1-v)}{f_T(F_T^{-1}(1-v))} + c \frac{\left(F_T^{-1}(1-v)\right)^2}{v \ln v} \right) dv = 0. \qquad (28)$$

Now, under the given hypothesis, the rest of the proof follows along the similar arguments of the proof of Theorem 1.   □

Theorem 4 *Let $T_1, \ldots, T_m$ be a random sample of size m as described in Theorem 3. Assume $c > 0$. Then, F_T belongs to the family of Pareto distributions if and only if*

$$n!\, \mathcal{E}_{n,w}^k(T_{1:m}) = \left(\frac{k^{n+1}}{mc} \right) E\left[T_{1:m}^2 \left(-\ln \bar{F}_{T_{1:m}}(T_{1:m}) \right)^n \left(\bar{F}_{T_{1:m}}(T_{1:m}) \right)^{k-1} \right], \quad (29)$$

for fixed n, k and for all $m = m_s$, $s \geq 1$ such that $\sum_{s=1}^\infty \frac{1}{m_s} = \infty$, where $T_{1:m}$ denotes the first order statistic.

Proof The proof follows from similar arguments of Theorems 2 and 3. Hence it is omitted.   □

The following theorem gives sufficient condition under which the random variables T_1^2 and T_2^2 belong to location and scale family of distributions.

Theorem 5 *Let T_1 and T_2 have CDFs F_{T_1} and F_{T_2}, respectively. Then, the random variables T_1^2 and T_2^2 belongs to the same family of distributions, but for a change in location and scale if*

$$\mathcal{E}_{n,w}^k(T_{11:m}) = c\, \mathcal{E}_{n,w}^k(T_{21:m}), \quad c > 0, \qquad (30)$$

for fixed n, k and for all $m = m_s$, $s \geq 1$ such that $\sum_{s=1}^\infty \frac{1}{m_s} = \infty$.

Proof From (30), after simplification we obtain

$$\int_0^1 v^{mk}(-\ln v)^n \left(\frac{F_{T_1}^{-1}(1-v)}{f_{T_1}(F_{T_1}^{-1}(1-v))} - c \frac{F_{T_2}^{-1}(1-v)}{f_{T_2}(F_{T_2}^{-1}(1-v))} \right) dv = 0. \qquad (31)$$

If (31) holds for $m = m_s, s \geq 1$ such that $\sum_{s=1}^\infty \frac{1}{m_s} = \infty$, then from Lemma 2 we get the differential equation of the form

$$F_{T_1}^{-1}(p) \frac{d F_{T_1}^{-1}(p)}{dp} - c F_{T_2}^{-1}(p) \frac{d F_{T_2}^{-1}(p)}{dp} = 0, \quad p \in (0, 1). \qquad (32)$$

Solving Eq. (32), we obtain $(F_{T_1}^{-1}(p))^2 = c(F_{T_2}^{-1}(p))^2 + b$, where $b > 0$. From this relation, the desired result follows. This completes the proof.   □

4 Empirical Approach

Let $\{T_1, \ldots, T_m\}$ be a set of iid random variables drawn from a distribution with CDF F_T and SF $\bar{F}_T$. Denote the empirical survival function corresponding to $\bar{F}_T$ by $\tilde{\bar{F}}_m$. If $T_{(1)} \leq T_{(2)} \leq \ldots \leq T_{(m)}$ represent the order statistics of the sample $T_1, \ldots, T_m$, then

$$\tilde{\bar{F}}_m(u) = \begin{cases} 0, & u < T_{(1)}, \\ 1 - \frac{s}{m}, & T_{(s)} \leq u < T_{(s+1)}, \ s = 1, \ldots, m-1, \\ 1, & u \geq T_{(s+1)}. \end{cases}$$

Kayal [10] proposed an estimator for the weighted generalized CRE from empirical point of view. Here, we study an estimator for WECRE$_n$ given by (5). The empirical WECRE$_n$ is

$$\mathcal{E}_{n,w}^k\left(\tilde{\bar{F}}_m\right) = \frac{k^{n+1}}{n!} \int_0^\infty u \left[-\ln \tilde{\bar{F}}_m(u)\right]^n \left[\tilde{\bar{F}}_m(u)\right]^k du, \ n = 1, 2, \ldots, k \geq 1. \ (33)$$

Using $\tilde{\bar{F}}_m(u) = 1 - \frac{s}{m}$ for $s = 1, \ldots, m-1$, the Eq. (33) becomes

$$\mathcal{E}_{n,w}^k\left(\tilde{\bar{F}}_m\right) = \frac{k^{n+1}}{n!} \sum_{s=1}^{m-1} \int_{T_{(s)}}^{T_{(s+1)}} u \left[-\ln\left(1-\frac{s}{m}\right)\right]^n \left(1-\frac{s}{m}\right)^k du$$

$$= \frac{k^{n+1}}{n!} \sum_{s=1}^{m-1} U_s \left[-\ln\left(1-\frac{s}{m}\right)\right]^n \left(1-\frac{s}{m}\right)^k, \quad (34)$$

where $U_s = \left(T_{(s+1)}^2 - T_{(s)}^2\right)/2$. Making use of the binomial theorem, (34) reduces to

$$\mathcal{E}_{n,w}^k\left(\tilde{\bar{F}}_m\right) = \frac{k^{n+1}}{n!} \sum_{s=1}^{m-1} \sum_{l=0}^{n} (-1)^l \binom{n}{l} U_s \left(\frac{m-s}{m}\right)^k \left(\ln(m-s)\right)^l \left(\ln m\right)^{n-l} (35)$$

Example 6 Consider m iid observations $T_1, \ldots, T_m$ from a Weibull population. Assume that it has density

$$f_T(u; \gamma) = 2\gamma u \exp\{-\gamma u^2\}, \ u > 0, \gamma > 0. \quad (36)$$

It can be shown that $Y_s = T_s^2$, $s = 1, \ldots, m$ follow exponential distribution with mean γ^{-1}. Here, $2U_{s+1} = T_{(s+1)}^2 - T_{(s)}^2$, $s = 1, \ldots, m-1$ are independent. Denote $A_{ms} = m - s$. Further, it is easy to show that they have exponential distribution having average $(\gamma A_{ms})^{-1}$. For relevant detail, see [20]. So, for $n = 1, 2, \ldots$ and

Table 2 Computed values of $E(\mathcal{E}_{n,w}^k(\tilde{\bar{F}}_m))$ for the PDF considered in Example 6

γ	m	$E(\mathcal{E}_{1,w}^1(\tilde{\bar{F}}_m))$	$E(\mathcal{E}_{5,w}^1(\tilde{\bar{F}}_m))$	$E(\mathcal{E}_{5,w}^2(\tilde{\bar{F}}_m))$	$E(\mathcal{E}_{1,w}^5(\tilde{\bar{F}}_m))$	$E(\mathcal{E}_{10,w}^5(\tilde{\bar{F}}_m))$
1.0	10	0.396072	0.032876	0.260133	0.489629	0.450637
	25	0.449366	0.089407	0.391381	0.498334	0.497900
	50	0.471234	0.146352	0.447210	0.499583	0.499991
2.0	10	0.198036	0.016438	0.130067	0.244815	0.225318
	25	0.224683	0.044704	0.195691	0.249167	0.248950
	50	0.235617	0.073176	0.223605	0.249792	0.249995
3.0	10	0.132024	0.010958	0.086711	0.163210	0.150212
	25	0.149789	0.029820	0.130460	0.166311	0.165967
	50	0.157078	0.048784	0.149070	0.166528	0.166664

$k \geq 1$, from (34), we obtain

$$E\left(\mathcal{E}_{n,w}^k\left(\tilde{\bar{F}}_m\right)\right) = \left(\frac{k^{n+1}}{n!}\right) \sum_{s=1}^{m-1} (2\gamma A_{ms})^{-1} \left[-\ln\left(\frac{A_{ms}}{m}\right)\right]^n \left(\frac{A_{ms}}{m}\right)^k \quad (37)$$

and

$$Var\left(\mathcal{E}_{n,w}^k\left(\tilde{\bar{F}}_m\right)\right) = \left(\frac{k^{2n+2}}{(n!)^2}\right) \sum_{s=1}^{m-1} (2\gamma A_{ms})^{-2} \left[-\ln\left(\frac{A_{ms}}{m}\right)\right]^{2n} \left(\frac{A_{ms}}{m}\right)^{2k}. \quad (38)$$

We present numerical values of (37) and (38) of the empirical WECRE$_n$ for the distribution as in Example 6 in Tables 2 and 3.

Below, we study the asymptotic behaviour of the estimator proposed in this section.

Theorem 6 *Suppose* $X_1, \ldots, X_m$ *be iid observations from the Weibull distribution whose PDF is given by (36). Then, for* $n \in \{1, 2, \ldots\}$ *and* $k \geq 1$

$$\frac{\mathcal{E}_{n,w}^k\left(\tilde{\bar{F}}_m\right) - E\left[\mathcal{E}_{n,w}^k\left(\tilde{\bar{F}}_m\right)\right]}{\sqrt{Var\left(\mathcal{E}_{n,w}^k\left(\tilde{\bar{F}}_m\right)\right)}}$$

converges in law to the standard normal distribution when $m \to \infty$.

Proof Proof follows using similar argument as in Theorem 3.2 of [10]. Thus, it is omitted.   □

Table 3 Numerical values of $Var(\mathcal{E}^k_{n,w}(\tilde{\tilde{F}}_m))$ for the model as in Example 6

γ	m	$Var(\mathcal{E}^1_{1,w}(\widehat{\tilde{F}}_m))$	$Var(\mathcal{E}^1_{5,w}(\widehat{\tilde{F}}_m))$	$Var(\mathcal{E}^2_{5,w}(\widehat{\tilde{F}}_m))$	$Var(\mathcal{E}^5_{1,w}(\widehat{\tilde{F}}_m))$	$Var(\mathcal{E}^5_{10,w}(\widehat{\tilde{F}}_m))$
1.0	10	0.027778	0.000749	0.033571	0.042778	0.096472
	25	0.014759	0.002741	0.034618	0.017146	0.040497
	50	0.008315	0.007002	0.023416	0.008573	0.020133
2.0	10	0.006944	0.000187	0.008393	0.010695	0.024118
	25	0.003689	0.000919	0.008654	0.004286	0.010124
	50	0.002079	0.001750	0.005854	0.002143	0.005033
3.0	10	0.003086	0.000083	0.003730	0.004753	0.010719
	25	0.001640	0.000409	0.003846	0.001905	0.004500
	50	0.000924	0.000778	0.002602	0.000953	0.002237

5 Conclusion and Future Work

This chapter provides a useful study on a shift-dependent extended CRE of the k-th upper record. The residual version of the proposed measure is also considered. Various properties including the effect of affine transformation, bounds, stochastic orderings and aging properties have been studied. Further, we establish several characterizations based on the proposed measure. A nonparametric estimator is proposed. Finally, its asymptotic property is studied. Similar to the proposed measure, a shift-dependent version of the extended cumulative entropy for the k-th lower record and its past form can be proposed. In future, we will consider and study the properties of this measure.

References

1. Asadi M, Zohrevand Y (2007) On dynamic cumulative residual entropy. J Stat Plan Inference 137(6):1931–1941
2. Baratpour S (2010) Characterizations based on cumulative residual entropy of first-order statistics. Commun Stat-Theory Methods 39(20):3645–3651
3. Di Crescenzo A, Longobardi M (2006) On weighted residual and past entropies. Scientiae Mathematicae Japonicae 64:255–266
4. Dziubdziela W, Kopocinski B (1976) Limiting properties of the kth record values. Appl Math 2(15):187–190
5. Ebrahimi N (1996) How to measure uncertainty in the residual life time distribution. Sankhya Indian J Stat Ser A 58:48–56
6. Hwang JS, Lin GD (1984) On a generalized moment problem II. Proc Am Math Soc 91(4):577–580
7. Kamps U (1998) Characterizations of distributions by recurrence relations and identities for moments of order statistics. In: Handbook of Statistics, vol 16, pp 291–311
8. Kapodistria S, Psarrakos G (2012) Some extensions of the residual lifetime and its connection to the cumulative residual entropy. Probab Eng Inf Sci 26(1):129–146
9. Kayal S (2016) On generalized cumulative entropies. Probab Eng Inf Sci 30(4):640–662
10. Kayal S (2018) On weighted generalized cumulative residual entropy of order n. Method Comput Appl Probabi 20(2):487–503
11. Kayal S, Moharana R (2017a) On weighted cumulative residual entropy. J Stat Manag Syst 20(2):153–173
12. Kayal S, Moharana R (2017b) On weighted measures of cumulative entropy. Int J Math Stat 18(3):26–46
13. Mirali M, Baratpour S, Fakoor V (2017) On weighted cumulative residual entropy. Commun Stat-Theory Methods 46(6):2857–2869
14. Moharana R, Kayal S (2017) On weighted Kullback-Leibler divergence for doubly truncated random variables. RevStat (to appear)
15. Navarro J, Aguila YD, Asadi M (2010) Some new results on the cumulative residual entropy. J Stat Plan Inference 140(1):310–322
16. Navarro J, Psarrakos G (2017) Characterizations based on generalized cumulative residual entropy functions. Commun Stat-Theory Methods 46(3):1247–1260
17. Psarrakos G, Economou P (2017) On the generalized cumulative residual entropy weighted distributions. Commun Stat-Theory Methods 46(22):10914 10925
18. Psarrakos G, Navarro J (2013) Generalized cumulative residual entropy and record values. Metrika 76(5):623–640

19. Psarrakos G, Toomaj A (2017) On the generalized cumulative residual entropy with applications in actuarial science. J Comput Appl Math 309:186–199
20. Pyke R (1965) Spacings. J R Stat Soc Ser B (Methodol) 27:395–449
21. Rao M (2005) More on a new concept of entropy and information. J Theor Probab 18(4):967–981
22. Rao M, Chen Y, Vemuri BC, Wang F (2004) Cumulative residual entropy: a new measure of information. IEEE Trans Inf Theory 50(6):1220–1228
23. Shaked M, Shanthikumar JG (2007) Stochastic orders. Springer, New York
24. Shannon CE (1948) The mathematical theory of communication. Bell Syst Tech J 27:379–423
25. Tahmasebi S, Eskandarzadeh M, Jafari AA (2017) An extension of generalized cumulative residual entropy. J Stat Theory Appl 16(2):165–177

Impact of Mobility in IoT Devices for Healthcare

M. Arogia Victor Paul, T. Anil Sagar, S. Venkatesan
and Arbind Kumar Gupta

Abstract In recent years there is a huge expansion in the field of Internet of Things (IoT) and Mobility in transform a way for the innovative kinds of secure data information across the industrial sector in the current generation. Healthcare industries are innovative and are creating energetic integrated platforms that facilitate the aggregation of data from a broad range of wearable devices and the concerned applications that are connected to the medical devices. Such platforms also give out a foster real-time engagement with patients and also it enables organizations to generate actionable monitoring on the health of each individual patient. There are especially certain required standards for healthcare organizations that provide the end to end care to the elderly and disabled patients. To support healthy and safe living of elderly individuals or disabled patients, a system that can monitor their daily activities. In this chapter a secure mobile and IoT based platform can help in tracking out essential vitals, providing personalized tips, reminders and educational content on medical conditions in a proactive and timely manner. Also this chapter discusses about how mobility in IoT lead healthcare involves the convergence of mobile, cloud, devices and social media to enable patients, care takers, and healthcare providers to access secure data and improve the quality along with outcome of both health and social responsibility.

Keywords Mobility · IoT · Healthcare sector · Security issues · Technologies
Networking domain · Wireless networks

M. Arogia Victor Paul (✉) · T. Anil Sagar · S. Venkatesan · A. K. Gupta
Department of CSE, Dayananda Sagar College of Engineering, Bangalore, India
e-mail: victor15.rymec@gmail.com

T. Anil Sagar
e-mail: anilsagar001@gmail.com

S. Venkatesan
e-mail: selvamvenkatesan@gmail.com

A. K. Gupta
e-mail: arbind.gupta@gmail.com

© Springer International Publishing AG, part of Springer Nature 2019
S. Patnaik et al. (eds.), *Digital Business*, Lecture Notes on Data Engineering
and Communications Technologies 21, https://doi.org/10.1007/978-3-319-93940-7_11

1   Introduction

1.1   Research Description

Mobility in Internet of Things (IoT) devices gives an idea about the transferring secure data transmission in healthcare scenario that will guide the way to create a trend across the industries in the present generation. In this chapter healthcare industry at present gives more energetic growth among those improving its roots to a huge maximum benefits which impacts out of a new technological advancement in bringing out of many technical capabilities to evolve new possibilities of patient care in hospitality sector in a highline fashion. Healthcare companies are also creating energetic integrated platforms that will also facilitate the aggregation of data transmission from a broad range of wearable devices and the also it is concerned about the applications that are connected to the medical devices. Such integrated platforms will also gives out a real-time involvement with patients and also it enables the organizations to generate an actionable and absolute monitoring on the health of each and every individual patient.

Internet of Things (IoT) gives information about the pattern fashion which gives out well connection set, established the concept of about anyone, might be anything, at anytime, at anyplace, any kind of service, and also any network that are established to a network strategy. Here the IoT devices makes a trend into next-generation technologies that can be very powerful in impact about the complete business kinds of strategy for spectrum and this could be thought of interconnection which is an unique to identify smarter objects. So that this kind of wearable devices impact today's nature of internet infrastructure and its basic architecture along with many extended benefits. From this overall criteria there is a kind of benefits which are typically may include about many advances in the field of connectivity of these many kinds of transformations and acquiring new possibilities. So this might be chance of many different evolution services that goes tracking the way to connect a many unique kind of machine to machine scenarios in relating to the healthcare sector to bring out an innovation.

1.2   Extent & Objectives of the Research

This type of system in healthcare sector will be helpful to the future generations to emerge out with different possibilities. Therefore, the aim in introducing the term automation which is conceivable in nearly each and every fields of sector in nowadays. IoT which provides most acceptable solutions for very vast range of applications such as establishing a smart cities innovations, eliminating congestion of traffic issues, waste maintenance, security assurance, high requirement emergency healthcare services, transferring logistic materials, retails services, industrial maintenance control setup, and many more.

Internet of Things (IoT) has the capability in maintaining and which gives importance to a huge healthcare application such as mobile healthcare, maintaining body fitness, chronological disease management, along with maintaining healthcare for patients. There exist much complex issues for the treatment in medication at home premises, so that healthcare management requirement is another major important potential application in healthcare industry. There is a huge requirement of various bio-medical devices, different compact sensors, along with diagnostic fields in maintaining imaging devices can be viewed from very well equipped instruments as smart devices which constitute a well setup major part in IoT. IoT healthcare in future generations are vital in reducing cost to minimal, and increase life quality of the patient, enriching the major experience of the technician in meantime in next comings days. From this information nature of biomedical industry, IoT has huge potential to minimal its device inactive through a mobile monitoring system. Also in the field of Internet of Things (IoT) it can correctly identify various healthcare devices for their smooth working and in a continuous working operation.

1.3 Significance of the Research

This chapter provides the concept setting trends in Internet of Things (IoT) which is based on biomedical information research, which also satisfies various related issues that must be relatively completes the process, also it must uniquely represents way to transform secure data in healthcare technology in Internet of Things (IoT). In this kind of regard there exists a much importance to the society and human kind, in this chapter it contributes about by classification of many existing Internet of Things (IoT) management in some important information based healthcare network which will give out some studies into many different trends and in presenting very broad information about some basic summary of each aspects to the healthcare sector.

- To provide an extensive range of survey information report on Internet of Things (IoT) this is based on nature of healthcare services, its applications and also along with its management.
- To highlight various different industrial kinds of efforts to embrace it's in depth analysis in Internet of Things (IoT) which is in compatible to many healthcare products which has been produced in its nature and also its prototypes which related to it.
- To provide extensive insights into a field of security maintenance and also concentrating on privacy issues which surrounding the Internet of Things (IoT), hereby proposing a unique private and secure model analysis in depth.
- Major discussion about some core technologies that can gives modelling concept in healthcare technologies that is based on the nature of Internet of Things (IoT).
- Also by highlighting some various different kinds of policies and their special strategies that can make sense, which also proved to support many research adventures.

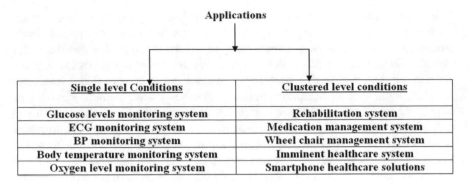

Fig. 1 Different applications in the field of healthcare

- Also to provide challenges in an open environment and in open issues that must be able to be addressed and to make Internet of things (IoT) based healthcare technologies which are robust in nature.

1.4 Applications of the Research

IoT based healthcare management is very much essential which is applied to a different variations in the nature, which include paediatric care, monitoring aged patients, taking care of different chronic viral diseases which makes difference in sense to mankind, and also in the maintaining many variety kind of management of private health information and fitness data in the systems, among others in many cases. For a better conceptual understanding of these extensive related topics, this chapter clearly and broadly categorizes the discussion in into many different kinds of applications. These applications are further broadly divided into two different broad categories which are called as single level condition and clustered level condition applications in the system of Healthcare industry applications.

At first a single level probably refers to and uniquely for a particular or a specific kind of disease in nature, whereas second category explains about cluster level which has a huge number of disease types, to its level conditions together as whole terms of nature to be reviewed. Below figure illustrates about this categorization to the applications in the field of healthcare industry. This gives out an inherently dynamic sort of information in nature status which can be easily made overcome by some additional services along with distinct features covering both single level conditions and clustered level condition solutions to the healthcare sector (Fig. 1).

2   Literature Survey

In this chapter it gives out evidence-based strategy policy with similar strategic technology keen to become a role model of all certain related case and that of practical initiations in similar fashion. Also, these kinds of strategic policies and some certain kind of regulations will play out important roles in converting the healthcare sector in future coming years. Though Internet of Things (IoT) on healthcare sector by means of some sort existing policies, electronic Health policy and there key goal for kind of policy conditions that are transforming globe. This section for healthcare discusses about some nations and their inbuilt organizing teams that works in the same conditions on both Internet of Things (IoT) along with electronic health strategic policy and their specific strategy.

(1) Indian government introduces a specific electronic health policy in between years 2000 to 2002 by means to make use of certain specific information in the field telecommunication technology for electronic health sector to form a comprehensive guidelines and its related recommendations for the country's secure information technology infrastructure in the field of healthcare sector in round 2003 and also in the creation of a telemedicine task work force in around 2005. These well maintained effort are boost that effects its conditions and they make use field of Internet of Things (IoT) in India's mobile healthcare management sector in progress. (2) Australian government in early around the year 2008, had then maintained and started an council called the Australian health ministers' advisory council in which it had developed a certain specific strategic kind of conditions to monitor out national collaboration in means of its terms of certain trends in healthcare and also by collaboration in electronic healthcare sector means series of national consultation by some initiatives which in terms include the concept of commonwealth, action state, along with some differentiated territory governments. (3) In Japanese Ministry of Internal Affairs and their respective communication system had developed concept of U-Japan kind of Policy early in the year 2004 and to accelerate the concept in realization of network to access the ubiquity nature. For this kind of trend the concept of cost savings to be very improved in clinical outcomes from a healthcare sector in information technology. Also it has been working for some means of real time recommendations for the electronic healthcare policies in the country. (4) In France around 2008, government had supported the make in kind creation of a concept of object naming service called root server for the country to enable trends in the advancement of the Information of Things. It also registered with the GS1 France, where every product is supposed to be maintained uniquely identified by using some global standards. By this way, the entire total consumers list will be convinced and that product data will be very accurate, well authenticated, and transform uniformities across the country. Also the government of France has working on improving the information technology infrastructure of hospitals in country, for the special purpose of the use of electronic Health, and solutions to the challenges in semantic way of interoperability. (5) Germany government in around 2003, had established its core electronic health activities by means in the protocols by governing the healthcare management sector. According to a very

High-Tech Strategy by around 2020 conditions, the existing industrial policy is very efficient and strategic in initiative and in achieving these long transforming goals. (6) South Korea had planned to expand its trend market for Information of Things (IoT) to $28.9 billion by end of the year 2020. Also the government had confirmed a strategy plan for developing an IoT services along with its products by establishing certain systems means by consisting of services, a platform for approvals, an integrated network, a kind of device, and related IT sectors. (7) In china around in the year 2010, China Ministry for Industry and IT had announced that it would encourage the technique for formulation by means of a plan to the IoT. Also these measures are expected to make and stimulate in the development of the Information of Things (IoT). (8) In USA, the Federal Trade Commission people had approached a policy and by some regulatory implications of the Information of Things (IoT). Also it is focused in two major areas, which are the state of prominence and also their respective features, for a non-consumer facing some related network devices and the question is about how these devices that are counter part of the Information of Things (IoT) that it can assured to have a proper security entry. (9) In European Union team has measured certain kind of policy related challenge that has to be addressed by some policy assurances from mid to long kind of terms with strategic perspective, by means of recommendations after assessing policy conditions options for stimulating the certain development of Information of Things (IoT) in the Europe continent. (10) Also World Health Organization (WHO) in both developing and developed countries, made mobile phones is used for public support. In around 2011, main aim was been taken to promote the use of mobile Health for the concept of tobacco control in the developing countries. However, their exists certain standard projects for the most mobile health in the growing countries usually a text messages like short message services that was been used for awareness and also it has been focus on certain diseases like human immunodeficiency virus (HIV), mean corpuscular hemoglobin (MCH), and malaria many other kind of diseases. Also to propose its focus to important countries to integrate electronic healthcare usage in their health information by means of transforming its related health for all.

3   Problem Statement for Impacts of Mobility in IoT Devices for Healthcare

In the certain improvement of a wireless healthcare sector where it offers interesting hidden challenges, and concepts like reliable transmission of data, mobile support to the node and also detection process for fast event, which empowers about the delivery of data in time, management of power system, computation of node in the network and in the terms of middleware stats. Further however, by deploying a certain new technologies in the field of healthcare without considering a security might makes privacy to the patient data which is complicated. Further, in wireless networks along with some sort of medical sensors which covers wide range of mobility healthcare sector applications, which as physiological kind of data monitoring system, along with activity kind of monitoring in the terms of healthcare in the campus, and also maintain location tracking for athlete in the tracks.

Consequently, this Wireless Multimedia Sensor Networks (WMSNs) shares their confident individual data with some physicians, some insurance companies, and also health-coach or also even with their family. Therefore there exists an chance for unauthorized collection of information and use the patient vital data by a high kind of adversaries which cause related risks to the patient, which can also make the patient data available to public which makes wireless electronic healthcare can offer support to patient monitoring in hospitality sector, but also their exists the physiological kind of certain data of an individual person which are very complex in nature, so that this kind of issues with respect to the security and privacy can become some of the major importance for the healthcare, especially when it comes to undertaking the concept of wireless electronic healthcare technology.

However, all these studies will discuss out about the specific related information for the above security related kind of problems, that we can make use of the concept of impact of mobility in IoT devices in healthcare has focused on security in terms of wireless medical sensor networks to explore secure wireless healthcare data monitoring applications and this will impact all issues will be in detailed and understand in the further in depth analysis in this chapter.

3.1 Difficulties Which Impact of Mobility in IOT Devices for Healthcare Sector

In this chapter there are many difficulties which are best explained by the IoT healthcare network which is very much important in the Internet of Things (IoT) for a related healthcare. It encourages a reasonable acceptance to the Internet of Things (IoT), which makes sense, also it facilitates the concept of data transmission and some sort of undertaking the related medical data, and where it make use of electronic healthcare related kind of query in the industry which relates to make an effort to bring out every possibility in it. However difficulties along with the network related issues will figure out concepts like

- Topology related issues like physical configurations links, application maintaining scenarios, some sort of activities, and also related use cases platforms.
- Platforms like library, framework, and environment.
- Architecture type of difficulties such as software organisation system as whole, hierarchical model which also reflects the system status,

And also with other minor all types of this issues that related to all kind of the impacts in healthcare (Fig. 2).

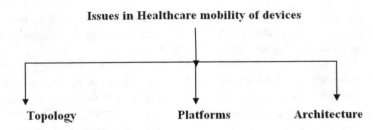

Fig. 2 Issues in healthcare mobility of devices

3.2 Existing System Scenarios

Data monitoring here right decision can be made by acquiring more information, which is accurate in nature.

Tracking applications in this computer can take a decision by means of the quality and reliability issues. Also here we get the exact transferring out data along with where it improves safety and quality of life.

Time management here it gives the amount of time management that saved in observing in related to the certain number of trips which had done that would be misuse in its nature of saving transfer rate.

Money effective here financial aspect is the very best advantage where it could be very much affordable to every mankind.

3.3 Disadvantages of Existing System

Compatible issues currently, their does not exists much standards for a acquiring and to monitor kind different sensors. Also Universal Serial Bus (USB) or a Bluetooth device which requires not too much difficulties that to do in the right aspects.

Complex issues here many aspects takes locate for the failure along with some complex nature of the system, in which two persons may receive same messages that over the data transmission are over and both of you may end up with the same messages.

Privacy issues it is very major kind of issue related to the Internet of Things (IoT). Here data should be encrypted and should be very secure enough before, that the data which as the information should not disclose to other which are not related.

Safety issues there might be a threat that certain software can be taken by some unknown personal and all the confidential information might be misplaced by means of certain techniques. So that the possibilities are never endless. So that our prescription being must be changed or might be our secure data transmission details being hacked and that could put us at risk.

3.4 Proposed System

An Internet of Things (IoT) in healthcare sector is not a complex issue but it might have continuation for the development. So that it is complicated to make out identification and to predict all the possibilities along with the certain threats and also with the association of the IoT in medical profession. So that there should be a need to have security aspects to be worked out in the security solutions for the apparent use to maintain the capability issues to monitor to transform to the requirement. To achieve this kind of security targets can be achieved by designing a very much security systems to reach our final destinations by the decisions by taking over the unnoticed problems available and properties that can be solved by the impact of experience along with certain knowledge factors that has be accounted.

3.5 Advantages of Proposed System

There exists much kind of emerging technologies for Information of Things (IoT) which is based on healthcare, so that it is complicate in finding out a certain kind of exemption list. So that here further discussion focuses on major many complex core technology that might have the inbuilt potential in order to make revolution the Internet of Things (IoT) based sectors.

Cloud based system gives concept of integration of cloud system computing into an Internet of Things (IoT) which is based on healthcare management technologies that can provide the facilities with some access to all the resources that are shared, by services offered based upon by over the request to the network over, and then by operating the executions to meet out in the need of various requirements.

Grid based system gives out the data insufficient the way of computation capabilities by means of medical data sensor nodes which can be further addressing by means of introducing grid based system computation which is well in maintain to the unique type of healthcare network. In this more accurate clustered computing can be addressed, which can be analysed as the major in the history in the field cloud system.

Big Data management with a huge amount of essential electronic health data which has generating with the electronic data medical sensors and to provide certain tools for increasing efficient in analysing the health diagnostics which very important and in monitoring future stages of medical healthcare.

Network connection ranging from one network for certain short-range communications like wireless local area network (WLAN) or wide area network (WAN) is to very long-range communications which are essential part of the physical infrastructure of the Internet of Things (IoT) based healthcare.

Ambient Intelligence technique is helpful for the end users, along with the customers, and also customers in a healthcare network in humans which are of crucial information.

Augmented monitoring being part of the Internet of Things (IoT) which is realistic in nature, which gives augmented monitoring in reality which plays a key role in healthcare industry. It is very helpful in performing surgery in the healthcare sectors. *Wireless Wearable devices* which make patient engagement and also to calculate certain evolution of crowd in the field of population in improving the health standards by facilities by means of embracing along with the medical data wearable sensors. This has many advantages by means of data information that are connected, healthcare target oriented communication systems and in gaming technologies.

4   System Security Standards

These Security requirements for Internet of Things (IoT) solutions for healthcare are very much related to the certain standards in the field of systems for communication technology which has many cases.

Confidential security standards ensure the condition that it cannot accessible in the medical field management for the unauthorized user, so that there is very much need of security standards with higher confidential criteria.

Integrity security standards ensure that there is need of medical data which are not altered by means of some the other related committees, so that there is need for the stored data standards that are very much compromised by means of efforts.

Authenticity security standards enable an IoT healthcare device which has the possibility in identifying peer to peer communications standards.

Available security standards ensure about there is a need for IoT in the healthcare sector that has to be in the range of authorized authority, where in which it deals with the denial of service guaranteed attacks.

Data acquired freshness standards include about the kind of data freshness and key freshness. Also in this kind of data acquired freshness it basically implies that there is a need of wide requirement of fresh data.

Non Repudiation security standards imply that the node cannot be denied by the way of sending the messages that are sent earlier in the range of destination that implies the security by non repudiation methods.

Authorization security standards ensures about the need of an authorized nodes which are accessible by means of the authorization of network services.

Resiliency security standard are integrated and connected by means of healthcare devices that are meant of compromised supervision, that there is a need of scheme that has security that should still protect to make the information safe by any attacks.

Fault Tolerance security standard is a security scheme that there is need for continuation that has to provide the security by means of presence by finding out any kind of fault involved in it.

Self Healing security standard requires medical data device information in IoT in Healthcare sector networks which has the ability to run out the need for the energy requirement, which enables the standard levels of security standard criteria involved

5 Security Healthcare Challenges

Security challenges give the idea about the IoT that should satisfy the following requirements that are not ensured by means of some of the security challenges which might be failed, so that here a novel approach had some of the measures that will address the concept of new challenges that are been taken into consideration by Internet of Things (IoT).

Computation challenge in IoT healthcare devices are embedded by using a very low running speed. Here central processing unit (CPU) in such kind of devices is not so much powerful in kind of speed categories. In addition to this devices that are not well designed to perform the computationally very cost effective in its standards. So that there is a need of security solution where it minimizes which consume resources and maintains high security performance in a challenging task.

Memory challenge gives the idea about Internet of Things (IoT) healthcare devices that should make use of very low memory. In such kind of things there is need of operating of system environment by means of the system software. So that there is huge requirement of low cost effective memory to execute a highly complicated security related protocols.

Energy challenges will gives out an idea about the healthcare networks which includes a healthcare device that should make use of effective limited resource battery backup. So that there is energy constraint requirement in IoT that is very much helpful in finding out in an aware of security challenge to the system.

Mobility limitations in general, that maintains the healthcare devices that not meant by means by a static, but also briefs about the knowledge of mobility in nature. Such devices are connected through the IoT, service providers. Also now a day there is a need of wearable devices that should make use of home based network. So that there is a need for requirement for the development of mobility constraint by means of some related algorithm which is a robust challenge.

Scalability challenges which gives out an idea regarding the IoT devices that gradually increased and means eventually, so that there is need for more number of devices to be connected for acquiring the network information globally in the particular system. So that there is need for designing a very high scalable issue need to be addressed by means of the security schemes, and there is no need of compromise to the security requirements.

Communications Media limitations which are meant for the well connection of both to the local and global networks that are through because of range of wide wireless contacts that make uses of such devices like Bluetooth, Zig-Bee and many others. So that it is very complex to find out a security protocol, which it can be make use of both use of wireless and wired channel devices.

Multiplication of many to many healthcare devices that should be make use of an IoT networks that makes an diverse effect in the nature of multiplication, by ranging by means of different types in the range of networks. So that there is a kind of open challenge in designing the limitations that will address some schemes, so that there is a need of simplest multiplication of connecting devices in the simplest form.

Dynamic Network Topology which has many limitations so that there is a need a chance to join an IoT healthcare device networks by means of anywhere and anytime. So that there is need of addition of network either by means of dynamic network topology which will be role in devising a network security model.

Tamper Resistant Package gives the idea about of limitations that give a security for a physical which is a major important aspect by means of Internet of Things (IoT), for healthcare devices. There is a need for a challenge that makes the attacker may tamper within the devices and this may extracts all the confidential secrets, or even can make modifications in the programs, or even can also manipulate the information. Tamper resistant packaging will gives limitations that will gives out an idea regarding such kind of attacks.

6 Mobility Issues

Internet of Things (IoT) paradigm will continue to evolve, in around in many additional IoT healthcare devices that also offer services that are in the expected range. Sp that when an attacker may try to attack the devices network by means of some security threats so that there is a need not to compromise in existing and future extension of IoT healthcare devices in the networks. Here there exists much kind of threats which are predictable, tangible and hard to predict. In this chapter need to address the security threats where this is applicable to both existing and future potential threats with reference to some of the key properties, by means of host, information in the type of data, and some network compromise of specific data.

Attacks based on host nature of properties
In this there exists a three different kind of attacks that can be launched by means on host nature of properties like the nature of software compromise, hardware compromise, and also by a user compromise. So there exist much vulnerability and many glitches that make the force the healthcare IoT devices to make a malfunction by means of attacks.

Attacks based on Information differences
In this there is a need of transferring and storing data that can be manipulated or been analysed by means of some unauthorized person. So that here he can input any kind of fake information and can also remove the information true nature in the healthcare devices, So that such attacks can might also have the following concepts of modification of data, interruption in the network, interception in the transferring of data, Fabrication of models, and replay of fake information and many more.

Attacks based on network properties In this attacking type it has two different kinds of forms, such as protocol and layer specific compromise attacks. In this there is a need for standard protocol compromise and network related protocol attacks, which can play vital role in the field of attacks in the network. To improve the kind of performance in IoT healthcare system care we should also need to address concept of security, by means of connectivity, and along with long nature of connections in

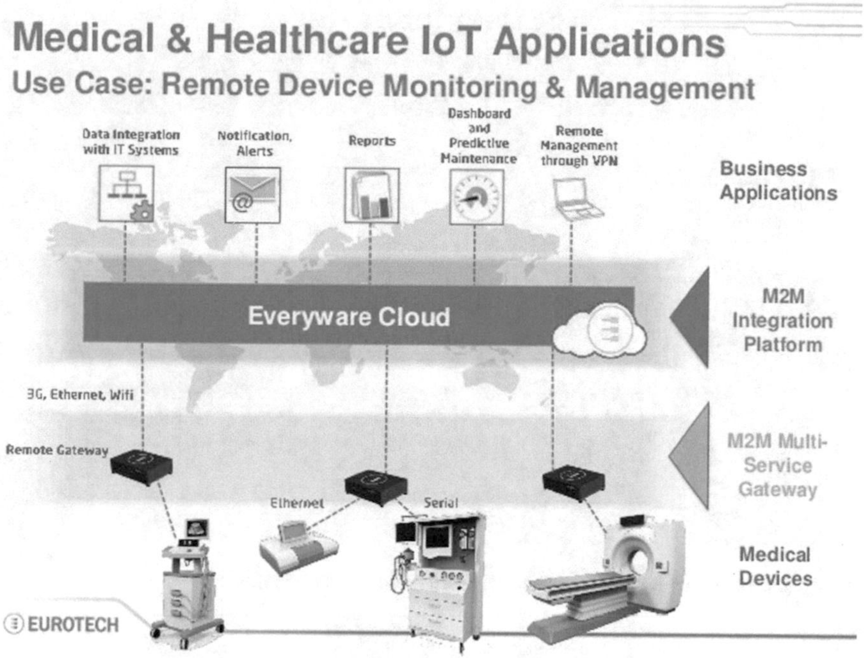

Fig. 3 Relationship between medical and healthcare IoT applications

the networks, which may vary different conditions in the environmental aspects, that should make use of security aspects and their techniques ensures in each and every protocol stack in the system.

7 System Design

See Fig. 3.

8 Performance Analysis

Pie-Chart
   See Fig. 4.
Graph
   See Fig. 5.

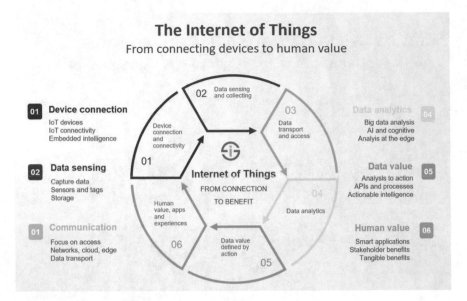

Fig. 4 Internet of Things (IoT) connecting devices to human benefits

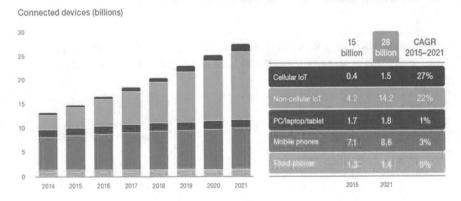

Fig. 5 Internet of Things (IoT) way of emerging its trends around the globe

9 Disadvantages in Traditional Method

Researchers are tend to be designing the required methodology and implementing the trend to involve in a depth analysis many IoT-based mobile medical healthcare services, maintenance and resolving services related to technological and architectural issues on it. Many challenges are needs to be solved with care. In this section both discovered and undiscovered issues near information technology healthcare services. *Standardization* there are numerous existing product manufacturing vendors and also new vendors are also joining to this group though there is a lack of following the rules and regulations of interfacing devices, this leads to interoperability issues. To resolve these issues a quick action is required. For example a group of experts can take care about standard policies about IoT mobile medical healthcare. In this we need to follow the topics like protocol layer stacks and communication network layers, physical, network and media access control layers, along with device interfaces. Here organizing many services like maintenance of electronic health and medical records is another major issue. Here administration has many types together with access management and healthcare expert registration. A variety of mobile health and electronic health associations and IoT researchers are able to work jointly, and existing standardization such as the Internet Protocol (IP) for alliance in orientation of smart objects the information technology along with innovation foundation, and many other forms an IoT technology functioning groups for the better standardization in the improvement of IoT based healthcare medical services in overcoming traditional issues.

IoT mobile Healthcare medical platforms can be a structural in designing an IoT based healthcare, because of more complicated nature and in the trend of making the common medical devices are available, by means of acquiring the real time operating of the systems with more useful. Also there should be a need of modification of platform computation with a run time libraries which was like a service oriented kind of a method that will be very much suitable for building the platform that can be broken by many different interfacing of packages. Here to construct a simplest and suitable platform, there is a need of service oriented architecture (SOA) that can be taken into consideration of services that was been in exploiting the usage under different application programming interface (API's). Also there is a need of specialized libraries, frameworks and many platforms that are need to be built to support the development of healthcare sector, for making use it in many ways. So that this concept will be helpful in classifying disease in many ways.

Cost Analysis method gives idea about low-cost IoT based healthcare technology is analysed by researchers but there is no evidence of comparative study on this so a cost analysis of an IoT is useful.

The App Development process setup, development, debugging and testing is the four fundamental steps on the android platform. Related methods are taken on other platforms also. An authorized medical expert must required to ensure the quality of healthcare app development. One of the crucial part of medical science application are recent advanced updates.

Technology Transition a smart technology environment can be developed by healthcare organizations by using sensors and IoT methods to modernize the traditional devices. Thus, creating an IoT system from an existing configuration will be a challenging task. Here there is a need for the proper method to be followed to overcome compatibility of backward issues and be aware of flexible kind of nature can be assured by means of machines in it.

Low Range Power Protocol in IoT healthcare development, there are many devices they behaviour is different to each other from their status of existing. To ensure service availability every communication layer will have an added power consumption challenge.

Network Approaches different kinds of patient, data, service centric approaches and architectures are three fundamental kinds of approaches in IoT for healthcare network. In the patient approach scheme, there is a need of healthcare systems are been divided according to the involvement in need of patients and their family members to consider for the medical treatment. In the data centric approach healthcare structure can generally be separated into kind of many objects based on the captured data. In a service approach scheme, healthcare structure is allocated by the means of assembly characteristics that they must provide in order to gain confidence of network reliability.

Scalability Issue in IoT healthcare in the field of networks, where it will make use of different kind of applications, maintenance of service and updating all the concerned databases. So that there is a need for requirement of additional information in the network which may get from the both individuals and a kind of healthcare organizations.

Continuous Healthcare Monitoring in most of the cases, where a patient is required to undergo a long-term monitoring so this method plays a vital role.

New Diseases and a kind of Disorders in the Smartphone which are being considered as one of the healthcare gadget responsible for cutting edge technology. Here all the applications which are formed by healthcare systems are going to be added every day, so that these kinds of categories are limited to certain specific list of diseases. There is a need for Research and Development activities for finding out a new kind of disorders and diseases that are very much important, in evolving out the new innovation technology aspects. So that early detection of this kind of rare deadly diseases can be found out in the essential task.

Identification multiple caregivers and multiple patients are in the healthcare environment and this requires a proper identification technique.

Business Model a new business model is required in the IoT healthcare business to educate doctors and nurses to learn about new technologies. Traditional system strategy which is a complicated kind of issue, because it involves many different kind of new requirements, many of them are like policies, new infrastructure systems, transformed organisational structures, operational process and many more.

Quality of Services (QoS) in Healthcare services reliability, maintainability and QoS are the most important parameters that are highly time sensitive. Here the system availability and a Vigorous nature plays a vital role in contributing the QoS because there is a need for maintaining the certain disaster management that can make the situations complicated in medical field.

Sensitive data Privacy protection in the field of safe precaution data, available from sensors along with various devices which are illegal in accessing the vital information. So that there is a need for high security standards, that are supposed to be introduced in the organizations, and applications. So that users can make use of it. Here the main challenge are avoiding threats, attacks prevention, and focusing on different vulnerabilities. From all this there is a need of optimal algorithms that are need to be collaborated by means of protection, reaction and detection of various kinds of services that can be summarized in further stages.

Resource Effective Routing IoT healthcare system should be designed to make the most of security performance while having less resource consumption due to resource constraints.

Physical Security all the devices in IoT based healthcare should be temper resistant packaging. So that there is a chance of unauthorised person may break the data and stole the healthcare related data, so that attacker may manipulate the vital data or even they can misplace the devices with some malicious networks.

Secure Routing here there might be chance of real time communication which are vital and proper in representing a proper routing format and forwarding the other protocols that well related to the healthcare network mainly because of attacks in capture of devices.

Data Transparency here in the IoT based cloud services can be adopted by means of some medical device data that are meant for data mainatainence of healthcare. So that there should be a need for transparent of data and need for life cycle development of personal data which can be traced and control of data can be takes place.

Security of Handling IoT Big Data might be chance of huge amount of healthcare data is generated by using a biomedical devices and sensors. So that it securely stores the data that was been captured, where it provides measures in handling, data transfer, mainatainence, and without compromising the integrity, confidentiality, and privacy issues that require much effort and secure attention of the system.

Mobility is one of the major issues in IoT, where this chapter majorly focus on. In this we can assure ability in IoT healthcare using the network support to the patients where they can be connected anytime and anywhere in mobile. This kind of feature is very helpful in accounting patient's environments dissimilar.

Edge Analytics in edge devices develop a kind of quality of entry devices and plays very prominent role in the IoT. So there is need for healthcare data analytics should be inspected at regular intervals and by this need to improvise the designers to optimise network architecture and maintain data traffic.

Ecological Impact in full phase scale will maintain the need for the healthcare needs a kind of biomedical sensors that set in maintaining devices. This should be considered as critical impacts on the environments, human health and users monitoring. For this reason there is a need for manufacturing of device, and there is a use of more devices and their proper maintenance.

10   Conclusion

This chapter briefs about the various features of IoT based healthcare sector technology that gives out about the various architectural network, which will be helpful in accessing the support to the platforms that has the IoT backbone and will facilitate medical data transmission and their reception of data. There is need for effective research and development, which will be made an effort in IoT driven healthcare services and their applications. This chapter gives detailed research activities with order of references to showcase the IoT sector that deals about the paediatric and elderly care of patients, there is a need for supervision of chronic disease, private health care and the fitness management. Also for the deeper insights need for the industry development and enabling technologies, which makes need of motivation and present trend to outlook how the recent technologies, which are ongoing progress in the various kind sensors, devices and internet applications, which are highly motivated and which are very cost less effective healthcare gadgets, and need of association of health services to limit the development of possible IoT based healthcare service for further development of technology related to healthcare. To better understand IoT security in healthcare, this chapter in overall gives in depth knowledge in a range of high security requirements and challenges, which reveals various kind of approaches in related to the problems in the area of healthcare models that can be moderate in the association of security risks.

11 Future Work

The discussion on the topic of healthcare for mobility aspects matters such as standardization of healthcare, type of networks, kind of business models, type of quality service, and related data protection that is very much expected that is supposed to make easy for the advancement of research in the IoT based healthcare service. This chapter performs about the inbuilt knowledge that provides in mobility electronic health care knowledge their policies and regulations for the better benefits of stakeholders which are in measuring the healthcare IoT technologies. This analysis and the results of this survey for the chapter is expected to be very much helpful in the field for research, engineers, policy makers and healthcare professionals altogether working in the area of IoT healthcare technology.

Bibliography

1. Jara, AJ, Belchi FJ, Gomez-Skarmeta AF, Santa J, Zamora-Izquierdo MA, Alcolea AF (2010) A pharmaceutical intelligent information system to detect allergies and adverse drugs reactions based on Internet of Things. In: Proceeding of the international conference on pervasive computing and communications workshops (PERCOM workshops), Mar/Apr 2010, pp 809–812
2. Rookie VM, Prasad R, Prasad NR (2011) A cooperative Internet of Things (IoT) for rural healthcare monitoring and control. In: Proceedings of the international conference on wireless communications, vehicular technology, information theory and aerospace & electronic systems (wireless VITAE) Feb/Mar 2011, pp 1–6
3. Awareness Day 2014 activities by program type. http://www.samhsa.gov/sites/default/files/ch ildren-awareness-dayactivities-by-program-2014.pdf, Accessed 7 Dec 2014

4. Vicini S, Sanna S, Rosi A, Bellini S (2012) An internet of things enabled interactive totem for children in a living lab setting. In: Proceedings of the international conference on engineering technology and innovation (ICE), Jun. 2012, pp 1–10
5. Höller J, Mulligan C, Tsiatsis V, Avesand S, Karnouskos S, Avesand S, Boyle D (2014) From Machine-to-Machine to the Internet of Things: Introduction to a New Age of Intelligence. Elsevier, Amsterdam, The Netherlands
6. Kortuem G, Sundramoorthy V, Fitton D, Kawsar F (2010) Smartobjects as building blocks for the internet of things. IEEE Internet Comput, 14(1):44–51
7. Lee J, Reyes BA, McManus DD, Mathias O, Chon KH (2013) 'Atrial fibrillation detection using an iPhone 4S'. IEEE Trans Biomed Eng 60(1):203–206
8. Chen N-C, Chu H-H, Wang K-C (2012) Listen-to-nose: a low-cost system to record nasal symptoms in daily life. . In: Proceeding of the ACM International Conference on Ubiquitous Computer, Sep. 2012, pp 590–591
9. Wadhawan T, Lancaster K , Zouridakis G, Yuan X, Ning S, Hu R (2011) Implementation of the 7-point checklist for melanoma detection on smart handheld devices. In: Proceedings of the Annual International Conference of the IEEE Engineering in Medicine and Biology Society, Aug/Sep 2011, pp 3180–3183
10. Guangnan Z, Penghui L (2012) 'IoT (Internet of Things) control system facing rehabilitation training of hemiplegic patients. Chinese Patent 202587045 U, 5 Dec 2012
11. Yue-Hong Y, Yi Z, Jie FY, Jian L, Wu F, Chao X (2014) Remote medical rehabilitation system in smart city. Chinese Patent 103488880 A, 1 Jan 2014
12. Tan L, Wang N (2010) Future internet: the internet of things. In: Proceedings of the international conference on advanced computer theory and engineering (ICACTE) Aug 2010, vol. 5, pp V5-376–V5-380
13. Chung W-Y, Jung S-J, Lee Y-D, A wireless sensor network compatible wearable u-healthcare monitoring system using integrated ECG, accelerometerandSpO2 (2008). In: Annual International Conference of the IEEE Engineering in Medicine and Biology Society (EMBS), Aug 2008, pp 1529–1532
14. Laranjo I, Santos A, Macedo J (2013) Internet of Things for medication control: E-health architecture and service implementation. Int J Rel Quality E-Healthcare 2(3):1–15
15. White PJF, Friesen MR, Podaima BW (2014) Algorithms for Smartphone and tablet image analysis for healthcare applications. IEEE Access 2:831–840
16. Liang S, Trinidad M, Hai S, Zilong Y (2011) Childhood autism language training system and Internet-of-Things-based centralized training center. Chinese Patent 102184661 A, 14 Sept 2011
17. Pang Z, Chen Q, Tian J (2014) Intelligent packaging and intelligent medicine box for medication management towards the Internet-of-Things. In: Proceedings of the 16th international conference on advanced communications technology (ICACT), Feb 2014, pp 352–360
18. Roomer K, Kellerer W, Mattern F, Ostermaier B, Fahrmair M (2010) Real-timesearchforreal-worldidentities: a survey. Proc IEEE 98(11):1887–1902
19. Guinard D, Wilde E, Trifa V (2010) A resource oriented architecture for the web of things. In: Proceedings of the Internet of things (IOT), Nov/Dec 2010, pp 1–8
20. J.Lee, K.H.Chon, O.Mathias, B.A.Reyes, and D.D.McManus (2012) Atrial fibrillation detection using a smartphone. In: Proceedings of the annual international conference of the IEEE engineering in medicine and biology society, Aug/Sep 2012, pp 1177–1180

Multiple Mobile Elements Based Energy Efficient Data Gathering Technique in Wireless Sensor Networks

Bhat Geetalaxmi Jairam and D. V. Ashoka

Abstract WSN applications primarily focus on data accumulation from the various sensor nodes spread across the environment. Many existing data gathering protocols work on the principle of using Cluster Head (CH) which is the designated node in a cluster for collecting data and Mobile Element (ME) which collects data from various CH's and deposits the data in the Base Station (BS). The proposed work on creation of an efficient data gathering technique in WSN, is the result of intense survey of existing technique in related framework and immense understating of the short coming of these existing protocols. The things that predominantly stand out from the survey performed are overflow of buffers at sensor nodes, visiting schedule of MEs, data fusion aspect and Idle listing concept, have not been well addressed. These limitations pave way for inception of novel data gathering technique for WSN. In this paper Energy Efficient Data Gathering Technique using Multiple Mobile Elements (EEDGMME) is introduced. Better efficiency in data gathering technique is achieved by data fusion at Cache Point (CP) which intends to reduce the instances of transmissions, visiting schedule for MEs to reduce buffer overflow and resultant data loss at various nodes of the network, Sleep-Awake duty cycling which prevents the instances of Idle listing and hence conserve the nodes energy. Practical simulation results prove the theoretical perspective of improved performance gains in comparison to the existing protocol Mobile Element based Energy-Efficient Data Gathering with Tour Length-Constrained in WSN (EEDG). Proposed technique EEDGMME provides better packet delivery ratio, lesser delay, reduced overhead, optimum energy consumption with decreased packet drop and hence enhances the network usability span.

B. G. Jairam (✉)
The National Institute of Engineering, Mysuru, Karnataka, India
e-mail: g.nettar@rediffmail.com

D. V. Ashoka
JSS Academy of Technical Education, Bengaluru, Karnataka, India
e-mail: ashok_d_v@rediffmail.com

1 Introduction

WSN is usually composed of nodes constituted as wireless, ad hoc hosts. Every node supports multi-hop routing protocol. WSNs are even self organizing. The sensor nodes are low cost devices equipped with processor, storage capacity, transceiver and battery power. Once these nodes are deployed, using short-range wireless communications, the nodes form a network. To monitor physical and/or environmental variations, sensor networks are deployed. Temperature variations, sound functions, vibrations, pressure measures, Motion pollutants, etc. are monitored, in specific area of interest, using WSNs. The data gathered by the sensor nodes spread across the network, is relayed to one or more sink nodes.

WSNs are implemented in areas where timely collection and processing of sensitive data is the main concern. As in a battlefield/warzone, sensor nodes are distributed over the concerned area and military grade information is sampled at regular intervals. This collected/sensed data is transmitted to the BS for related processing [1]. In a WSN deployment sensor nodes are constrained by limited available energy, hence it becomes necessary to conserve the sensor node energy in order to extend the network lifetime. Energy consumption can be reduced by eliminating redundancy in collected data, reducing the transmission instances by making use of data fusion techniques [2].

Major issue of concern for data gathering in WSN are as mentioned below:

Sink being a static node performs as a gateway between WSN and user application. With multi-hop transmission, the sensor nodes have an option of sending the sampled data to the static sink. Bottleneck congestion is created near the sink since accumulated data from all nodes reach the sink [3].

The overloaded node can hinder network performance, might as well stop data transmission towards the sink, thereby reducing network lifetime. At the same time, non uniform distribution of sensor nodes and sink not being at a proper gathering position, can create energy imbalance issues in the network [4].

In WSN implementation, efficient energy utilization is a major requirement. The sensor network usually has large number of nodes, operating over an extended period of time. These nodes energy resources have to be maintained efficiently. Since the process of data gathering consumes large amounts of energy, it creates, need of efficient communication strategies being designed, implemented precisely and efficiently [5].

In battlefield monitoring applications, where time elapsed from collecting data to the time the collected data is processed, becomes a dominant factor of consideration—the emphasis should be on minimal data transmission time. Distance between data-collection points and data-processing points, makes this delay-sensitive task relatively difficult [6].

The specific application of sensors, as in military purpose, deployment have additional requirements like reliability, network survivability, real-time operation which impose additional pressure on maintaining energy efficiency [7].

The extensive study of existing data gathering protocols focus on two major criteria, which are energy conservation and network lifetime. The efficient use of power is an essential aspect in maximizing network lifetime [8].

As the wireless connectivity is ad hoc in nature, no guarantee of reliable forwarding of sampled data from sensor nodes to sink can be anticipated. Even it may cause data loss, energy wastage, reduction in data collection by sink and increased energy depletion at the sensor nodes [9].

Other issues to be handled in increasing usefulness of network during gathering of data in WSN are as specified:

- Path from sensor to sink is to be opted for, depending upon topology of the network and associated profit.
- In order to conserve energy, if same path is used by sensor nodes to transmit data across the network, link failure may result.
- Reduction in delivery ratio due to link failures, results in increased retransmission probability as well as end-to-end transmission delay and utilization of energy.

2 Related Work

Several research in recent years have considered the use of mobile elements for data gathering in wireless sensor networks.

- Tinybee: Data Gathering System developed by Ota et al. [10] makes use of mobile agents called "Tinybees" which are dispatched by a mobile server to collect data from the designated area. With respect to transmission delay and energy consumption, this protocol has better performance results but with the lack of visiting schedule for the Tinybees, buffer over flow at sensor node may happen.
- Zhao et al. [11] developed Mobile Data Gathering With SDMA (MDG-SDMA), uses special mobile collectors called SenCar which operate like mobile base stations, collecting information from predetermined set of sensor nodes. SenCar having multiple antennas, achieve uniform energy consumption via a single-hop transmission to the base station. Again the visiting schedule of these SenCars was not part of the proposal which may result in buffer overflow at the sensor nodes.
- A Cost Minimization Algorithm for Mobile Data Gathering in WSN proposed by Zhao et al. [12] attains optimum data control for sensors along with provision of sojourn time to the mobile collector. Drawback is visiting schedule for the mobile collector is not considered thereby leaving a chance of buffer overflow at the sensor nodes.
- Kinoshita et al. [13] in Enhanced Environmental Energy-Harvesting Framework (EEHF) proposed a framework for prolonging of lifetime of WSN. Framework includes data gathering scheme by estimating obtained environmental energy, thus helping sensors to make efficient use of environmental energy for betterment of the network lifetime. EEHF scheme uses two methods, Enhanced-EEHF and clustering method to get the desired work done. In every round of E-EEHF, expected

environmental energy is predicted for the future which, as a result, improves the estimation accuracy. Thus helping in setting the estimation interval. The clustering mode assumes that every sensor has rechargeable and non rechargeable battery. Whenever possible, energy from the rechargeable battery can be used to extend the network lifetime. EEHF based data accumulation scheme is a proposal. It is yet to be experimentally ascertained.

- Exploiting Mobility for Energy Efficient Data Collection (EMEEDC) method proposed by Jain et al. [14], has 3-tier architecture consisting of Access Points, Mobile Ubiquitous LAN Extensions and Sensors. The MULEs carry data from sensor to access point, thus sensors need to transmit data over shorter distance requiring lesser transmission power. Responsibility lies on the sensor to find nearby MULEs to offload its data. To aid in this discovery, MULE continuously transmits discovery message to detect a nearby sensor. The architecture assumes that a MULE will eventually reach an access point and at least one MULE reaches a sensor. To overcome the limitations, which arise from the above assumption, MULE-to-MULE communication can also be implemented. Various issues like visiting schedule of MULEs, reliability of data gathering, communication channels, network-layer, end-to-end connectivity and Sleep-Awake Duty Cycle have not been addressed.

- In Energy-Conserving Data Gathering by Mobile Mules in a Spatially Separated Wireless Sensor Network (ECDGMM) technique proposed by Wu et al. [15], Sensor network is divided into isolated subnetworks. The mobile MULEs collects data from these possibly distant subnetworks. The problem is modeled as a dual objective problem namely Energy-Constrained MULE TSP. The ECM-TSP aims to achieve minimal MULE traversal paths. Optimum energy consumption in all sensors is also maintained so as not to surpass a pre-defined threshold. Unresolved issue of the above described method being buffer overflow during data collection since traversal path is determined but visiting schedule is not implemented.

- In Minimum-Path Data-Gathering (MPDG) technique proposed by Wu et al. [16], sensor network is spatially divided into several subnetworks. Mobile mule visits these subnetworks for sensed data collection in a well-organized manner. Occasionally WSN are naturally separated into isolated subnetworks due to some physical constraints, resulting in cost-effective subdivision. Coordination amongst these isolated subnetworks has to be maintained. In a spatially separated WSN, primary problem of data gathering is mainly constrained by latency of the mobile data collector, mule and delay in uploading data from each subnetwork of WSN. The path taken by mule to traverse the subnetworks, thus becomes a point of concern. MPDG, a generalization of the Euclidean TSP, tries to resolve the problem. With the Mules traversal path knowledge, nodes proactively calculate transmission/reception with the ferry. The efforts are to compute optimized ferry route so that data transmission latency is optimized and transmission time of each node is also met. By conducting multiple ferries even packet drops probability can be attempted to be minimized. The visiting schedule of each mule traversal path has not been considered for inclusion. This becomes a possible loop-hole in the implementation.

- Energy-management is an important concern for prolonging WSN network life time. As major consumption of power occurs during communication of collected data, transceiver action should be optimized for prolonging the network lifetime. There is a survey done in [17, 18] on the existing energy management schemes which clearly says that keeping nodes in the awake mode when there is no data to be sensed, consumes energy. In WSN, sensor nodes in the field transmit their measured values to the BS in a many-to-one communication pattern. Since measured values get updated in small sizes, update interval is maintained higher. All the time period between successive listening intervals is not utilized for data gathering. Keeping the sensor node awake for unused time periods would be an energy-draining exercise. This idle listening should be avoided for conserving energy.

As stated by Bhat et al. [19], with the help of IEEE 802.11 technology private wireless network can be set up by achieving good throughput. S-MAC is one of the special MAC protocol integrated within IEEE 802.11 MAC protocol suit. S-MAC provides adaptive listing concept [20–23]. There is another special MAC protocol called IEEE 802.15.4 which also provides adaptive listing mechanism. This feature is possible in IEEE 802.15.4 only during beacon-enabled mode. The drawback is if the duty cycle is not set correctly it may incur transmission latency [24–26]. Another option being D-MAC which is best suited for flat topology. Unlike S-MAC, D-MAC doesn't suffer from data forwarding interruption problem [27, 28]. This provides explicit duty cycle adjustment mechanism [29]. As stated by Anastasi et al. [30], in a WSN with MEs, motion of ME can be controlled by allowing nodes to sleep and awake based on the ME schedule and proximity.

3 Practical Understanding of Unresolved Issues

State-of-the-art survey of data gathering protocols in WSN has been carried out, focusing on the following criteria:

- Necessity of clustering and Cluster Head creation
- Need of Mobile Elements
- Scheduling, dispatch of Mobile Element using tour optimization techniques
- Correlation between Data gathering and Data Aggregation
- Comparison of existing Data Gathering Techniques highlighting respective drawbacks.

The surveyed papers exemplify the use of MEs for Data Gathering Techniques providing good improvement in the measured values of metrics like energy consumption and network lifetime. The surveyed techniques have considered the concepts of ME usage for data gathering. Observation made during survey of the existing research makes it evident that metrics not taken into proper consideration by certain data gathering protocols in existence are *overflow of buffers at sensor nodes* and *visiting schedule of MEs* [31].

In order to estimate the practical limitations of the existing Multiple Mobile Elements based data gathering protocols, two existing protocols EEDG [32] and IAR [33] have been implemented and compared depending upon various simulation scenarios having different set of nodes by authors [34].

Different sensor network scenarios have been established depending upon number of nodes (20, 40, 60, and 80, 100). For each scenario readings are taken to measure five analytical metrics of performance—Delay, Packet Drop, Packet Delivery Ratio, Energy Consumption and Overhead. Statistics obtained from the comparison clearly indicates that EEDG technique implementation is more efficient than IAR technique implementation. With the comparison it also gets evident that EEDG has certain limitations as stated below:

- EEDG does not consider inclusion of *Data Fusion Concept*. Using Data Fusion Concept, number of packet transmissions in the system could have been further more reduced.
- *Idle listening* feature has not been dealt with in EEDG. During the Idle-state of the nodes if the respective nodes are temporarily deferred from any activity, then the overall reduction in the energy consumption of the sensors across network is achieved. Thus positively affecting extension of network lifetime.

3.1 Motivation

Distributed data aggregation protocols are predominantly used in WSN for gathering and transmission of sensor data. If data aggregation concept is not used, data transmission from every sensor node to the sink will take place. In addition, it will also relay the packets of the downstream nodes. Since most of the time, a tree topology rooted at the sink is used to collect data, all packets are delivered to the sink node via its immediate neighbours. This situation results in faster depletion of energy in immediate neighbours of the sink, as compared with energy levels of other nodes of the network. Nonetheless, when the MEs have to operate in larger sensing areas, the MEs may skip visiting certain far-off nodes in order to save energy. If MEs are bound to visit all nodes in larger sensing areas, this may results in increased data gathering latency.

Existing researches, like Distributed Intelligent Data-Gathering Algorithm, have tried to deal with the above mentioned problem by incorporating cluster based architecture in WSN. A central node in each cluster is considered as Cache Point. MEs are made to visit only CPs, while other sensor nodes of the cluster transmit their data to designated CP. In such architecture, efficient tours for MEs are computed to visit CPs with minimized communication distance.

However, the recent work encompasses following challenges:

- While selecting CPs, residual energy of the nodes are not considered, which would have more impact on improving lifetime of sensor networks.
- Data aggregation concept is not incorporated at CPs which causes more number of transmissions in the sensor networks.
- MEs are always in passive open listen mode even if certain CPs doesn't have any data to transmit.

These challenges have to be specifically addressed to improve efficiency of the data gathering technique [32, 38].

Afore mentioned findings about the existing protocols, is the main motivation behind proposing a novel energy—efficient approach for data gathering.

4 Proposed Solution

Objectives of this research work is to implement—A novel technique of data gathering in energy efficient way using Mobile Elements to improve lifetime of WSN. The scope of the proposed data gathering technique is to achieve reduction of energy consumption, prevention of buffer overflow at CPs, improving network lifetime, by implementing the proposed technique in different practical scenarios.

To address the formulated challenges in the previous section, authors have proposed *"Energy Efficient Data Gathering Technique Using Multiple Mobile Elements in Wireless Sensor Networks (EEDGMME)"* which accomplishes task at hand by:

- Selection of CPs based on residual energy and incorporation of Data Fusion aspect at CPs to reduce energy consumption.
- Inclusion of Sleep-Awake Duty Cycle technique to ensure reduction of power consumption.
- Establishing visiting schedule for MEs based on earliest deadline of CPs to reduce buffer overflow.

Figure 1 clearly indicates issues of data gathering and proposed solutions to solve these issues.

4.1 Overview of the Proposed Work EEDGMME

To achieve optimum results for the proposed technique, after the node deployment, the node having utmost residual energy is chosen as CP for every cluster. MEs are deployed to perform the data collection from the selected CPs. Every ME is assigned visiting schedule based on the earliest deadline of CPs. Number of MEs required for the network is decided based on Round Trip Time (RTT). Number of nodes shared by an individual ME, is decided based upon number of hops node is farther

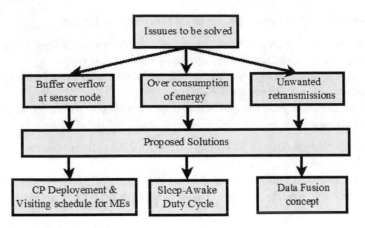

Fig. 1 Issues and proposed solutions

from a ME. Data fusion technique is applied at the CPs to reduce number of packet transmissions. In order to prevent energy wastage, the MEs go into sleep mode when there is minimal data collection job. The MEs after collecting data from all CPs in the predetermined path, deliver data to the sink.

The EEDGMME protocol overcomes limitation of existing EEDG protocol and performs the required job of data gathering using following techniques:

- Estimation of residual energy
- CP Deployment
- Deployment of MEs
- Visiting Schedule of the MEs
- Data Fusion at CP
- Sleep-Awake Duty Cycle implementation.

Figure 2 shows overall flow diagram of the proposed work.

4.2   Detailed Explanation of Inclusive Algorithms

This section includes detailed explanation of the algorithms incorporated in the proposed work EEDGMME.

4.2.1   Deployment of Sensors and Formulation of Clusters

In the practical scenario, sensor nodes are deployed by patching the *Naval Research Laboratory* (NRL) sensor extensions to the standard NS2 package and clusters are formed using relative geographical positions.

Fig. 2 Flow diagram of the
proposed work EEDGMME

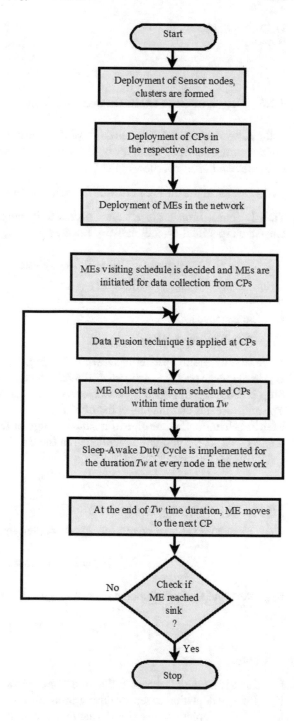

Fig. 3 "hello" Packet
Format

4.2.2 Deployment of CP in Respective Clusters

In the network scenarios, CPs are deployed in each cluster to collect information from the sensor nodes. MEs visit CPs by following a predetermined visiting schedule so as to prevent buffer overflow at the CPs.

Every sensor in the cluster performs the below mentioned steps [35]:

Step 1: Sensor broadcasts a "hello" packet to its neighbouring sensors of the same cluster, as specified symbolically by Eq. 4.1:

$$N_i \rightarrow Ne_i : Hello \tag{4.1}$$

where,

N_i ith sensor node
Ne_i Neighbouring sensors

Format of "hello" packet is shown in the Fig. 3. The residual energy parameter E_{res} in the hello packet is estimated as mentioned in Eq. 4.4.

Step 2: On receipt of "hello" packet, every node acknowledges and identifies itself. The node creates and maintains list of NeL_i.

Step 3: Depending upon value of residual energy of the node E_{res}, node N_i is configured as a CP, if the condition mentioned in the Eq. 4.2 is met:

$$E_{res} > E_{th} \tag{4.2}$$

where,

E_{th} Threshold value of node's energy, computed using following Eq. 4.3:

$$E_{th} = Initial\ Energy/2 \tag{4.3}$$

E_{res} Residual Energy of node N_i is estimated using Eq. 4.4 [36]:

$$E_{res} = E_i - (E_{tx} + E_{rx}) \tag{4.4}$$

where,

E_i Energy of the ith sensor at the time of initialization
E_{tx} Energy depletion at sensor during transmission
E_{rx} Energy depletion at sensor during data reception

Hence, the node with the residual energy greater than half of its initial energy gets configured as a CP. In comparison with the residual energy of all CPs, CPs with higher amount of residual energy send their collected data to farthest MEs and the ones with lesser amount of residual energy send their collected data to nearer ME.

4.2.3 Deployment of MEs

In larger networks with widely scattered sensor nodes, MEs move around specific areas and collect information from various sensors. MEs are deployed in the equally split network area called clusters. Deployment is done in a way that MEs will serve all nodes of the clusters uniformly. While deploying MEs following criteria has to be considered [37]:

- MEs count
- Nodes shared by the MEs

Estimating Number of MEs

Parameters required to calculate ME count:

G   Number of sensors deployed in area of $l \times l$ units
V   Speed at which the ME moves
T_{cache} Time needed for the CP's cache to be filled. Deadline of CP decides T_{cache}
T_{serve} Time needed to empty/serve the cache of the CP

Estimation of number of MEs needed is accomplished in the following steps:

Step1: Initialize G
Step2: Compute RTT using Eq. 4.5

$$RTT = 2 \times (l/V) + (G \times T_{serve}) \qquad (4.5)$$

where,

RTT   Round Trip Time
(l/V) Inter-cluster movement time of ME
$(G \times T_{serve})$ Time taken by ME to collect data from CPs of the corresponding cluster

Step3: Calculate MEs count for the network.
    The number of MEs required for data collection is selected based on the following condition:

    if $(RTT \leq T_{cache})$
                One ME is enough for data gathering
        else
                (RTT/ T_{cache}) number of MEs are required for data gathering

Table 1 Hop count of CPs from each ME

CPs	Hops away from ME_1	Hops away from ME_2
N1	1	5
N2	2	4
N3	3	3
N4	4	2
N5	5	1

    end if

Note: The ME does not perform any data collection in its reverse direction.

CPs Shared by the MEs

The sharing of the MEs by the CPs depends on the hop count of the CPs. Hop count is count of number of hops CP is farther from a ME.

Consider the following example.

Let N_1, N_2, N_3, N_4, N_5 be the deployed CPs, ME_1 and ME_2 be the Mobile Elements. From Table 1 it can be deduced that

- N1 is served by ME1
- N2 is served by ME1
- N4 is served by ME2
- N_5 is served by ME2

Whereas N3 can be served by either ME1 or ME2 since N3 is equidistant from both ME1 and ME2. The tie between selection of ME1 or ME2 is randomly resolved in favour of either ME1 or ME2. The ME which wins the tie, corresponding CP is added to its visiting schedule.

4.2.4 Data Fusion Technique at CP

The data fusion technique is applied at the CP when an incoming packet has the same destination as any previous packet waiting at CP to get transmitted. The CP will wait for some w seconds before sending the incoming data packets in order to potentially fuse new incoming data packets to make aggregated data packets.

In the wake of size of incoming data packets being greater than space remaining in CP's cache, incoming packet is split into smaller packets. These smaller chunks are considered as individual potential packets for data fusion. When CP's cache gets filled up, the already fused packets is transmitted out and remaining un-fused packets are taken for further fusion events, along with new incoming packets. The steps involved in the data fusion are given in Algorithm 4.1.

Algorithm 4.1. Data Fusion at CP

Let CP be the cache point, $\{fp\}$ be the fused packet, L be size of fused packet, L_{max} be the maximum allowable size of fused packet, w be the waiting time, Sp be the available space in CP, Pi be the incoming packets.

> **Input:** Incoming sensor data packets
> **Output:** Outgoing data packets from CP
> 1. $L=0$, $Sp = L_{max}$
> 2. arrival of Pi^{th} packet
> 3. **if** size of $Pi <= Sp$ **then**
> 4. { add Pi to $\{fp\}$
> $L = L + $ size of Pi
> $Sp = L_{max} - L$
>   start timer w }
> 5. **else**
> 6.   { fragment Pi
>   Consider each fragment as next arriving packet
>   go to step 3 }
> 7. **end if**
> 8. **if** timer w expires **then**
> 9.   transmit $\{fp\}$, go to Step 1
> 10. **else**
> 11.   { next packet arrives
> 12.   **if** size of packet $<= Sp$ **then**
> 13.   { attach packet to $\{fp\}$
> $L = L + $ size of packet
> $Sp = L_{max} - L$, Repeat step 4 till timer w expires}
> 14. **else**
> 15. { fragment packet, consider each fragment as next arriving packet, go to step 8 }
> 16. **end if** }
> 17. **end if**

4.2.5 Visiting Schedule of ME

The visiting schedule or trajectory of each ME is created by computing shortest route from every ME to CP, meeting the deadline constraints of the CP to reduce buffer overflow.

Deadlines at different CPs are computed using Eq. 4.6:

$$\text{Deadline} = \text{Current Time} + \text{Buffer Overflow Time} \qquad (4.6)$$

The Earliest Deadline First (EDF) algorithm is applied to know which CP to visit next in the visiting schedule. The step involved in computation of EDF are mentioned in Algorithm 4.2.

Algorithm 4.2. Earliest Deadline computation

Let *cost_am[1..n][1..n]* be cost adjacency matrix, *bot[1..n]* be CP's buffer overflow time, *dl[1..n]* be deadline of CPs, *sn* be start node, *ct* be current time, *cn* be current node.

> **Input:** *cost_am[1..n][1..n], bot[1..n], sn*
> **Output:** *dl[1..n]*
> 1. Initialization:
> $ct = 0, cn = sn, dl[1..n] = bot[1..n]$
> 2. select (node i $\neq$ cn) which has earliest deadline
> 3. **if** *dl* [i] $< ct + cost\_am$ [cn] [i] **then**
> 4. { Omit node i from visiting schedule
>   go to step 8 }
> 5. **else**
> 6. { $ct = ct + cost\_am$ [cn] [i]
> $cn = i$
> $dl[i] = ct + bot$ [i]
>   }
> 7. **end if**
> 8. **if** node is left unvisited **then**
> 9. go to step 2
> 10. **end if**

The visiting schedule of each ME is summarized in Algorithm 4.3.

Algorithm 4.3. Visiting Schedule of ME

Let $ME_1, ME_2, \ldots ME_n$ be the Mobile Elements, $C_1, C_2, \ldots, C_k$ be the clusters, CP_1, $CP_2, \ldots, CP_k$ be the Cache Points of the clusters, $D_1, D_2, \ldots, D_k$ be the deadlines of the CPs.

> **Input:** Mobile elements (ME)
> **Output:** Visiting schedule and path

1. Partition the network into clusters
2. The MEs are deployed initially
3. **for each** ME_i, i $= 1, 2, 3, \ldots, n$
4. **for each** CP_j, j $= 1, 2, 3, \ldots, k$
5. Let $p1_i$ be the initial position of Me_i
6. Find set of paths from ME_i to CP_j
7. Let {P} be the set of paths from ME_i to CP_j
8. Find the visiting time V_t of each path in {P}
9. **for each** path P_y in {P}
10. **if** P_y is shortest and $V_t(P_y) < D_j$), **then**
11. Choose the path P_y as the visiting path
12. **end if**
13. **end for**

14. Let p2$_i$ be the position of ME$_i$ after collecting data from CP$_j$
15. **end for**
16. **end for**

4.2.6   Sleep-Awake Duty Cycle

In order to ensure the reduction of power during idle state, a duty cycle technique is applied. The periodic duty cycle scheduling of SMAC [38, 39] which is included in the IEEE 802.11 MAC protocol. Sensor nodes operate in listen/sleep phases depending upon pre-determined time duration. Each slot/frame of time is partitioned to have listen" time and "sleep" time. Slot "listen" is decided depending upon MAC layer parameter—size of contention window and bandwidth of channel. Duty Cycle is calculated using Eq. 4.7.

$$Duty\,Cycle = T_{ON}/\,Tw \qquad (4.7)$$

where,

T_{ON} Time for which node is ON in "listen" mode
Tw Total time in slot/frame.

The steps involved in the duty-cycle technique are presented in Algorithm 4.4.

Algorithm 4.4. Sleep-Awake Duty Cycle

Let S$_{min}$ be the minimum threshold value for the size of gathered data in the CP.

Input: Sensor nodes and ME
Output: Sleep-Awake duty cycle
1. To identify neighbours, all nodes broadcast SYNC packet periodically with hop count $= 1$
2. Initialize   all nodes are in AWAKE mode
3.   **if** nodes reaches end of listen slot **then**
4.   {   Node enters SLEEP mode for the current slot to be awakened in the next slot go to Step 18 }
5. **else**
6.   {
7.   **if** node is an ME **then**
8.   {
9.   **if** Size of gathered data in CP $<$ S$_{min,}$ **then**
10.   { ME enters SLEEP mode for the current slot to be awakened in the next slot go to Step 18 }
11.   **else**
12.   { ME remains in AWAKE mode for the current slot and continues data collection go to Step 18 }
13.   **end if**

14. **else**
15.   { Node remains in AWAKE mode for the current slot and continues data collection go to Step 3 }
16.   **end if**
17. **end if**
18. End

5 Simulation Experiments

For the proposed data gathering technique EEDGMME, performance evaluation using simulation, has been performed and the acquired results are analysed. The analysis of the result is based on comparison with the existing EEDG protocol. Packet Delivery Ratio, Packet Drop, Energy Consumption, Delay, Overhead are the metrics used for comparison. Static scenarios have been considered for statistical comparison amongst EEDGMME and EEDG techniques.

5.1 Simulation Model for EEDG and EEDGMME Techniques

NS2 [40], an all-in-one network simulator tool, is put in use for simulating the proposed technique. Sensor network is created and implemented using a flat grid of 500×500 m area. For different scenarios with different number of nodes, energy model has been ascertained to estimate energy consumption. Transmission range has been set to 250 m and CBR traffic is maintained. EEDGMME and EEDG protocols have been simulated in static environments. EEDGMME protocol's performance is compared with EEDG technique.

For performance analysis, five different scenarios are considered depending upon number of nodes (60, 80, 120, 150, 200) implemented in WSN. For each scenario readings are taken to measure five analytical metrics of performance—Delay, Packet Drop, Packet Delivery Ratio, Energy Consumption and Overhead. Various setting and values of different parameters of the simulation are summarized in Table 2.

5.2 Performance Metrics

EEDGMME and EEDG techniques are analysed depending upon the specified metrics of performance:

- Packet Delivery Ratio—It is a parameter which computes number of packets received at sink in comparison to the number of packets sent by the nodes.

Table 2 Simulation settings and parameters

Parameters	Values
Number of participating nodes (N)	60, 80, 120, 150, 200
Simulation area (A)	500 × 500 m
Range of transmission (R)	250 m
MAC	IEEE 802.11
Simulation duration	50 s
Source of traffic	CBR
Size of packet	500 byte
Initial energy	20.1 J
Transmitter	0.660 W
Receiver	0.395 W

- Overhead—Overall amount of control information propagating in the system in comparison to amount of data transmission from sensors to sink.
- Packet Drop—Average number of packets dropped during data transmission.
- Energy Consumption—Overall energy spent by sensors.
- Packet Delay—Total transmission time of data packets.

5.3 Performance Analysis of EEDGMME and EEDG Techniques

The simulation results bring out some important characteristic differences between the data gathering protocols. The various measurements done in the simulation implementations are compared with respective values corresponding to EEDGMME and EEDG techniques.

Figures 4, 5, 6, 7 and 8 show the graphs of individual performance metric plotted, based on the statistics obtained after simulation.

Figure 4 depicts the graphical representation of various values of overhead measurements versus different number of nodes, for EEDGMME and EEDG implementations in static environment. As number of nodes increases, overall overhead incurred in the proposed work EEDGMME is comparatively less than that in existing EEDG technique.

Figure 5 depicts the graphical representation of various values of Energy Consumption of EEDGMME and EEDG technique for various number of nodes. As number of nodes increases, overall energy consumed by proposed work EEDGMME is comparatively less than that in existing EEDG techniques.

Figure 6 depicts the comparative graphs of EEDGMME and EEDG implementations where different values of Packet Drop are plotted against various number of nodes in the different scenarios. As number of nodes increases, overall packet

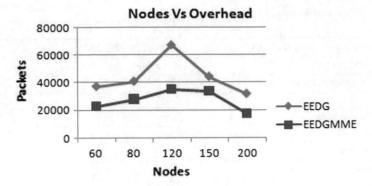

Fig. 4 Nodes versus overhead in static environment

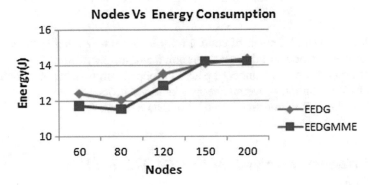

Fig. 5 Nodes versus energy consumption in static environment

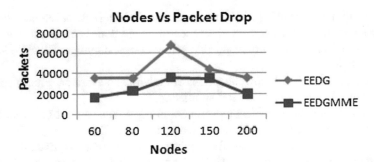

Fig. 6 Nodes versus packet drop in static environment

drop occurred using proposed work EEDGMME is comparatively less than that in existing EEDG techniques.

Figure 7 depicts the comparative graphs of EEDGMME and EEDG implementations where Packet Delay measurements are plotted against various number of nodes in the different scenarios. As number of nodes increases, overall delay incurred by

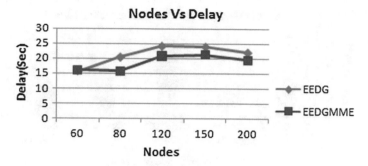

Fig. 7 Nodes versus delay in static environment

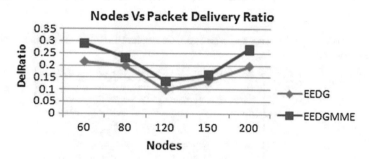

Fig. 8 Nodes versus packet delivery ratio in static environment

proposed work EEDGMME is comparatively less than that in existing EEDG techniques.

Figure 8 depicts the comparative graphs of EEDGMME and EEDG implementations where Packet Delivery Ratios are plotted against various number of nodes in the different scenarios. As number of nodes increases, overall Packet Delivery Ratio achieved by proposed work EEDGMME is comparatively more than that in existing EEDG techniques.

Figure 9 shows the average percentage of individual performance metrics of EEDGMME and EEDG technique for different network sizes in static environment. Figure 9 clearly indicates that the average percentage of overhead and delay incurred, energy consumed, packet drop occurred is less and delivery ratio obtained is more with respect to EEDGMME as compared with EEDG technique.

5.4 Comparative Performance Analysis

Table 3 shows comparison between performance analysis of EEDGMME and EEDG protocols depending upon on different number of nodes implemented in static environment. Table 3 includes the performance metrics—Overhead, Energy Consump-

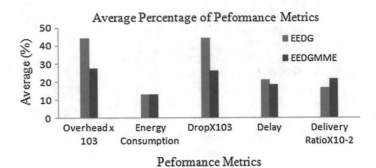

Fig. 9 Average percentage of performance metrics comparison between EEDG and EEDGMME in static environment

Table 3 Comparative performance analysis of EEDGMME and EEDG for various number of nodes in static environment

● Efficient			● Comparatively inefficient	
Performance Metrics	EEDGMME (Average %)	EEDG (Average %)	Percentage Increase/Decrease of EEDGMME compared to EEDG	Efficient Protocol
Overhead	27.265	**44.2588**	38% Decrease	EEDGMME
Energy Consumption	12.88473	**13.275544**	3% Decrease	EEDGMME
Packet Drop	26.2904	**44.199**	40% Decrease	EEDGMME
Delay	18.63372	**21.23776**	12% Decrease	EEDGMME
Packet Delivery Ratio	21.5165	**16.7677**	22% Increase	EEDGMME

tion, Packet Drop, Delay, Packet Delivery Ratio depending on which comparison is evaluated.

The fourth column in the Table 3 indicates the Percentage Increase/Decrease in the values of concerned Performance Metrics of EEDGMME in comparison with that of EEDG static simulation scenarios.

In static environment, EEDGMME technique attains 38% less overhead, 3% less energy consumption, 40% less packet drop, 12% less delay and 22% more packet delivery ratio as compared to EEDG technique. EEDGMME technique hence proves to be more efficient than the existing technique EEDG. Thus EEDGMME helps in prolonged network lifetime.

In case of EEDGMME better results are achieved for the metrics Overhead, Energy Consumption, Packet Drop, and Delay as compared to EEDG findings. Packet Delivery Ratio is found to be comparatively more and hence better, with EEDGMME as compared with EEDG results.

Hence it is proved that for the respective static scenarios simulated, EEDGMME has overall better performance than EEDG.

6 Conclusion and Future Work

In this proposed work, proficient data gathering technique for WSN is investigated. To enhance network lifetime and to optimize energy consumption, special algorithms have been explained and implemented in various simulation scenarios. The proposed technique EEDGMME has been put to test in order to prove that it indeed achieves better performance results. Performance of EEDGMME and existing protocol EEDG were compared in static and as well as in dynamic environment, with different network sizes. The simulation results were appropriately tabulated. In conclusion, it is affirmed that in static environment, the proposed work EEDGMME fares comparatively superior than the existing protocol EEDG. Since proposed work can save energy and enhance life time, it is well suited for the applications where battery power is the primary consideration. However some aspects of this technique still need to be improved.

The scope for the future work includes following:

1. Handling failure of MEs—While traversing the network according to the visiting schedule, if the ME fails then data loss will be resulting in the network. Since, the deadline of the concerned CPs would have exhausted waiting for the ME to arrive for data collection.
2. Decreasing further consumption of power—To achieve optimized consumption of power in EEDGMME, Sleep-Awake Duty Cycle technique has been implemented. To achieve even better optimization of power consumption, IEEE802.15.4 protocol can be incorporated. With the use of beacon frames, which are an integral part of IEEE 802.15.4 protocol, proper co-ordination between CPs and the visiting ME can be established for data collection.
3. Extending the work to HetNet—By making use of sensors with enhanced computational prowess, upgraded memory and different communication potentials, the protocol can be implemented for Heterogeneous Networks.

Acknowledgements This work is sponsored and supported by grant from Vision Group on Science and Technology (VGST), Govt. of Karnataka (GRD-128). The authors wish to thank Dr. S. Ananth Raj, Consultant, VGST, and Prof. G. L. Shekar, NIE., for their encouragement in pursuing this research work.

References

1. Cui T, Chen L, Ho T, Low SH, Andrew LL (2007) Opportunistic source coding for data gathering in wireless sensor networks. In: IEEE international conference on mobile adhoc and sensor systems, 2007. MASS 2007. IEEE, pp 1–11
2. Ye Z, Abouzeid AA, Ai J (2007) Optimal policies for distributed data aggregation in wireless sensor networks. In: 26th IEEE international conference on computer communications. INFOCOM 2007. IEEE, pp 1676–1684
3. Puthal D, Sahoo B, Sharma S (2012) Dynamic model for efficient data collection in wireless sensor networks with mobile sink 1

4. Bi Y et al (2006) A power graded data gathering mechanism for wireless sensor networks. Acta Automatica Sinica 32(6):881
5. Yu Y, Krishnamachari B, Prasanna VK (2004) Energy-latency tradeoffs for data gathering in wireless sensor networks. In: Twenty-third annual joint conference of the IEEE computer and communications societies, vol 1. INFOCOM 2004. IEEE
6. Caillouet C, Li X, Razafindralambo T (2011) A multi-objective approach for data collection in wireless sensor networks. In: Ad-hoc, mobile, and wireless networks. Springer, Berlin, Heidelberg, pp 220–233
7. Soyturk M, Cicibas H, Unal O (2010) Real-time data acquisition in wireless sensor networks. INTECH Open Access Publisher
8. Ramanan K, Baburaj E (2010) Data gathering algorithms for wireless sensor networks: a survey. Int J Ad hoc Sensor Ubiquitous Comput (IJASUC) 1:102–114
9. Lu M, Wu J (2008) Utility-based data-gathering in wireless sensor networks with unstable links. In: Distributed computing and networking. Springer, Berlin, Heidelberg, pp 13-24
10. Ota K, Dong M, Li X (2009) TinyBee: mobile-agent-based data gathering system in wireless sensor networks. In: IEEE international conference on networking, architecture, and storage, 2009. NAS 2009. IEEE, pp 24–31
11. Zhao M, Ma M, Yang Y (2008) Mobile data gathering with space-division multiple access in wireless sensor networks. In: The 27th conference on computer communications. INFOCOM 2008. IEEE
12. Zhao M, Gong D, Yang Y (2010) A cost minimization algorithm for mobile data gathering in wireless sensor networks. In: The 7th IEEE international conference on mobile ad-hoc and sensor systems (IEEE MASS 2010). IEEE, pp 322–331
13. Kinoshita K, Okazaki T, Tode H, Murakami K (2008) A data gathering scheme for environmental energy-based wireless sensor networks. In: 2008 5th IEEE consumer communications and networking conference. IEEE, pp 719–723
14. Jain S et al (2006) Exploiting mobility for energy efficient data collection in wireless sensor networks. Mobile Netw Appl 11(3):327–339
15. Wu F-J, Tseng Y-C (2013) Energy-conserving data gathering by mobile mules in a spatially separated wireless sensor network. Wirel Commun Mob Comput 13(15):1369–1385
16. Wu F-J, Huang C-F, Tseng Y-C (2009) Data gathering by mobile mules in a spatially separated wireless sensor network. In: 2009 tenth international conference on mobile data management: systems, services and middleware, pp 293–298. IEEE
17. Pandey GK, Singh AP (2015) Energy conservation and efficient data collection in WSN-ME: a survey. Indian J Sci Technol 8(17)
18. Huang P et al (2013) The evolution of MAC protocols in wireless sensor networks: a survey. IEEE Commun Surv Tutor 15(1):101–120
19. Jayram BG, Ashoka DV (2015) Qualified breakdown of Wi-Fi with and without standard in wireless networks-onus on throughput, Int J Adv Res Comput Commun Eng 4:19–21
20. Demirkol I, Ersoy C, Alagoz F (2006) MAC protocols for wireless sensor networks: a survey. IEEE Commun Mag 44(4):115–121
21. Network, High PERformance Local Area. Mobile communications, Chapter 7: Wireless LANs. Diss. University of Karlsruhe (1996)
22. Kim TH, Choi S (2006) Priority-based delay mitigation for event-monitoring IEEE 802.15. 4 LR-WPANs. IEEE Commun Lett 10(3):213–215
23. Molisch AF et al (2004) IEEE 802.15. 4a channel model-final report. IEEE P802 15.04: 0662
24. Bhar J (2015) A Mac protocol implementation for wireless sensor network. J Comput Netw Commun 2015:1
25. Hurtado-López J, Casilari E (2013) An adaptive algorithm to optimize the dynamics of IEEE 802.15.4 networks. In: International conference on mobile networks and management. Springer International Publishing
26. Stanislowski D et al (2014) Adaptive synchronization in IEEE802. 15.4 e networks. IEEE Trans Ind Inform 10(1):795–802

27. Jayram BG, Ashoka DV (2012) Comparison of MAC layer protocols for wireless sensor networks. In: Proceeding of international conference on evolutionary trends in information technology

28. Jayram BG, Ashoka DV (2012) Performance improvement of DMAC for data gathering in wireless sensor networks. In: Proceedings of international conference on global research. Sharing innovative thoughts, gaining memorable knowledge, pp 61–68

29. Jayram BG, Ashoka DV (2014) MAC layer protocols for WSN—comparison and performance improvement strategy. Int J Eng Res 3(4):217—220

30. Anastasi G, Conti M, Di Francesco M (2009) An adaptive sleep strategy for energy conservation in wireless sensor networks. Technical Report DII-TR-2009-03. http://info.iet.unipi.it/~anastasi/papers/DII-TR-2009-03.pdf

31. Jayram BG, Ashoka DV (2013) Merits and demerits of existing energy efficient data gathering techniques for wireless sensor networks. Int J Comput Appl 66(9)

32. Almi'ani K, Viglas A, Libman L (2010) Energy-efficient data gathering with tour length-constrained mobile elements in wireless sensor networks. In: 2010 IEEE 35th conference on local computer networks (LCN), pp 582–589

33. Kim J-W, In J-S, Hur K, Kim J-W, Eom D-S (2010) An intelligent agent-based routing structure for mobile sinks in WSNs. IEEE Trans Consumer Electron 56(4):2310–2316

34. Jayram BG, Ashoka DV (2016) Comparative analysis of data gathering protocols with multiple mobile elements for wireless sensor network. In: Proceedings of first international conference on information and communication technology for intelligent systems, vol 1. Springer International Publishing, pp 281—290

35. Kumar KA, Ribeiro VJ (2009) REEF: a reliable and energy efficient framework for wireless sensor networks. In: 2009 First international communication systems and networks and workshops. IEEE, pp 1–9

36. Narayana G, Akkalakshmi M, Damodaram A (2016) Energy efficient polynomial based group key management protocol for secure group communications in MANET. Int J Appl Eng Res 11(9):6701–6705

37. Jea D, Somasundara A, Srivastava M (2005) Multiple controlled mobile elements (data mules) for data collection in sensor networks. In: International conference on distributed computing in sensor systems. Springer, Berlin, Heidelberg, pp 244–257

38. Ye W, Heidemann J, Estrin D (2004) Medium access control with coordinated adaptive sleeping for wireless sensor networks. IEEE/ACM Trans Netw 12(3):493–506

39. Halfbrodt H-C (2010) MAC protocols for wireless sensor networks. Institute of Computer Science Freie Universität Berlin, Germany, Jan 2010

40. Pradhan S, Sharma K, Dhakar JS, Parmar M (2016) Cluster head rotation in wireless sensor network: a simplified approach. group 4(1)

Part IV
Information Management
& Social Media

Online Social Communities

Anuradha Goswami and Ajey Kumar

Abstract Social communities helps people to interact and engage both socially and cognitively. Direct and indirect relationships are built through these interactions with the members in the community. The members may be known, unknown, friends or relatives. With the invent of Web 2.0, Internet users were empowered with the ability to interact, share content and collaborate over the dynamic web pages. The evolution of technologies has helped the transition from offline to Online Social Communities (ONSC). Social interaction also got transformed within communities encouraging both offline and online connect. In this chapter, a comprehensive review of ONSC is done. First, the ONSC dimensions-user relationships, social interaction, common interests and virtual environments along with OSNC features-purpose, membership, rules and terms, user generated content, users-benefits and types are discussed in detail. Types of communities for ONSC are given based of three different school of thoughts. Second, the importance of ONSC management—user acquisition, user engagement and user retention, across different user participation life cycle stages—initial, growth, matured and decline, which is further mapped to user membership involvement phases, is given. Various perspectives, recommender systems and user engagement tools for ONSC are also included. A survey of ONSC structure detection techniques of disjoint and overlapping communities for both static and dynamic network structures is done. Finally, the applications of ONSCs across various domains are explained. The summary of popular ONSC available till date and supported businesses flourishing based on ONSCs are explicitly mentioned.

Keywords Web 2.0 · Social communities · Online social communities · Online social community management · User participation life cycle · Disjoint communities · Overlapping communities · Domains

A. Goswami · A. Kumar (✉)
Symbiosis Centre for Information Technology (SCIT), Symbiosis International (Deemed University) (SIU), Rajiv Gandhi Infotech Park, Hinjewadi, Pune 411057, Maharashtra, India
e-mail: ajeykumar@scit.edu

A. Goswami
e-mail: anuradha@scit.edu

© Springer International Publishing AG, part of Springer Nature 2019   289
S. Patnaik et al. (eds.), *Digital Business*, Lecture Notes on Data Engineering
and Communications Technologies 21, https://doi.org/10.1007/978-3-319-93940-7_13

1 Introduction

The word '*community*' has been traditionally anchored in local and neighbourhood interactions which was synonymous to social cohesion. The concept of community usually used to connect people socially and cognitively, considering individuals homogeneous in nature, broadly termed as *groups* [1–4]. The intensity or velocity of interactions or transactions between the community members has got increased over the years. Since 1970s, plethora of studies documented a shift from the local interaction to long distance one within the communities. These communications happen through engagement around different disciplines, functions, information needs, professions, etc. Few of the members are known to each other, some are unknown and the rest of the friends and relatives live elsewhere but contributing to the same community structure [5–8]. Relationships are built through these communications, both direct and indirect, which in turn create communities, broadly conceptualized as *Social Communities* as shown in Fig. 1.

Traditional researches have provided couple of definitions for communities and few are given as follows:

- *A group of people living in the same defined area, sharing the same basic values, organization and interests is a community* [9].

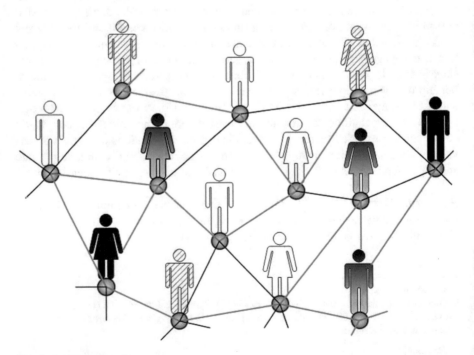

Fig. 1 Social communities

- *An informally organized social entity which is characterized by a sense of identity* [10].
- *A population which is geographically focused but which also exists as a discrete social entity, with a local collective identity and corporate purpose* [11].

People in community-based societies function in a discrete work group be it an organization, household in a neighbourhood, or as members of the kinship groups. These groups often have restrictions for inclusion and have a definite hierarchical structure for e.g. parents and children, supervisors and employees, organizational executives and members, etc. [12]. These structures form the core of the network that can be defined between individuals which form the vertices and connected more densely with each other within the group structure than with the individuals of other groups or communities. Thus, communities become a small representative of the whole network, which reveals much similar characteristics of the whole network graph or structure. Therefore, comprehending the whole network requires examining one or few communities of the same. This makes the study of different types of communities, community structure detection significant for researchers.

The concept and technology of Web 2.0 has revolutionized the way people communicate with each other throughout the world. Internet users were empowered with the ability to interact, share content and collaborate over the dynamic web pages. A complete evolution happened in the type of communication networks, where traditional static networks got replaced by its dynamic form. Dynamic networks like online social networks, wireless, sensor, mobile networks etc., which functions through distributed protocols and where nodes and edges appear or disappear over time, is globally trending now.

Another perspective is the driving force that makes the users contribute to all such social communities. Social research conveys it is the social capital which empowers the communities and creates the urge amongst users to participate and be members of the social communities [13, 14]. This social capital is generated both by the independent users as well as through interactions with other users of the network platform. Users connect through message exchange, creating friendships, extending support, and writing comments, likings or ratings. All these combine to form the source for social capital. Social capital formation in social communities lead to social and functional benefits, like information exchange, knowledge support, etc. [15, 16]. For community operators, it is very important to conduct a proper management of the social communities, so that active user participation is ensured, which further adds onto the social capital. It's a mandate for the community operators to understand the user behaviour of their community, user engagement procedures, so that they can develop measures to build up a long-term relationship with the users.

The rest of the chapter is organized as follows: Sect. 2, describes the types of social communities. Section 3, gives a brief review of the evolution of technology which has supported different types of social communities to prevail over time. Section 4, focus on the online social communities, detailing its important dimensions, features, pre-requisites, types of possible memberships and different types of online social communities. A proper online social community management throughout the life-

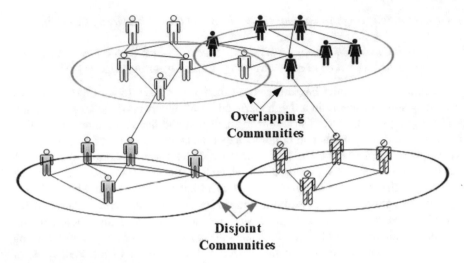

Fig. 2 Offline communities

cycle of the community users is discussed in Sect. 5. This section also involves a detailed study on the user acquisition, user engagement and user retention factors along with the recommender systems and user engagement tools used by the community operators. Section 6, elaborates different online community structure detection algorithms, considering both disjoint and overlapping communities which are very common in nowadays. Various application of online social communities in different business domains are described and summarized in Sect. 7. Lastly, Sect. 8, concludes the chapter.

2   Types of Social Communities

The individual members of the groups or social communities can prefer to inter-act with each other both offline and online, leading to the formation of *Offline Social Community (OFSC) and Online Social Community (ONSC)* respectively. For instance, there are many communities who usually conduct their activities in a dis-tributed fashion in an organizational perspective, where members are encouraged to meet offline, face-to-face for meeting [17], without any online communication into existence as shown in Fig. 2.

For example, Wikimania [18], an annual meeting, held every year since 2005, of contributors to the Wikimedia Foundation's wiki projects is an OFSC. Local Wikimania-type events occur frequently, which allows Wikipedians community members from different cities and countries to meet and discuss their online technology-mediated collaborations. eBay conducts an annual eBay Live! Event to promote networking among its distributed vendors offline. Similarly, regular offline events are conducted by World of Warcraft [19] for their members to socialize on meeting face to face. Also, through different conferences and conventions, the geographical variation amongst the interested community members gets blurred with the evolution of time for OFSC. The offline face-to-face meeting of users also result in more trust among each other through creation of relationships which in turn result in more dedication or involvement in getting engaged online and produce more relevant and valid content. There are several researches who have clarified why offline meetings are required to enhance the online social community relationships. People can be better connected through offline interaction which facilitate communication [20].

A study by Parks and Floyd [21] has shown that offline interaction leads the users to communicate online through email, phone, etc. In a study by [22] on *Metafilter* (community weblog) [23], offline gatherings are shown to be important for online discussion community. The online users can socialize more through these offline gatherings. Oldenburg [24] studied three places in which users spend their time.

- *First place is home*, where people feel comfortable the most and hold significant power over those who are visiting.
- *The second place is work place*, where people feel comfortable accompanied with some behavioural and occupational restrictions and expectations.
- *Third place is the social settings*, where people meet each other and discussion happens over topics imposed by external identities.

Studies by Putnam [13] and Steinkuehler and Williams [25] proposed that these social settings are in decline, which are getting replaced by virtual worlds of online social settings on communities as shown in Fig. 3. Computer aided communication may remove the visual traditional comfort of interaction but sometimes this is convenient for the persons who are less comfortable in gatherings [26, 27]. Also, online communication limits on the amount of information which gets transferred within the limited period of time [28]. This lead to the desire in developing relationships offline where information transfer and engagement is much more efficient. Also, offline meetings could help to socialize new members of an online community. Bryant et al. [29] have studied how Wikipedia editors have become Wikipedians through offline meetings in adopting the hallmarks of central participations.

An ONSC, on the other hand, are virtual groups, supported by Internet technology and are guided by norms and policies [30]. Kim [31], defines online community as a '*group of people with a common purpose, interest, or activity, who get to know each other better over time*'. ONSC's are a part of virtual space where individuals feel they are a part of it, by interacting on a common topic of interest. The various components depending on which online communities differ are—purpose (e.g. education, business etc.), the software environment support (e.g. bulletin board, chat, instant messaging, online social network etc.), size of community, duration of their existence and stage of their life-cycle, member's culture (e.g. international, national, local, political influence, religion influence etc.) and governance structure accompanied by respective rules and norms [32].

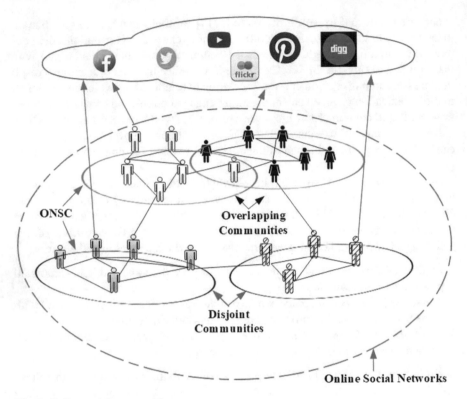

Fig. 3 Online social communities

3 Evolution of Technology for ONSC

The first and frequently used communication technology on Internet [33] was developed by ARPANET in 1972. Initially, it was a point to point technology where one person was able to send email just to one another person. But, in 1975, listservers were designed to enable one user to send an email to multiple receivers. The two ways listservers namely *trickle through* and *digests* were used. *Trickle through* send the messages as they are received whereas *digests* presents the messages in chronological order or receipt. Systems with improved GUI came into existence in mid 1980s.

The technology of bulletin board was established mimicking the physical bulletin board. People keep messaging on this board related to a specific topic of choice and the thread of discussion continues. The first message on a topic initiates a thread and later on responses get stacked on it. Over the years, this technology has evolved a lot with the addition of search engines, list of emoticons, user profiles and web pages and graphical two dimensional pictures.

Both the technologies mentioned are asynchronous communication technology as it does not require the partners to be present at the same time. There can be a lag of days, months or years between one reading and replying the messages or expressing his personal view on the same. Typically, later synchronous solutions like chat systems, instant messaging or texting platforms were developed that require the presence of both the partners at the same time to justify the involvement in the community discussion.

Nowadays, the invent of Online Social Networks incorporates user profiles, platform for discussion between users and formation of communities to facilitate discussions and knowledge sharing, belong to both synchronous and asynchronous communication technology.

ARPANET was followed by the invent of World Wide Web by Tim Berners Lee in 1991 and was released by CERN (the European Organization of Nuclear Research). This development was followed by widespread use of different web sites, development of different online communities on OSN platform emerged with various forms of communications software. Graphical, three-dimensional environments such as Palace and later ActiveWorlds made its space. Highly sophisticated gaming software's also changed the complete scenario. MP3 technology innovation also impacted the concept of community. It showed how communities can form around a particular technology with respect to the distribution, sharing and also stealing music. The open source era has also effect on the community building, for example Slashdot, Internet telephone, streaming video, photographs, sound, voice, web cam, blogs, wikis, etc. are all enhancing the way online communities collaborate or communicate. With the advent of mobiles, the concept of communities has now shifted onto mobile apps where people can always remain connected through OSN platforms.

4 Online Social Community

The most recent forms of ONSC are social networking sites which are web-based services. ONSC allows people to construct a public or semi-public profile in a closed system, define a list of other users to connect with, and view and traverse other user's list of connections and the ones made in the system [34].

Previous researchers have studied what helps communities to grow and what motivates contributors, but the reasons that people create new communities in the first place remain unclear. According to different researchers, the most important dimensions of ONSC are-*user relationships*, *social interaction*, *common interests*, and *virtual environments*. Each of them are discussed as below:

- *User Relationships*: The social relationship between members of a ONSC becomes the central aspect and emphasized by [35], as virtual communities which shows social aggregation on web and form personal relationships. Finally, these relationships build the social context for interaction between the users.
- *Social Interaction*: A closely knit people in ONSC interact and communicate with each other in the community, exchanging ideas, information and knowledge. This

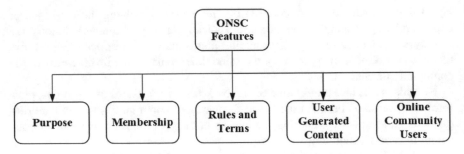

Fig. 4 ONSC features

is an ongoing, continuous process which forms and maintains relationships [36]. Not all members are actively engaged in exchange process or content production process [37]. So, interaction can happen both in active or passive way from the users which not always include constant activities of all members, but can also happen through only consumption of information.

- *Common Interests*: The users of ONSC are always facilitated by a specific orientation and topic of the community involved so that they can share their topic of interests and feelings. This helps the people of common interests to come closer and form relationships through ONSC [38, 39]. The purpose of the community and the user's interests give a longevity in the existence of ONSC [30].

- *Virtual Environment*: The term coined as 'online' refers to connection which takes place in online domain. Social ties can happen both online and offline, but in context of ONSC, both get intertwined with each other [40]. People can meet and interact offline too, once they have got an online connect. Thus, the online community offers an interface for users meeting virtually. The web-based service and technology facilitates these relationships and interactions, and accommodates the virtual community of people [30, 34].

A study in [41], highlighted the results of a survey of over 300 founders of new communities on the online wiki hosting site www.Wikia.com. The research analysed the motivations and goals of wiki creators, finding that founders have diverse reasons for starting wikis and diverse ways of defining their success. Many founders see their communities as occupying narrow topics, and neither seek nor expect a large group of contributors. The study also found that founders with differing goals, approach community building differently. A future scope for community platform designers are *to create interfaces that support the diverse goals of founders more effectively*.

4.1   Pre-requisite for ONSC

Every online community should adhere to some 'must have' features. These features are summarized in Fig. 4 and detailed as follows [42]:

(a) **Purpose**: The purpose of an ONSC can be explained through *three C's—Content, Community* and *Commerce*:

- *Content* perspective determines from where to obtain the user generated content. It can be in form of long posts, blogs, microblogs, etc., if it is a text. Otherwise, videos, images, and photos, are also a part of user generated content which can be obtained from the ONSC platforms.
- *Community* perspective defines the tools used by the ONSC to maintain relationships, perform business, satisfy customers, etc. Plethora of tools which can be used by the ONSC for proper maintenance of communities and retention of users are discussion boards, directories, Newswires, etc.
- *Commerce* involves planning of the revenue model for ONSC. This could be through advertising, third party apps, etc.

Once a construction of communities is done, it should be communicated to the members of the ONSC, to let the first time visitors know about what the community is all about, the aim and objective of the same and the rules and regulations to follow in order to opt for a membership. According to [30], *"Communities who have clearly stated goals appear to attract people with similar goals and who are often like each other, this creates a stable community in which there is less hostility"*.

(b) **Membership**: Every ONSC should provide an opportunity to the users to register and become member of the same through a well-defined procedure. This gives a belongingness to the member users with a sense of identity and also offers a scope to the community manager to know who the users are. The identity can be in the form of real name of the user or some unique username. This username acts as an identification to the user for his every messages, posted in ONSC in form of chats, tweets, posts, comments, etc. All the users in the community can know each other through this unique identification of username. The non-members are not allowed to get all premium facilities of the ONSC as compared to the members. The ONSC should have a proper procedure of registration for the users where complete information on benefits should be conveyed properly to the interested users followed by filling up a form by the user, detailing his/her personal/professional information. At the same time, the details required should be limited by the ONSC to avoid boredom among users to fill the form. This registration process should be followed by a 'welcome email' from the ONSC manager, detailing the benefits as a member, so that the member could feel comfortable and feel the belongingness towards the ONSC as a member. With evolution of time, ONSC also encourage the user to keep on updating his profile so that their public profiles also get updated, enriching the 'who's who' directory.

If people belong to multiple online communities, their joint membership can influence the survival of each of the communities to which they belong. Communities with many joint memberships may struggle to get enough of their members' time and attention, but find it easy to import best practices from other communities. A study by [43], researched on the effects of membership overlap on the survival of online communities. By analysing the historical data of 5673 Wikia communities, the study found that higher levels of membership overlap are positively associated with higher survival rates of online communities. Furthermore, it is beneficial for young communities to have shared members who play a central role in other mature communities. The contributions of the study are two-fold. Theoretically, by examining the impact of membership overlap on the survival of online communities, they identified an important mechanism underlying the success of online communities. Practically, the findings of this study may guide community creators on how to effectively manage their members, and tool designers on how to support this task [43].

According to [38], the typical membership development process from a community operator perspective consists of four stages:

- *Attract Members*: Community operators need to get attention for the online community. So, potential customers or users need to be convinced to try the service by becoming member or registering for the ONSC.
- *Promote Active Participation*: Second stage involves to make the users participate more actively, spending more time in the platform.
- *Increase Loyalty*: This stage facilitate relationships to other members and to the community operator. The goal should be no member should leave or lost due to decreased level of interest. Therefore, user retention is of high priority.
- *Generate Profit*: The member create revenue for the ONSC, in case the community is an commercially oriented operator. For example, through advertising or fees.

(c) **Rules and Terms**: Every ONSC should define the legal and the social boundaries within which the users are expected to remain through document of rules, regulations and terms provided by the community. They are described as follows:

  - *Terms of Use of Document*: This document should include the legal issues such as privacy, copyright, intellectual property, etc. Also, the community managers should adhere to the format of the Terms of Use documents to ensure the inclusion of all clauses of mandate. The visibility of this document should be 24/7, so that the users are aware of the terms at ease. Every user intends to get registered with a ONSC should agree with the terms by signing against the statement of agreement provided by the community manager on behalf of the ONSC.
  - *Rules and Regulations*: This refers to a set of rules and guidelines published by ONSC which should be adhered to by the users. It should also contain the penalty which the users have to undergo, if they break the rules under any circumstances. The most common platform which invites trouble for the ONSC members are on discussion boards through offensive comments which sometimes can be interpreted as libellous. ONSC can have a check on this, through proper policing on the posts which are getting posted on the discussion board platforms, which might not be always preferred from the user point of view. Sometimes peer pressure against offensive comments can also play positively to withdraw a negative statement. If nothing works,

ONSC can deliberately remove the offensive material from the discussion platform, thus stopping any further dispute among the existing members of the community. As a last resort solution, the authority can also replace the offensive statement by an Editor's posts explaining the reasons of removal of the said material from the discussion portal. It is also advisable on behalf of the members to update the community rules from time to time.

(d) *User Generated Content*: The user generated content of an ONSC play a significant role in retaining users to the community platform. This content can be in form of text, images, videos, photos, etc. The generating platforms can be from submitted articles, posts on discussion boards, transcripts from online events and directory entries. Two types of ONSC interaction is possible: *asynchronous* and *synchronous*.

- *Asynchronous Interaction*: Interaction between users which happens over time where replies to 'posted' messages can be made by the other users anytime over following days or weeks, is an asynchronous interaction. Any discussion board platform, where members discuss on a specific topic or in general, involves users in this type of interaction. The community members join for discussion are normally like-minded with similar interest. Thus, ensuring quality of content discussed, where users can gain knowledge overtime through active participation.
- *Synchronous Interaction*: This type of interaction normally happens in a chat room environment where users post comments and get replies in real time. The quality of content of chat rooms are normally poor if not regulated by a regulatory body. So, chat rooms can yield some relevant information if a focus is given to the same. This can be achieved through scheduled online discussion events with 'guest speakers' and 'moderators' with a definite focus and scope in mind of the community managers.

(e) *Online Community Users*: There are two user dimensions which should be adhered to, by the community managers for ONSC. Firstly, the benefits the users will get on registering as a member for the ONSC. Secondly, the types of users the ONSC platform is targeting.

i. *User Benefits*: The value of participating or registering in an ONSC comes from the level of satisfaction or fulfilment of specific needs of the user. According to the study of Hagel and Armstrong [44], the different types of communities are formed out of varying user needs such as relationship building, out of interest, for transaction and out of fantasy. It is very evident from this study that the members essentially leverage profit from—Social Support and Relationships, Information exchange, Economic benefits and Entertainment. Each of them are described as follows:

- *Social Support and Relationships*: The member needs can be catered to by connecting socially with other community members, finding new friends and like-minded people. Also, social support is offered by these people when experiencing personal problem in lives. For example, the existing ONSC on illness and for severe diseases like cancer, or critical heart patients, provides a platform to their members to share their emotional feelings and hopes regarding their sickness, exchange experience of pain with other similar kind of members in the community.

- *Information Exchange*: One of the greatest advantage of being a part of ONSC is to get facilitated by the information posted/provided by the other members of the community on topics of interest which can solve many problems of specific users. In this respect, the knowledge sharing and problem solving communities play a very significant role [45, 46]. So, the knowledge and information exchange which happens through the social capital lying with other connected members, actually benefit the users of ONSC.

- *Economic Benefits*: User decisions are taken by exchanging information and experiences from other existing members of the ONSC, always result in economic benefits. For example, getting a better buying deal for a product, eventually ending up with a buying decision is very much influenced by the peer group in the community. Word-Of-Mouth (WOM), product recommendation and rating platforms are some of the activities in ONSC which benefits the member's buying decision very much. Among this, as WOM is the most trustworthy and risk reducing marketing communication channels which is very much beneficial for the users to take proper decision making for a product buy [47, 48].

- *Entertainment*: ONSC provides a very good environment from where the members can enjoy taking roles of virtual characters, play games together or just posts and consumes text, photos and videos to acquire fun.

  The above motivates users to join the ONSC like information exchange, friendship etc. are also supported by [49]. The study also showed that the reason differs with the type of the communities. Also, the need of participation for users might change with the purpose of the ONSC. Another study by [50, 51] determined factors like purposive value of ONSC as a form of information needs, self-discovery, social-enhancement, maintaining interpersonal connectivity and entertainment value as the main objectives for using ONSC. Hence, social factor, information, entertainment and extrinsic factors like economic or status oriented benefits are considered to be the most dominant benefits which enhances the stickiness of members to their respective ONSC [50–54].

ii. *User Types*: Members of ONSC can be categorized into two types—based on their motivation and based on their pattern of participation.

- *Members based on Motivation*: According to the study of [55], members can be segmented to four different clusters, based on their motivation to participate in ONSC. The motivational factors, as mentioned in the study, are platform assistance, venting negative feelings, concern for other consumers, extraversion/positive self-enhancement, social benefits, economic incentives, helping the company and advice seeking [55]. The four clusters are given as:

- *Self-Interested Helpers*: These type of users are strongly driven by the economic incentives.
- *Multiple Motive Consumers*: These members are motivated by large number of different factors.
- *Consumer Advocates*: These type of members are motivated mainly by the concern for others.
- *True Altruists*: These type of users are strongly motivated by helping other consumers as well as other companies.

    Several motivations may work for the clusters at different levels at the same time. Also, different clusters show separate levels of participation. For example, according to [55], multiple-motive users score highest on all dimensions, showing the highest contribution activity and visit frequency on the ONSC platform. It is misleading for the community operators to depend on this division of clusters. So, mostly the operator does the segmentation of users on their own, with the changing motivations which differs across different ONSCs.

- *Members based on Participation*: This refers to the categorization of members with respect to differing levels of participation. The first level of categorization of users of ONSC can be between *members who are registered users* and *non-members who are non-registered users*. The only aim of ONSC should be to convert non-members into members and retain existing members to maintain sufficient number of users on their respective platforms.

    Members in ONSC can become *inactive members* if they stop posting/using the platform over a long period of time without de-registering. The inactivity of users is often based on their last activities depending on the number of months they have not logged in [56]. *Active members* are regular users, who log in regularly and use the platform. Active members are further divided into *lurkers* and *posters*.

- *Lurkers* visit and use the community by reading posts, spends significant amount of time on the ONSC but do not post messages [57].
- *Posters* actively invests time and effort in ONSC, posts content and actively take part in discussions with other community members (2006).

The share of lurkers to posters depends on the type of ONSC, but according to studies, there are higher number of lurkers in many ONSC than the posters [58]. Though the lurkers are not active posters, their contributions can be measured through increased page impressions and consequently higher advertising revenues, every time they login the ONSC site. According to [57], posters are further divided into two groups: *frequent posters* and *infrequent posters*. *Frequent posters* show active participation and *infrequent posters* are passive participants but contribute to and interact on the platform in irregular intervals. Studies demonstrated that the two poster groups differ in their levels of trust and motivation towards the ONSC. Another study by [59], classified users with respect to the participation factors based on their frequency of visits, duration of visits, retrieved information, supplied information and discussed information. This study proposed six different user clusters:

- *Core members* represent the most active users within ONSC,
- *Conversationalists* make frequent, but short visits, produce high degree of engagement and discussion,
- *Informationalists* show high participation in information retrieval and sharing, low on discussion, visit frequency and duration,
- *Hobbyists* visit frequently the ONSC platform, visit for longer time but are low on information retrieval, sharing and discussion,
- *Functionalists* are high on information retrieval but low on participation with respect to visit frequency, duration, information sharing and discussion,
- *Opportunists* scores are far below the scores on all other five clusters with respect to all the parameters.

A categorization of user's is also done in [60], depending on the frequency by which users perform 20 different activities on the platform. The different types of users are:

- *Introvert users* who are least active users, who use ONSC mainly for emails,
- *Novel users* occasionally contribute to the ONSC platform, through communicating with friends, sharing comments and messages but spend more time than introverts,
- *Versatile users* perform many activities, but occasionally and through variety of platforms other than ONSC, and
- *Expert communicators* who perform a great variety of activities with a high frequency, being the most active group of all the existing ones.

From all these studies, it is inferred that lurkers or the users who contribute the least, form the largest part of users in the community. In contrast, the most active users with highest level of contribution make the smaller proportions of the ONSC platform.

4.2   Types of ONSC

There are different research disciplines who proposed various classification schemes for ONSC. Every scheme differs in criteria of differentiation. Armstrong and Hagel [61], proposed one of the first typologies for ONSC on purpose for which they are organized. There are four types of communities designed towards customer-oriented communities:

- *Communities of Interest* are formed based on some shared interest, expertise, or passion.

- *Communities of Relationship* are developed by individuals with a need to come together and share life experiences.
- *Communities of Transaction* facilitate economic exchanges through information sharing related to those transactions, and
- *Communities of Fantasy* provide people with the opportunity to develop environments and personalities in imaginary world of fantasy.

Primarily, in ONSC, most online communities mainly focus on any one of these four types of communities.

Another approach studied in [62], suggested two communities as *member-initiated* and *organization-sponsored communities*. Additionally, with respect to the relationship orientation, member-initiated communities can be social or professional, while organization-sponsored communities can be commercial, non-profit and government communities. The problem with all these typologies are the type of communities are not always strictly disjoint in nature. A more comprehensive and brief framework of different dimensions of communities are given in [63] and are summarized as follows:

- *Initiator*: With respect to the initiator of a community which mostly can be an organization, different communities feasible are companies, non-profit organizations and government organizations [62]. Again, member or private persons can also initiate communities based on their interests and ideas.
- *Living Environment*: Communities can be of professional or job-related and centered around the private or social life of its members [62]. Professional communities could be of different professions, can include different business networks like LinkedIn etc. focussed on connecting business contacts.
- *Commercial Orientation*: Communities can be differentiated with respect to their commercial orientation i.e. whether commercially driven or not. Many OSNC are commercially having the main motive to achieve profit. Non-commercial communities include like member-initiated brand communities, established based on the passion of a specific brand like car or bikes etc.
- *Community Function*: This is with respect to the overall objective of the community. When online community is a core business like Facebook, the community need to reach some economic objective. On the other hand, online communities as a support function are only part of the overall business model and support the core business. For example, eBay where community supports eBay users in using the auction website.
- *User Segment*: Communities can be consumer-oriented and business-to-business (B2B). In B2B case, a firm initiated social network can be established for its customers, providing a knowledge sharing platform for its users. For consumer oriented communities, it can be related to a company's products and brands which offers social networking to all of the people of the society.
- *Content Focus*: Theme oriented communities which varies along demographics, geographies and topics [38]. Demographics can be with age or gender, geographics can be by country, state, region etc., topics can have an elaborate range from sports, brands, health etc.

- *Needs*: Armstrong and Hagel classified communities according to user needs into four groups: information needs (communities of interest), social needs (communities of relationship building), economic needs (communities of transaction), and self-explanation and entertainment needs (communities of fantasy).
- *Access*: ONSC can be either open to registration to everyone or exclusive for people who receive invitations from other members or operators.

The different types of online communities as suggested by [42], in general can be given as follows:

- *Communities of Practice (CoP)*: Communities formed from members belonging to the same profession or job are termed as CoPs. For example, online community of accountants, doctors etc., all are CoPs.
- *Communities of Circumstance (CoC)*: Communities which are formed around a circumstance are CoCs. For instance, the community of Red Cross who serve people during circumstances of both natural and manmade disasters can be termed as CoC.
- *Communities of Purpose (CoPu)*: Communities where the members share a common perspective or objective to perform being in the community. The missing people squad community which normally gets formed during natural disasters to keep track of the missing people and inform the family members accordingly belong to this group of community.
- *Communities of Interest (CoI)*: Members of community who shares an interest being in the group are called CoI. For example, TripAdvisor which is a community of people who love travelling, share experiences etc.

The above communities are not necessarily disjoint. Sometimes, a particular community may get counted for more than one variation of online community.

5   Online Social Community Management

A positive attitude of the users towards ONSC, further increases the user participation. This positive attitude is one of the key element to keep the communities alive and attractive. It motivates the community managers for proper ONSC management to understand what aggravates the user's participation and positive perceptions towards ONSC. As a part of ONSC management, the community operators should ensure that enough new and actual content is provided on the community website. Additionally, frequent interaction and contribution is important and much needed to build a stronger relationship to the community and keep the community attractive for both lurkers and posters.

A proper ONSC management include community management to deal with users in different phases of their life cycle [63], given as *user acquisition*, *user engagement*, and *user retention*. The different stages across the relationship between user membership development and user participation life cycle where the ONSC operators can step in, is illustrated in Fig. 5. At the initial stage of user participation

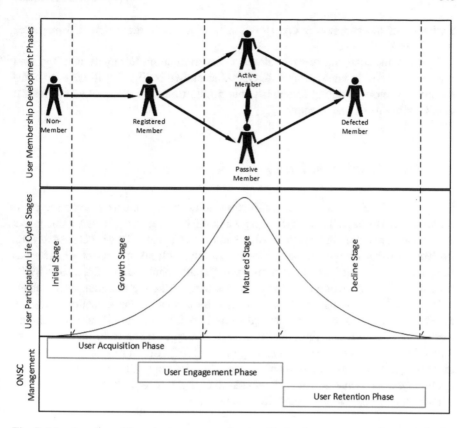

Fig. 5 Mapping of user life cycle stages to user membership development process for proper ONSC management

period, every user is a non-member of an ONSC. Eventually, depending on a proper ONSC management through using of effective marketing channels, the user acquisition phase focus on making a strong member base. This facilitates in compelling a non-member to become a registered member which also initiates the growth phase in the user participation life cycle.

After a user gets registered in an ONSC site, the challenge of the community operators as a part of ONSC management is to engage them and to promote the users activity on the ONSC platform. This initiates the matured stage of the user participation where users can be active members, who are frequent posters of messages, thus adding to the intellectual property of the ONSC site. Otherwise, the users can be a passive member, who just browse the ONSC platform without any active participation. Challenge of the community operators here are to convert as many passive members to active members to retain them to the site, avoiding user defection.

At the end, with the initiation of the decline phase of the user participation life cycle, the user tends to be a defected member. In some cases, users will unsubscribe and in many cases, users simply do not log in ever to the ONSC site. Here, the challenge of the community operators are how to retain these defected members back to the ONSC site.

Thus, if the mapping of user roles in the participation life cycle with the user membership development process in ONSC as shown in Fig. 5, will help the community operators to define different focus points to build a robust and successful ONSC with proper management.

5.1 User Acquisition Factor

Every ONSC requires to maintain sufficient number of members to keep the community active through the interaction between the users. In order to attain this, users must be acquired both in growth as well as in matured phase. This is because as time evolves, existing users stop using the community with no production of information, and new users join in the community with additional content. So, every ONSC should continue attracting new users through new marketing communication channels [64–66] to keep the community alive. Interpersonal communication skills are an effective means of acquiring new customers to the community sight. Traditional research has shown that the effectiveness of WOM is much more than the other marketing channels like personal selling channel [67]. So, in this stage of community management, the community managers should acquire valuable insights on how communication channels differ and what kind of users, in terms of their attitudes and behaviour towards online community, can be attracted.

5.2 User Engagement Factor

It is a mandate for the ONSC to retain their individual users, thus maintaining a minimum number of users in their site [68] through encouraging them to participate actively [69]. According to the studies, it is a challenge to retain the users in ONSC and maintain a constant rate of users. Some of the ONSC attract people at the initiation period of the community and others do it at any point of time during the ONSC's lifecycle [70]. The main factors on which the sustainability of online communities depends on are quality content generation and continuous increase in the number of users around this content over the time. From the initial stage through the community life cycle, various challenges exist. For example, when a community gets formed or launched, the most important challenge is to motivate potential users to join the community as members and contribute through valuable content. Once the community start gaining sufficient number of user memberships and also user generated content, the challenge is to maintain sufficient stream of user generated content over the long run, and ensuring the relevance and validity of the content.

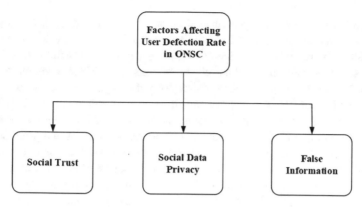

Fig. 6 Parameters affecting user defection rate

5.3 User Retention Factors

The usage of ONSC by the registered users continues until the perceived benefits of the users from the membership are higher than the cost incurred. There are plethora of ONSC platforms existing nowadays who compete with each other with varying topics, geographies, etc., on user retention. Though the number of ONSC platforms are on rise, but according to [71], a social-media fatigue trend is on for the users. So, it is very much required for the community managers to retain their users, thus maintaining the crowd at their community intact. In order to achieve this, they need to know the parameters responsible for user retention.

Past research lacks empirical evidence of factors that influence the user defections in ONSC. A study in [72], emphasized the importance of social influence on the user adoption and retention behaviour in ONSC. This happens to be the first study to find out the impact of dynamic social structure of users on their retention. The user's position in the network, the configuration of the user's current network and the participation and engagement in the community are tested to understand the reasons for user defection. The study also found that these effects can change over time. According to this study, parameters that can affect the user defection rate from ONSC platforms are summarized in Fig. 6. The details of each parameters are as follows:

- *Social Trust*: The anonymity and lack of face-to-face interactions between users in ONSC has made trust relevant for users [39]. If the user cannot trust the information retrieved or the sources of information, the users will lose the relevance of ONSC for exchanging information. Thus, trust in online context plays a vital role in the retention of users by ONSC [73].

  A study in [74], defined trust as a *"willingness to reply on an exchange partner in whom one has confidence"*. Studies proposed different types of trust which are adapted in relationships for ONSC. A study in [45] found the interpersonal trust which is related to the trustworthiness and honesty of all members of the ONSC, positively affects knowledge contribution and collection. The identification based trust was demonstrated in [75], which is defined as the members trust due to emotional interaction among themselves. A group of studies in [76], define trust on ONSC as a multidimensional construct consisting of the elements of ability, benevolence and integrity, which collectively has a positive effect on the active participation and desire to retrieve and share information among users [77].

Trust between a pair of users is an important piece of information for users in an online community (such as electronic commerce websites and product review websites) where users may rely on information to make decisions. In a study by [78], the problem of predicting whether a user trusts another user was addressed. Most prior work infers unknown trust ratings from known trust ratings. The effectiveness of this approach depends on the connectivity of the known web of trust and can be quite poor when the connectivity is very sparse which is often the case in an online community. In this paper, the authors proposed a classification approach to address the trust prediction problem. A taxonomy was developed to obtain an extensive set of relevant features derived from user attributes and user interactions in an online community. As a test case, the study applied the approach to data collected from *Epinions*, a large product review community that supports various types of interactions as well as a web of trust that can be used for training and evaluation. Empirical results show that the trust among users can be effectively predicted using pre-trained classifiers. Overall, measurement of trust in ONSC is mainly directed towards the community or the group of members as a whole. That is why, recent studies are based on generalized trust which affect user behaviour in ONSC [79].

• *Social Data Privacy*: Users are typically connected to friends, acquaintances and family through ONSC with a perception that the network provide secure, private and trusted platform for online interaction [80]. But, ONSC has raised serious stakes for privacy protection because of the overwhelming amount of user data which would not have got exposed otherwise. Also, nowadays, ONSC leads to detailed profiles which are not comfortable for the users from multiple social spheres [81]. This unpredictable and unwanted disclosure of user information results in dire consequences later. News like a case of teacher suspended for posting gun photos [82] or employee fired for commenting on her salary compare with that of her boss [83] on Facebook are some of the examples. Additionally, user's privacy is also getting breached by both intentional events like Facebook Beacon controversy [84] and unintentional events like publishing of anonymized social data used for de-anonymization and inference attacks [85]. High volume of user's personal data, which is disclosed due to lack of technical awareness of users or due to the lack of proper sophisticated privacy tools provided by ONSC, have attracted organizations like GNIP [86]. GNIP aggregate and sell user's data including his profile details, network structure, feeds, etc.

  The privacy and security issues of ONSC which are responsible for increasing user defection rates are leakages and linkages of user information and content which can be related to information disclosure threats [87]. The number of entities who are involved with this information disclosure are as follows:

- *Leakages to Strangers*: Users can bring in enhanced risk while interacting with other users, especially unknown users or strangers or mere acquaintances. Also, some of these strangers may not be humans also (e.g. social robots [88]) or may be crowdsourcing workers strolling and interacting with users for mischievous purposes [89]. So, the challenge for the ONSC is to protect users and their information from other users.
- *Leakages to Third-party Social Applications*: Users may interact with third-party provided social applications for better functionality, linked to their profiles. ONSC provides an user interface to the application developers of the third-party to access user information to smoothen the process of interaction between users and external applications. For many undesirable purposes, malicious applications can collect user's private data [90]. So, ONSC should protect users from these malicious third-party applications which is a challenging task.
- *Leakages to ONSC*: ONSC facilitates users to interact to other users and third party applications, in exchange for full control over the user's information that is on the ONSC platform. Every ONSC explicitly mention this in their Terms of Service documents which user's must agree and read first. In reality, very few users understand and actually go through the extent of this exchange [91] and most users do not have a choice or option. This exploitation is actually a breach of trust by the ONSC of user's personal information. ONSC should provide proper solution to the users to lessen this exploitation so that the users can retain their trust on them.
- *Linkages by Aggregators*: Professional data aggregators, through large scale distributed data crawlers, collect publicly available user profile details by exploiting social network provided APIs. Large databases of profiles and links are then sold by the aggregators to the companies of different domains like insurance, credit-agencies etc. who build on decisions applying analytics to this databases [92]. Crawling user's data from multiple sites and domains and further linking them increases profile accuracy. This profiles can be then used to apply public surveillance where curious bodies like government can monitor different individuals in public through variety of media [93].

  ONSC should overcome all these challenges as far as possible through proper privacy controls and tools so that users feel the interest to engage and retain with the community for a longer period of time.

- *Negative/False/Controversial Information*: A single post of rumor/negative information in ONSC can sometimes spread beyond anyone's control. A rumor about two explosions in the White House is a perfect example of how a single tweet out of more than 9000 tweeted in the same seconds spreads and causes real damage [94]. Users share information based on different types of needs through ONSC platform. One of these need is also to verify negative/false/controversial information. This requires the community managers of ONSC to have proper tools/solutions so that negative/false/controversial information is detected very early if their diffusion life cycle. This feature can also help the ONSC platform to retain more users over time.

5.4   ONSC and User's Perspective

The total user engagement and user retention problem have two perspectives: First, *ONSC perspective*-identifies proper controls with ONSC to attract users to their site and understands the requirement of users with respect to content which will motivate them to contribute and participate. Secondly-*user's perspective* denotes the activities of interest for the user's and their take away from the ONSC membership, either in the form of cash or kind.

- *Research work on ONSC perspective*: A study done in [95], has looked at the problem of relevant content discovery on a large social network site deployed inside IBM. Beehive site was studied which consisted of over 50,000 registered users with 10k–15k unique visitors each month and 50% of users actively adding to the site content. It was observed that users come to the site to socialize and connect with co-workers [96] through sharing photos [97], lists [98] and events [96]. With the increasing trend of this content sharing, it was difficult for the participants to find relevant, interesting and valid data which needs to be filtered out of the available content. There are different controls to function in this place for example search, social tagging, most-viewed content, page highlighting, etc. but no mechanism exists to highlight content according to importance or quality at the online community level. This study designed a system that addressed the current challenge of content discovery in Beehive whose goal was to increase activity and social interaction around more diverse set of content. This system was designed to select a group of users, in the form of rotating panel who are eligible to rate or promote a content, by tagging the text as 'honey' [95]. The implementation of this system observed an increase in the user participation in promotions, boost in the social interaction and attracted more attentive users towards content.
  A study in [99], focussed on participation and feedback elements [100–103] for user retention by ONSC. The study examined these elements in an online-peer production community *Everything2*. The study discovered that high rates of drop outs of users from the site is observed if feedback and participation mechanism are not worked out. The administrators of the community should invest in means to provide positive feedback and strongly discourage negative feedback from the users at their early stage of joining the ONSC to keep the users with the community. On the other hand, to discourage incompetent and inefficient writers, the feedback mechanism can be used to serve as a social filter.
  Several factors were examined for blogger retention in an online social blogger community called *Blogster* [104]. The study researched on the variables which can predict high retention for users in ONSC. The predictor variables of retention have been characterized into five categories:
  - *Network Metrics Specific Variables* (clustering coefficient, degree, betweenness, closeness, pagerank and communicability centrality),
  - *User Activity Specific Variables* (number of posts, number of comments, number of photos, network or community age),

- *User Physiology Oriented Variables* (Age, gender),
- *Interactional Data* (blog traffic, other user's comments) and
- *Relational Parameters* (social tie strength and friend's retention).

The study tried a multiple regression model to predict retention, utilizing activity specific variables. The result showed that these variables have a very high predictive power for anticipating user retention. Also, the male and aged bloggers, who normally occupy central hub positions, have low clustering coefficient in the ONSC of blogging and have friends with higher retention, are more retained in ONSC than others.

A study in [105], proposed the student retention factors for the first time in the *Massive Open Online Courses* (MOOC) platform. The author tried to understand the reason behind the increasing number of drop outs [106, 107] from a MOOC course. The study identified patterns of motivation, affect and activity among MOOC students, through interview, which includes an understanding of withdrawal or failure to complete the course. This is followed by a qualitative study by using quantitative method to incorporate strategies on platform and course designs in order to benefit the MOOC community, enhancing student retention and decreasing the number of drop outs.

In another study [108], several characteristics were studied through which students engage with MOOCs. It has also explored methods for increasing the amount of student level activity in these platforms. Two major student activities are considered in characterizing the predominant styles of engagement: *viewing a lecture* and *handing in an assignment for credit*. Different patterns of student's behaviour towards these styles was studied to find out the extent to which their overall activity is balanced between these two modes of engagement. The styles of engagement suggested by the study with respect to the user activities are:

- *Viewers*, who primarily watch lectures, handling in few assignments,
- *Solvers*, who primarily handled assignments for a grade, viewing few of the lectures,
- *All-rounders*, who balance he watching of lectures with the handling in of assignments.

The styles of engagement based on the downloading content behaviour of students who may or may not actually look into it are:

- *Collectors*, who download lectures, handling in few assignments, if any. Unlike Viewers, they may or may not be watching the lectures actually.
- *Bystanders*, who registered for the course but their total activity is below a very low threshold.

To change the pattern of student engagement in MOOC courses, a system of badges was designed which comes in the form of incentives for student's activity and contribution. The study showed that even a small amount of variation in badge presentation can have an effect on the student activity. Some of the future directions suggested to enhance student's engagement were: designing predictive models of student behaviour and grades, personalization and recommendation mechanisms to help increase user engagement and learning, identifying student behaviour effectively and developing methods that automatically recognize students requiring help or lacking in understanding of concepts, understanding and facilitating students for use of forums and discussion boards and exploring more badges.

In the healthcare domain, *SOcial FAmily* (SOFA) model was proposed in [109]. SOFA is an online social networking system, built to enhance user engagement and motivating families, instead of individual users, to adopt a healthy lifestyle through proper static education information on diet, exercise and healthy tips provided in this site. The study identified that persuasive applications can increase user engagement. Persuasive applications are defined as interactive computing systems which are designed to change people's attributes or behaviours [110]. The study intends to find out whether the users engage themselves only for the social networking component or for both social networking intent and interest in acquiring underlying knowledge. The study followed the work of [111] for a quantitative formulation of user engagement. User performs 14 main activities while interacting in SOFA. Some of the activities are *login, view blog entries, view profile, update profile, provide food preference information, write wall messages, complete quiz*, etc. The number of times a user performs an activity i is denoted by V_i, and it varies with the importance of the activity. So, weight W_i are introduced per activity to describe the significance of an activity. Important actions are those which are most likely to initiate further engagement through more contribution of new content than consumption. Thus, the overall measurement of an individual's engagement "$User_{Engagement}$" is given as:

$$User_{engagement} = \sum_{i=1}^{14} W_i * V_i$$

This proposed model also has two representations of profile: one for each family member who is a part of the family network and other is each family has a single profile. The study shows that individual profile representation increase user engagement with the system compared with shared family profile.

Many ONSC are now capitalizing on their users for face-to-face interactions to motivate and encourage their participation. Also, ONSC like Meetup and Foursquare promote offline connections [112]. These online-offline model of communities are called as *hybrid communities* which was first described in [113] as a 'hybrid people and technology-based phenomenon as an electronic-to-face community'. Several studies have proved that hybrid online-offline connections between users can increase community development [114], and enhance user engagement in political and social movements [113, 115]. This issue of engagement and online-offline transition is critical in case of new users to the ONSC. This is because newcomers join with new ideas, focus, audience and workforce. Any lack of interest and trustworthiness shown from by the newcomers with regards to the ONSC can pose serious challenge to communities with respect to user retention and attraction [116]. Sometime, the hybrid nature of the ONSC put extra burden on the newcomers as they need to join offline interactions along with online one. The study in [112], found out the factor which influence the newcomers to join an offline event along with joining ONSC. Also, the study researched on finding the retention factor of newcomer in hybrid communities. The study was conducted on www.Meetup.com, gathering data for 2 years. In order to detail the new user's behaviour towards hybrid set up, the author came up with a lifecycle of members in Meetup groups. Through the lifecycle, it was relevant that the first offline event is an essential milestone for new user which influence their continued participation in both online and offline interactions, thus enhancing user engagement.

- *Research work on User's Perspective*: The participation of users in ONSC can be in many forms like commenting in discussion systems, writing articles, posting pictures, etc. A power law distribution of participation prevails in ONSC platforms which means that majority of contributions are mainly done by a minority of users [117]. Researchers in [118] compares two theories of user motivation in the context of participation in ONSC.
  - First theory is *Uses and gratification* theory [119, 120], which states regarding what motivates individual users to consume media based on their own anticipation of what will they receive after doing so.
  - Second theory is *Organizational commitment theory* [121], predicts that more the affinity a user feels with an organization, better will be their contribution. The study produces a comparative brative study between individual motivation against social motivation amongst both registered and anonymous users of a ONSC through the above theories. The study finally specifies how individual perceptions regarding their current and future activity on a site is affected by the individual and organizational motivation, resulting difference in both anonymous and registered users. Also, the user contribution does not always depend on the ease of use of the site but the reason may lie in some social or cognitive factors too.

    Plethora of studies are done to understand the reasons for users to participate and contribute in online communities. Some users contribute to enhance their professional and personal reputation [122, 123], few users participate with a wish to help others [123, 124] and some people wish to gain strong social connections [124]. There are also users who participate just to gain the information resource available at the community level [125], especially for open-source communities [122]. In a study [126], informal leaders in a ONSC limits their participation to the community in relation to their daily work. So, this type of participation ensures an effective and direct personal benefit to the users on participation. The informal leaders also make other users to contribute valuable information to the ONSC. At the same time, support was less for the participants who contribute for the sake of reputation enhancement [122]. The study also confirmed that there is no evidence that the contribution of informal leaders depends on their individual performance goals and reviews. So, this gives an open scope for the ONSC to give opportunity to their informal leaders or users to contribute in such a way so that it can urge other users to contribute as well.

The dynamics of user departure from ONSC and collaboration networks was studied in [127] from the perspective of local and global network structure. The study analysed the predictive power of the local neighbourhoods in determining the behaviour of the nodes along with the global changes in the network topology. The study measured the probability of arrival and departure of users as a function of the activity of their friends, thus quantifying the dynamics of the local neighbourhood of the users, at the local level. There are three important finding of this study: *firstly*, there is a strong clustered effect in the timing of the departure among friends which is not significant for arrival case; *secondly*, though both the number and fractions of neighbourhood activity/inactivity is correlated to the probability of departure of user, the fraction of inactive friends in the local neighbourhood can predict the probability of departure better than any other factor; *thirdly*, after a certain period of time the network loses its capability to predict the departure probability of a user, as significant fraction of friends has departed the ONSC already. The study also showed that at the peripheral region of the network, the users are having the tendency to depart the communities over the period of time as the network evolves, whereas the internal network core remains dense and gets populated over time.

User's motivation to get engaged in ONSC is also affected by the existing leadership roles in the concerned networks. Traditional research on leadership in online communities has consistently focused on the small set of people occupying leadership roles. In a study by [128], a model was proposed of shared leadership, which hypothesized that leadership behaviours come from members at all levels, not simply from people in high-level leadership positions. Although, every member can exhibit some leadership behaviour, different types of leadership behaviour performed by different types of leaders may not be equally effective. The study further investigates how distinct types of leadership behaviours (*transactional*, *aversive*, *directive* and *person-focused*) and *the legitimacy of the people* who deliver them (people in formal leadership positions or not) influence the contributions that other participants make in the context of Wikipedia. After using propensity score matching to control for potential pre-existing differences among those who were and were not targets of leadership behaviours, the study found that:

- *Leadership behaviours* performed by members at all levels significantly influenced other members' motivation.
- *Transactional leadership* and *Person-focused leadership* were effective in motivating others to contribute more, whereas *aversive leadership* decreased other contributors' motivations; and
- *Legitimate leaders* were in general more influential than regular peer leaders.

5.5 Recommender Systems from ONSC

Traditionally, social connections in an ONSC was not considered for modeling of a recommender system. But, studies like [129, 130] have proved the importance of influencers in ONSC for the development of a robust recommender system in a field like product marketing. Also, the integration of ONSCs and OSNs also improve the performance of the recommendation systems, which in turn increases the engagement of users in the ONSC platform [131].

An efficient exchange recommendation service platform with frequent updates on the listed items through new data structures to maintain promising exchange pairs for each user was proposed in [132]. The item exchange through online gaming platform or social network web sites is another popular behaviour and supported widely in most of the ONSCs for user engagement [132]. Item exchange is a process in which each user in the system is supposed to list some items which is no longer need by him/her along with the list of item which he needs or seeking for. Given the values of all items, an exchange between two online users is only feasible if,

- the unneeded list contains some items which other user needs, and
- the exchange items value collectively calculates to the same total value for both the participating users.

For example, Shede [133] is a quick growing internet-based product exchange platform in China, who performs millions of transactions every year. A huge gap exists in this exchange market between the increasing demands and the techniques supporting automatic exchange pairing.

The RecSysChallenge [134], is a traditional competition among Recommender Systems researchers. The 2014 edition is used to predict the extent of user engagement through tweets related to movies. This is done in three steps: the two lists are created using binary classification methods to split the tweets into one group having user engagement measuring to zero and the other group where user's engagement is not equal to zero. In the next step, each list is sorted through regression method followed by a concatenation of the two lists and sorting of the tweets. Naïve Bayes was used as a classifier and linear regression was used as a technique.

5.6 ONSC User Engagement Tools

Nowadays, there is an increasing emphasis on ONSC to use a mix of both online and offline tools/practices which can help to reach their audience/customers. But, with so many software systems and offline tools available, it is very challenging for the ONSC to decide how best to engage the users to its platform. Some of the tools/practices (both online and offline) which ONSC opt for user engagement are given as follows [42]:

- *Content Management Systems* (CMS): It is a computer application that supports the creation and modification of digital content. This application supports multiple users in a collaborative environment like ONSC. The CMS of ONSCs are also dynamically generated digital content, where the webpage content is served on real-time from a dynamic database than being statically hard-coded. A good CMS should provide a simple web interface where only editorial staff are given access to add, delete or modify content. The other benefit of a good CMS is it allows automatic content addition from a third party news feed or the content shared between websites.
- *Discussion Boards:* Discussion boards or threaded discussions are one of the most commonly used tool in ONSC. These boards allow asynchronous discussion to happen over a period of time. During launch of ONSC, it is very challenging to predict on what topic the members will initiate their discussions. Therefore, it is better to start with highly focussed discussion groups where topics/keywords will be in sync with the topic of the focus of the group. Sometime, the ONSC can encourage the discussion groups through posting questions to the community and clearly 'sign-posting' their existence on the homepage. To make a successful discussion board, the ONSC is required to encourage them from time to time, especially at the onset of the discussion board. The ONSC should ensure that through the discussion board, new users join in and posts of interests keep floating in the platform. Thus, resulting in a discussion board vibrant with content and new trending topics.
- *Scheduled Events*: An already-reputed ONSC can reach out even more people of a larger community through scheduling events like seminars, conferences, etc., in auditoriums. The ONSC members space is divided into two parts: first, the '*stage area*' consisting of the guest speaker and a host, second, the '*auditorium area*' containing attendees and member of staff who is present there to greet their guests. The question and answer session through discussion happens with a question from attendees to a guest speaker via host and replies also follow the same way backward to the attendee. Thus, this follows a more structured way of managing large number of users performing discussions is a large ONSC platform.
- *Newswires*: Posting of regular newsletters or newswires is an extraordinary way to attract members to a ONSC, keep regular contact with the members involved, and also to retain users to the ONSC site. A Community manager or an Editor should get involved in sending messages/articles on their own providing only with links back to the full article on the ONSC site.
- *Polls and Surveys*: This tool is used in ONSC for generating a feedback from the community members. Polls consists of a single question with a Yes/No option and is used for collecting member's opinion regarding facts. Surveys normally contains questionnaires and is used to find out information about the members of the community involved. Polls can be used frequently for generating conclusion but surveys should be used with proper plan and should have a focussed aim/objective. Brief surveys are always supported by the users with clearly mentioned goals to be achieved and benefits reserved for the users.

- *Directories*: This refers to the user generated content which can be of high value for the users who intend to gain information from the ONSC. Directories provide searchable lists of public member profiles (who's who), lists of professional products and services along with the suppliers (supplier's directories), etc. These directories should be updated over time, thus becoming a valuable community resource.

6 ONSC Structure Detection

A *network community* is a group of people who communicate and collaborate over networks, strengthening their identity and goals [135]. The *community networks* are a special case of a network community in which a physical social community coextends with the network community [135]. Online Social Networks (OSN) resembles online community networks, with ONSC as an online network community embedded into it (see Fig. 3). In OSN, individuals network with each other through friendships or other kinds of relationships. These networks form communities (ONSC) where subsets of vertices forming communities/groups have dense vertex-vertex connections revealing similar interests or properties, but between groups/communities where interests differ, the connections go sparse [136]. The communities in OSN can be *disjoint* or *overlapping*. In fact, in OSN, communities do overlap each other [137]. For example, in OSN like Facebook, Twitter, a user who favours art, movies or music can become an active member of all these communities of interests in OSN. The nodes or individuals that are present in more than one community play a key role as intermediator and also can predict the dynamic behaviours of the individual nodes in the network.

Detection of these communities are much significant in case of both non-overlapping and overlapping communities. Let us consider a case of worm containment problem in OSN [138, 139]. Nowadays, all OSN's like Facebook, Twitter, LinkedIn are available as openAPIs on dynamic networks like different mobile devices. This openness will make it far easier for the malicious software's like worms or viruses to propagate from one device to another. Under these circumstances, a possible way of controlling the spread could be to send patches to control worms to few critical users and then let them redistribute the same to others. Smaller the number of critical users, better the control [140]. But the open problem is which critical users to choose. This is where the community structure, both overlapping and non-overlapping comes into picture through plethora of detection techniques are available for both static as well as dynamic networks as shown in Fig. 7. Use of a right kind of algorithm is essential to obtain best results. If a network has disjoint communities, an overlapping algorithm will not help and similarly for overlapping communities, only overlapping algorithm will be the right kind of algorithm.

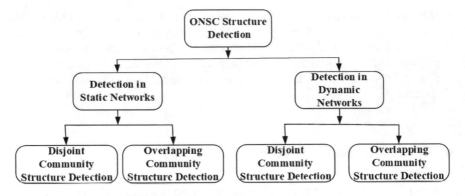

Fig. 7 ONSC structure detection techniques

6.1   Detection of Communities in Static Networks

In static networks, the communities are static and does not change their structure overtime. There are several effective methods proposed by researchers to detect both disjoint and overlapping communities in static networks [141], discussed as follows.

6.1.1   Disjoint Community Structure Detection

A beginning of a new era in the field of community detection was done through the popular divisive algorithm (GN) proposed by Girvan and Newman [136]. The main focus of GN algorithm was to achieve community detection through edge between-ness of an edge, which is the number of shortest paths that run between pairs of vertices and pass through that edge. According to the algorithm, communities in a network are connected by few intergroup edges through which all shortest paths between the communities should necessarily pass. So, these inter community connecting edges will have high edge-betweenness. Therefore, GN algorithm finds out edges with high edge-betweenness and then remove them iteratively to generate communities. The disadvantage is the high time complexity of the procedure which limits its use only for small networks. Also, the final number of communities detected is not declared.

GN algorithm was modified by Tyler et al. [142] and Wilkinson and Huber-man [143] through improvement in the time complexity by using Monte Carlo method to calculate edge-betweenness. According to Monte Carlo method, the edge-betweenness was calculated through some chosen random limited number of edges and also a stopping criterion was used to end the process. The sampling of edge betweenness mandates statistical errors and also the partitions are different for different choices of center of vertices. An improvement to the time complexity problem of GN algorithm was also tackled by Newman and Girvan [144] through his fast agglomerative hierarchical clustering algorithm which was also suitable for networks with millions of nodes. This algorithm started with single node resembling single community and then iteratively merge pairs of communities together, based on the modularity measure. The modularity is a measure of a cluster and depends on a notion that communities do not occur by random change. Following the work of Newman, many studies [145, 146 etc.] were done for community detection in networks through optimization of modularity technique.

Radhicchi et al. [147], proposed an algorithm which includes iterative removal of edges for community detection. But, unlike GN, instead of edge betweenness, the study used a local measure to recalculate the edge clustering coefficient to decide on edge removal. The local measure reduces the time complexity. The edges with low clustering coefficient are proposed to remove iteratively. For the remaining number of edges, the clustering coefficient is recalculated and the process continues. The main disadvantage of this process is that it assumes the presence of cliques or triangles in the networks which is not necessarily the case always in real-world networks.

6.1.2 Overlapping Community Structure Detection

Classical graph clustering methods mentioned above are not suitable for managing the presence of nodes/individuals in two or more communities at a time. Therefore, novel community detection algorithms were proposed for static networks as forceful assumptions of non-overlapping community nodes in static networks, can result in erroneous underlying community structures [141, 148].

Study of overlapping communities has received much attention only in the last couple of years with respect to both time and efficiency of the algorithm. OSN are naturally characterized by multiple community memberships. Based on the study done in [149], the overlapping community detection algorithms in static networks are categorized into five classes:

- *Clique Percolation*: In graph theory, a clique in a undirected graph is a subgraph of its vertices so that every two vertices in the subgraph are connected by an edge. The very first work on detecting overlapping communities using *Clique Percolation Algorithm* (CPM) was done by Palla et al. [86]. This is an extension of the GN algorithm to detect overlapping communities. The basic assumption of CPM is the formation of cliques through internal edges of communities where density is more. The study worked with k-cliques which represents complete graph with k-vertices. Two k-cliques are said to be adjacent if they share $k-1$ vertices. The possibility of the presence of a vertex in multiple k-cliques simultaneously makes the overlapping communities. CPM assumes large number of cliques as fewer cliques fails to detect meaningful covers. A software application known as *CFinder* was designed in [150], to detect overlapping communities on biological networks using the concept of k-clique.

  The drawback of CPM is it does not consider directed graphs. Later, in study by [86, 151], the author proposed a restricted version of CPM for directed networks, denoted as *Clique Percolation Method with directed cliques* (CPMd). In this work, only directed k-cliques are used for the identification of overlapping communities where directed k-cliques are defined as complete subgraphs of size k vertices in which an ordering can be made such that between any pair of nodes, there is a directed link pointing from the node with higher order towards the node with a lower one. Additionally, the link weights in a graph provides even better and deeper understanding of the overlapping structure of the weighted network. In this case, the CPM removes the links having lesser weight than a fixed weight threshold and search for communities considering the remaining links as unweighted. Farkas et al. [151] extended the concept of unweighted CPM through *Clique Percolation Method with weights* (CPMw) which take account of link weights through subgraph intensity used by Onncla et al. [152] in his search algorithm. The subgraph intensity is geometric mean of its link weights. A k-clique is included into a community if its calculated subgraph intensity is greater than a fixed threshold value. Therefore, for communities formed using CPM, links have weights higher than the link weight threshold. And for CPMw, the communities often contain links weaker than the intensity threshold. In assortative networks [153], where strong links to connect to strong links, both the algorithms are similar. But, for dis-assortative networks, the algorithms strongly differ.

EAGLE (agglomerativEhierarchicAlclusterinG based on maximaLcliquE) was proposed in [154] to discover both hierarchical and overlapping community structure of networks. EAGLE deals with maximal cliques instead of the set of individual vertices and use agglomerative framework. A maximal clique is defined as a clique which is not a subset of any other cliques. The final outcome is a hierarchy of overlapping communities which reveals a full community structure network. This is a computationally expensive algorithm. Another fast community detection algorithm called *Sequential Clique Percolation* method (SCP) was proposed in [155], designed for weighted as well as unweighted networks for a chosen size of clique. The algorithm inserts links to the network and keeps track on the merging community structure. If the links are inserted in decreasing order of weight, the algorithm is capable of detecting k-clique communities at chosen threshold levels in a single run and also produces a dendogram representing hierarchical community structure simultaneously. The method is suitable for dense weighted networks containing hierarchical communities where weight-based threshold of either the links or cliques formed by the process is necessary for getting a meaningful information about the structure. SCP is faster than CPM as far as the computational time is concerned.

- *Line Graph and Link Based Algorithms*: There are existing research studies which shows that links of a network can take part in detecting communities instead of nodes. In case of links, a node is called overlapping if links connected to it are a part of two or more communities. Ahn et al. [156] successfully identified overlapping communities and its hierarchy by identifying groups as links instead of nodes. The study used a hierarchical clustering with a similarity between links to build a dendogram. Each leaf in the dendogram is a link from the original network and branches represent the link communities. Link communities are extracted from this dendogram by cutting it at different thresholds at multiple levels. Each node inherits all memberships of its links and thus can belong to multiple overlapping communities. A study done in [157], used map equation to partition the network into communities. The output of their work is a map that simplifies and highlights the network structure and their relationships. Overlapping communities are detected by following the information flow through random walks. In a work done in [158], the line graph was extended to clique graph where cliques of a given order form the node in a weighted graph.

- *Fuzzy Based Algorithms*: In non-fuzzy overlapping communities, each vertex/individual belongs to one or more communities with equal strength i.e. either the individual belongs to a community or does not. But, with fuzzy overlapping case, though the individual can belong to more than one community, but the strength of its membership to each community varies. This membership is defined in different ways by different studies. Authors in [159], expressed this as a belonging coefficient which defines how a given node is distributed between communities. A method was proposed by combining spectral mapping, fuzzy clustering and optimization of a quality function by [160] which converts a network to $(k-1)$ dimensional Euclidean space to generate k communities using fuzzy c-means algorithm. In another study [161], the presence of a vertex in multiple communities at the same time is determined by exact numerical membership degrees in the presence of uncertainty in the data. This algorithm identifies outlier vertices that do not belong to any community, bridge vertices that belong to more than one single community and vertices belong to their own community through a new measure which is based on membership degrees. This algorithm is also capable of computing the centrality of a vertex with respect to its dominant community. This can also discover the fuzzy community structure of different real world networks including OSN with high accuracy. The limitation of this algorithm is that it works excellent with small datasets. But for large networks, the assumption that all vertices are connected to all other vertices does not hold good.

Disjoint detection methods were coupled with local optimization techniques to find the overlapping communities by Wang et al. [162]. Firstly, a partition is made through disjoint algorithm. Then communities attempt to remove or add vertices computing fitness scores on a community. The difference or normalized variances between these two scores form a fuzzy membership vector of the node. A Bayesian *Non-negative Matrix Factorization* (NMF), a probabilistic approach was used for extracting overlapping communities from a network by [163]. The model showed that how the degree of participation of two individuals in different communities can be a latent generator of the expected number of interactions between them. This study shows that NMF is capable of capturing the membership of a node in multiple communities as well as quantification of the strength with which the individual is connected to the communities. The entropy of the node membership distribution can be used to identify the core nodes in one community as well as the bridge nodes in two or more communities. The study also used mean entropy of membership distribution to quantify the degree of fuzziness in the network, thus reflecting on the clarity of the community structure.

- *Local Expansion and Optimization*: A study by Gregory [164] proposed an overlapping version of GN disjoint community detection algorithm called *Cluster-Overlap Newman Girvan Algorithm* (CONGA). It is done by splitting a vertex into two vertices repeatedly during divisive clustering process. This algorithm opted both split-betweenness and the conventional edge-betweenness concept for the same. Split-betweenness is the total number of shortest path passing through the imaginary edge. The edge-betweenness of edges is calculated first and then the split-betweenness of vertices followed by the removal of the edge having maximum edge-betweenness value. The vertex which is selected for split should also have maximum split betweenness. This process of edge removal and vertex splitting continues until all edges are removed. New communities are formed with every iteration under CONGA. The computational efficiency depends on the number of vertices in the network and also on how easily the network breaks into separate communities. To optimize the speed of the algorithm, Gregory in another study [165] proposed CONGO (CONGA Optimized) where the author used local betweenness to discover small diameter communities in large networks which is more effective. Another algorithm proposed by the same author was a transformation algorithm known as *Peacock*. Peacock is comprised on two phases for overlapping community detection. One phase consists of calculation of split betweenness of all vertices, then choosing the vertex having maximum score for the same. The chosen vertex is then split into two and again the process is repeated. The iteration continue until the split betweenness score becomes sufficiently small. In the second phase, for each split vertex, a new edge is added between the vertices. Lancichinetti et al. [148], proposed a method to detect overlapping and hierarchical structure communities using local optimization of a fitness function. A local structure is defined consisting of nodes in a community along with an extended neighbourhood of the nodes. The distribution of nodes in different communities is determined after finding the fitness value through local optimization, which automatically give rise to overlapping communities. A set of two axioms was formulated in [166] for nodes to be eligible to be a part of a community. The axioms are namely connectedness and local optimality. Connectedness in a community ensures the presence of a connected subgraph in the network. No path from one node to another node should be through any external node in a complete subgraph which defines a community. The axiom of local optimality demands that density of a community cannot be improved through addition or removal of a single node. To obtain these, a scan is performed in the communities in the order of the increasing node degree for all the nodes. Once the scanning is over, if the connectivity check examines multiple connected components, only the component with highest density is retained and the scanning process is repeated. The limitation of this algorithm is production of high number of overlapping communities which can be effectively managed by merging highly similar communities.

A *Detecting Overlapping Community Algorithm* (DOCA) proposed in [137], mainly consists of three phases. In the *first phase*, grouping of nodes is done to form local community through internal density calculation. The internal density should satisfy the connectedness function which depends on the size of the local communities and followed by multiple communities of different sizes. So, in this phase, small communities having less than four nodes are left out. In *second phase*, the dense community structure identified along with outliers are identified in bottom up manner. Common dense substructures should be merged in this phase. In the *third phase*, the unlabelled left out nodes are tried out to fit in some communities or classify them as outliers based on their connectivity structures. The advantage of DOCA is, it requires only local network topology knowledge to detect overlapping communities.

Order Statistics and Local Optimization Method (OSLAM) [167] is the first algorithm of its kind which deals with communities from networks having edge directions, edge weights, overlapping communities, hierarchy and community dynamics. The algorithm locally optimizes the statistical significance of the communities which are defined with respect to global null model [168]. This algorithm consists of three phases: In the *first phase*, it looks for significant clusters or communities till convergence. *Second phase*, analyses the set of clusters, and try to find out the internal structure presence or possible union of two or more clusters. In the *last phase*, it identifies the hierarchical structure of the communities. This is a very powerful algorithm for directed graphs and for detection of overlapping communities. Limitation of the algorithm includes lots of iterations performed for accuracy which increases the complexity of the algorithm.

- *Agent and Dynamic Based Algorithms*: A *Label Propagation Algorithm* (LPA) [169], is an agent based iterative method that detects community structure using the network topology only. Most of the algorithms discussed require a prior information on the number and size of communities to detect communities. But, for large complex networks where the network is huge and heterogeneous in nature, it is extremely difficult to detect communities with the existing algorithms. This gap is fulfilled by LPA which is a near-linear time algorithm. The algorithm propagates labels for vertices in the network and then detect communities using the label propagation. In this technique, labels are flooded but it is expected that the nodes in a community will be enclosed in a connected group expressed as local trapping of nodes. Community detection is done by tracking the node label propagation through its neighbours until end of community is reached. The end of community is flagged by dropping of the number of outgoing edges from a community beyond a certain threshold value.

Later, Gregory [170] extended the concept to nodes having multiple labels through *Community Overlap Label PRopagation Algorithm* (COPRA). A game theoretic framework was developed by [171] to detect overlapping communities. The communities are detected in a form of game played by selfish agents on the social network. Every agent has a gain and loss function, used to express the utility of the agent which they want to maximize. A Nash equilibrium of this game represents the community structure. The work in [172], detected communities in email objects using virtual ants through ant colony based algorithms. The algorithm allows the virtual ants to travel over the grid of email objects, detecting similar objects to form communities. The virtual ants evaluate an email object and identify its community through random walk on the grid. Swarm intelligence was also used in [173] to detect overlapping communities combining the individual view of community, instead of considering a global view of the network. Friendship groups of the individuals were used by the study to detect smaller groups in the network. Every friendship group in this study is assigned a unique identifier and alongside non-propagating nodes within each group were also found out through a decentralized control.

6.2   Detection of Communities in Dynamic Networks

Real world OSN are not always necessarily static. Most of the OSN like Twitter, Facebook, expand in size with the increasing number of user over time and thus lead to the formation of dynamic networks [174]. There are growing body of researchers studying on the analysis of communities and their temporal evolution in dynamic networks [175–181].

One of the key characteristic of studying social interactions in OSN is to incorporate the continual change in the network during analysis. Most of the traditional analysis of OSN are typically static where temporal information on interactions and evolution of network structure over time was not taken into account [182]. All static OSN research considered independent snapshots of graphs at different intervals or one aggregated network over the time period. But, for OSNs, a more dynamic behaviour is observed over the time where new nodes/edges can join in or get removed from the OSN. Hence, an emerging area of research interest is community detection, both disjoint and overlapping, should be considered in dynamic networks.

6.2.1   Disjoint Community Structure Detection

In a study by Nguyen et al. [23], *Quick Community Adaptation* (QCA) was proposed to detect disjoint communities. This is an adaptive modularity based method for identifying and tracing community structure for dynamic OSN. This approach has the power of efficiently and quickly update on network communities using structures from previous network snapshots and tracing the evolution of community structure over time. QCA was tested on Enron dataset, citation network and Facebook. Authors in [183], proposed a community detection methods based on contradicting the network topology and the topology-based propinquity, where *propinquity* is the probability of a pair of nodes involved in a community. Another study done in [184] presented a parameter-free methodology for detecting clusters on time-evolving graphs which is based on mutual information and entropy functions of Information theory. A work carried in [185], proposed a distributed method for community detection using modularity as a measure instead of objective function. *Dynamic Stochastic Block Model* (DSBM) model was proposed in [186] is a dynamic extension of a very traditional static model *Stochastic Block Model* (SBM) [187]. DSBM detects non-overlapping communities in OSN with constant number of nodes and communities over time and also require the number of communities a prior. An *AFFECT* algorithm was proposed in [188] to discover noise-free network from the observed network and its history. From the noise-free network, static non-overlapping community detection methods are used at each snapshot.

6.2.2 Overlapping Community Structure Detection

An innovative method for detecting overlapping communities on dynamic networks was proposed based on k-clique percolation technique [189]. But it is a time consuming process for large scale networks. Authors in [175], proposed *FacetNet framework* which analyses communities in dynamic networks based on the optimization of snapshot costs. This framework guarantees to converge to a local optimal solution but its convergence speed is slow and its input asks for the number of networks communities which is not a practice. The study used a stochastic block model for generating communities and a probabilistic model based on the Dirichlet distribution for capturing the community evolutions. The study also introduced two concepts: *Community Net* and *Evolution Net* to interpret the community level interaction and transition respectively. *Stream_Group* proposed in [190], is an incremental method to solve community mining and detect the change points in weighted dynamic graphs. A two-step procedure is followed to discover community structure of weighted directed graph in one time-slice: The *first step* constructs compact communities according to each node's single compactness which indicates the degree of a node belonging to a community in terms of graph's relevance matrix. The *second step* merges compact communities along the direction of maximum increment of the modularity. Apart from these studies, a particle and density based clustering method for dynamic networks was proposed in [191], based on extended modularity and concept of nano community and l-quasi clique by clique. Other than these, iLCD [192] find the overlapping network communities by adding edges and then merging similar ones. This framework does not perform so well with dynamic behaviours of the network which undergoes removal of edges/nodes. Finally, a two phase framework *Adaptive Finding Overlapping Community Structure* (AFOCS) was proposed in [140], for detecting, updating and tracing the evolution of overlapping communities in dynamic mobile networks. This two frame network first identifies basic network communities with FOCS and then employs AFOCS to adaptively update these structures as the network evolves. *Dynamic Label Propagation Algorithm* (DLPA) [193], detects both overlapping and non-overlapping community structures in large-scale network. This study also defines the confidence of a node to its neighbours, thus incorporating the importance of a node's neighbour into account. Also the study defined an inflation operator to control the overlapping rate. A weighted mean between the importance of a node in the previous snapshot and the current snapshot is calculated to find the consistent communities over time. In a recent research done in [186], a method called *Dynamic Bayesian Overlapping Community Detector* (DBOCD) was proposed that is able to discover consistent overlapping communities and their numbers simultaneously, in polynomial time by using the *Recurrent Chinese Restaurant Process*

(RCRP) as a prior knowledge, and the adjacency matrix as observations in every time snapshots of network. In this study, a generative model for a static network based on link communities was explained and in the second step, the generative model was extended for dynamic networks. The time complexity was reduced by replacing common node partitioning by link partitioning. *Dynamic Structural Clustering Algorithm* (DSCAN) was proposed in [194] for dynamic OSN that are continually changing their structure. The SCAN algorithm was first proposed for static networks. Later, DSCAN improved SCAN by allowing to update a local structure in less time than it would to run SCAN on the entire network.

Another very recent study done in [195], introduced a new paradigm of *hidden community structure*. Real world networks contain sparse community structure showing individuals belonging to weaker communities like group of medical patients that see each other at the doctor's office and communicate infrequently or high school alumni having infrequent contacts. These are hidden communities as compared to the dense or strong denser community structure like family, close friends, etc. Hidden community structure can be regarded as a special type of overlapping communities. A novel approach called *HIdden Community Detection* (HICODE), was proposed by the study for identifying hidden structures on OSN. The study formally introduced the concept of hidden communities through a hiddenness value of a community. HICODE was presented to detect both dominant and the hidden structure. Validated through several real world datasets, the study showed that higher the hiddenness value of a community, it is more difficult for an algorithm to locate the community. The study also showed that using any disjoint community algorithm as the base, HICODE as a tool is capable of finding overlapping communities.

7 Applications

There are plethora of personal as well as business platforms where ONSC have got their illustrative benefits. The list can be categorized as follows:

- *Personal Use*: There exist a strong research based work, showing how online interaction impact individual's well-being [196]. Existing studies in research [197] has shown that individuals supplement face-to-face interaction through joining ONSCs. Individuals prefer to use these social communities to prevent families, friends from becoming too fragmented by providing them social support in form of suggestions, comments, recommendations, control etc., regardless of their physical location [198]. Also, individuals having depleted offline social resources, can achieve huge benefit through involvement and making relationships in ONSCs [199]. In addition, acquiring social capital is important for the individuals, which can be heavily fulfilled through participation in ONSCs.
- *Business*: The advantages of development of communities for businesses are manifold. Communities help the companies to enhance customer relations as consumers get an opportunity to know about the environmental and social credentials of the

products and services they buy and the activities of companies in the sourcing and production of the products the company sell. Businesses can also improve on the risk management of their services, thus enhancing their reputation through communities. Social communities also effect employee relations through attracting conversations and retaining talent, making them motivated towards work. Engaging with customers make businesses able to keep close to emerging market trends and thus resulting in innovating new products or models.

- *News*: News, whether it is good or bad, true or false, always travels faster, but with the advent of social media and ONSC, the spread of news has become faster than ever, both locally and globally. Once a news posted online, the journey of the news starts by getting shared among readers through their ONSC of friends, colleagues and family, thus resulting in the news reach to exponential number of social community users.

- *Recommender System*: On internet platform, the number of choices are humungous [200]. There is a need to filter, prioritize and efficiently deliver the relevant information to avoid overload and also to facilitate the customers with personalized content and services. The ONSC provide rich data that can be used for recommender systems [201]. Some important type of data, supporting in getting a recommender systems constructed, is the relationship between users, preferences of users and the review of the users through their ONSC.

- *Education and Research*: ONSC has transformed the way teachers, students and researchers synthesize information across subjects, experiences, papers and incorporate various inquiries [202]. Educators can foster critical learning spaces encouraging students to enhance their capacities of analysis, imagination, critical synthesis, creative expression, self-awareness and intentionality through ONSC. Also, for the researchers, ONSC has become an important platform to increase the visibility of their papers, thus in turn increasing the number of citations of their publications.

- *Healthcare*: There are various ways in which healthcare is influenced through ONSC [77]. An existing ONSC boosts the amount of communication and cooperation among patients through exchange of information pertaining to similar problems. At the same time, professionals can share their experiences over care and treatment, thus resulting in better health decisions. A feedback can also be provided through these community relationships regarding disease and treatment given for the same. This can further generate a recommender systems helping new patients to consult with the doctor and list of preferable clinics to diagnose from. The community discussion happens without any bound on time and space limitation, increasing the number of users infinitely over a period of time.

- *Marketing*: The online community concept results in higher win rates through sales enablement, improved marketing campaign development, effectiveness and amplification, creates customer's affinity towards product/service, provides better natural search engine and social media optimization for portraying news on brand and products/services, facilitates in real-time market research and consumer intelligence on products and enhances channel presence, web sales and referrals.

- *Customer Service*: The presence of ONSC results in decreased volume of inbound support calls, improve customer self-service or first time resolution, increased customer retention and satisfaction, customer support and engagement and enhance customer service agent collaboration and knowledge sharing.

- *Product Development and Customer Intelligence*: ONSC largely contributes in consumer driven product innovation and insights during development of products. Increase in feedback and ideas from consumers and knowledge retention also helps in developing new products by businesses.

In Table 1, we tried to summarize the popular ONSC available across the globe in various domains along with their purpose. The list is exhaustive, only a glimpse is given to have a mind-mapping of the theory with the existing ONSC. Hence, better understanding of this chapter.

8 Conclusion

In this chapter, a robust definition of social communities is provided along with its importance in knowledge sharing among users. Two types of social communities were explored—online and offline social communities. The role of technology in the transformation of offline social communities to online social communities with its evolution from ARPANET to Web 2.0 was discussed. Different pre-requisites of online social communities acts as a decision making factor for community operators while creating an ONSC. These factors include purpose, membership of users, terms and conditions, user generated content type and the types of users to mention a few. This study also discussed about the requirement of online social communities management through proper handling of participation of users in the platform. For proper handling of users, understanding of inherent network structure-static, dynamic, offline and online is very important. Overall, the community operator need to take proper actions and should continuously improve on all the three phases of the membership development process—user acquisition, user engagement and user retention phases, for a proper ONSC management. They need to attract new users, activate the users to participate more actively and retain the users. The effectiveness of this procedure will be achieved if the community operators can actually understand the position of the users in the network and the structure of the network- static or dynamic, overlapping or disjoint. The ONSC structure detection techniques done by the researchers are also surveyed for the benefit of the readers. ONSC has lots of applications for personal as well as in business domains. The popular ones were considered from the existing ocean of ONSCs. Marketing, Healthcare, Education, etc., are known areas as of now, but sky is the limit in future.

Table 1 Domain-wise list of ONSC

Community services	Domain	Purpose
23snaps www.23snaps.com	Personal use	Allow parents to save photos, videos, measurements and stories of their children to a digital journal and share those uploads privately with their family and friends
aNobii www.anobii.com	Personal use	Allow individual reader to catalogue their books and rate, review and discuss them with other readers
ASKfm www.ask.fm	Personal use	Global social networking site where users create profiles and can send each other questions
Badoo www.badoo.com	Personal use	Contacting nearby people, search users from different parts of the world, encountering other users having similar choice and video chat
LiveJournal www.Livejournal.com	Blog hosting	Social networking service where users can keep a blog, journal or diary. Sending text messages, to-do-list feature, express lane, voice post and extra storage space are also some of the added features
Instagram www.instagram.com	Personal use	Allows users to share pictures, videos either publicly or privately to pre-approved followers. Addition of digital filters to images, add locations through geotags and hashtags to posts, linking photos to content are some important purpose of users
MixBit www.mixbit.com	Video sharing	Users can create dynamic shared videos
Musical.ly www.musical.ly	Video creation, messaging and live broadcasting	Browsing through popular musers, content, trending songs and sounds and hashtags are some of the takeaways
Pinterest www.pinterest.com	Visual discovery, collection and storage tool	Discovering information on the WWW, mainly using images and on a shorter scale, GIFs and videos
Slidely www.slide.ly	Video creation service and video slideshows	Image based social networks and cloud-based video creation service
Snapchat www.snapchat.com	Photo sharing, instant messaging, video chat and multimedia	Allows the users to create multimedia messages called snaps, which can consist of a photo or a short video, and can be edited to include filters and effects, text captions and drawings
Tumblr www.tumblr.com	Microblogging	Allows users to post multimedia and other content to a short-form blog. Users can follow other users' blogs

(continued)

Table 1 (continued)

Community services	Domain	Purpose
Foursquare www.foursquare.com	Recommender system	Provides personalized recommendations of places to go to near a user's current location, based on previous browsing history, purchases or check-in history
Untappd www.untappd.com	Recommender system	Allow the users to check into beers as they drink them, and share these check-ins, their locations with their friends, to rate the beers, earn badges, share pictures of their beers, see others choice of beers and suggests similar beverages
Readgeek www.readgeek.com	Recommender system	Online book recommendations engine and social cataloguing service. Allows users to search for books matching their individual preference making use of several algorithms
TV time www.tvtime.com	TV tracker social network	Online service centered exclusively on television series. Enables the users to get information and comment on the shows they follow, along with new episode release dates, user reactions etc.
Academia.edu www.academia.edu	Education and Research	Platform for sharing research papers, monitor their impact, and follow the research in a particular field
ResearchGate www.researchgate.net	Education and Research	Social network service for scientists and researchers to share papers, ask and answer questions and find collaborators
Meta www.meta.com	Education and Research	Produce real-time updates from PubMed and other sources covering the biomedical sciences. By browsing through papers and learning from user behaviour, the service pinpoints key pieces of research and provides relevant search results. Visualizations about the field of research by organizing papers by their date of publication and citation count is provided, facilitating the users to quickly identify key historical papers
Brainly www.brainly.co	Education	Operates a group of social learning networks for students and educators. Gamification is also included into the service as a form of motivational points and encourages users to engage in the online community through Q&A sessions among students

(continued)

Table 1 (continued)

Community services	Domain	Purpose
Edmodo www.edmodo.com	Education	Offers a communication collaboration, coaching platform to K-12 schools and teachers. Teachers can share content, distribute quizzes, assignments and manage communication with students, colleagues, and parents
Moodle www.moodle.org	Education	Free and open source learning management tool, used for blended learning, distance education, flipped classroom and other e-learning projects in schools, universities, workplaces and other sectors
Branchout www.branchout.com	Human Resource	Social networking application designed for finding jobs, networking professionally, and recruiting employees
LinkedIn www.linkedin.com	Human Resource	Business and employment oriented social networking service, used for professional networking, including employers posting jobs and job seekers posting their CVs
IBM connections https://www.ibm.com/us-cn/marketplace/ibm-connections	Business—Corporate collaboration	It's a Web 2.0 enterprise social software application developed to provide online social networking tools for employees. Microblogging, profile creation, communities, media gallery, blogs writing etc. are all components of it
Solaborate www.solaborate.com	Business	Social and collaboration platform dedicated to professionals and companies to connect, collaborate, discover opportunities and create an ecosystem around products and services. It also incorporates real-time analytics, file-sharing and document organization
Viadeo www.viadeo.com	Business	Professional social network whose members include business owners, entrepreneurs and managers
Xing www.xing.com	Business	Open business club which provides a career oriented social networking site for enabling a small world network for professionals. The capabilities are personal profiles, groups, discussion forums, event coordination and other common social community features
IdeaPlane www.ideaplane.com	Business	An enterprise social networking platform targeted at companies in heavily regulated industries like financial services. The platform is tailored to keep companies compliant with the regulations set forth by the organizations

(continued)

Table 1 (continued)

Community services	Domain	Purpose
Yammer www.yammer.com	Business collaboration	Freemium enterprise social networking service for private communication within organizations
PatientsLikeme www.patientslikeme.com	Healthcare	Provide a platform for patients to connect each other, having the same disease or conditions, tracking and sharing their experiences. This process generate data regarding real-world nature of disease which can form as a basis for more than 100 peer-reviewed scientific studies
HealthBoards www.healthboards.com	Healthcare	Health topics are given in individual message broads. Each message board topic typically consists of thousands of discussion threads, each related to a specific question, comment, or response initiated by the community member. This makes it one of the renowned patient health support platform and referred to as online health community
MedHelp www.medhelp.org	Healthcare	Pioneered in the field of consumer health information and communities on the Internet. Offers a series of personal health applications which includes personal health records and trackers, weight management tools, exercise, diabetes, sleep, mood, pain, ovulation and pregnancy conditions check
WebMD www.webmd.com	Healthcare	Online publisher of news and information related to human health and well-being. Information regarding drugs is also available here
HeathUnlocked www.healthunlocked.com	Healthcare	Uses health specific artificial intelligence to support patients to manage their own health through relevant recommendations and tailored health content, information and services to patients. Peer support is also available where various health conditions are discussed and actively engage patients with their proper healthcare
Moz www.moz.com	Marketing	SEO (Search Engine Optimization) consulting company and expanding their software to include tools for social media and brand reach. This community intend to initiate a novel way to do marketing where customers are earned rather than bought

(continued)

Table 1 (continued)

Community services	Domain	Purpose
MyCopyblogger www.mycopyblogger.com	Marketing	The users are encouraged to write blogs and also to share knowledge. The comments section of the posts are often featured with intelligent analysis of content marketing trends and actionable tips which helps the blogger manifolds
Reddit www.reddit.com	Marketing	Houses helpful and honest online marketing community where marketer can find latest and most relevant digital news and questions. The questions can be from digital catalogs, to ecommerce, and to the best SEO tools
GrowthHackers www.growthhackers.com	Marketing	Users who are interested at driving growth through their marketing belongs to this community. Free discussions with concerned individuals or groups, networking with professional digital marketers to decide on the way to use the online resources for better customer base are some of the keys which drives users in this community

References

1. Hillery GJ (1955) Definitions of community: areas of agreement. Rural Sociol. 20:111–122
2. Wellman B (2001) The persistence and transformation of community: from neighbourhood groups to social networks. Law Commission of Canada, Ottawa, p 101
3. Wellman B (2002) Little boxes, globalization, and networked individualism. In: Tanabe M, van den Besselaar P, Ishida T (eds) Digital cities II: computational and sociological approaches. Springer, Berlin (In press)
4. Wellman B, Leighton B (1979) Networks, neighborhoods and communities. Urban Aff Q 14:363–390
5. Fischer C (1982) To dwell among friends. University of California Press, Berkeley, CA
6. Wellman B (1979) The community question. Am J Sociol 84:1201–1231
7. Wellman B (1999) The network community. In: Wellman B (ed) Networks in the global village. Westview, Boulder, CO, pp 1–48
8. Wellman B (ed) (1999) Networks in the global village. Westview, Boulder, CO
9. Rifkin SB, Muller F, Bichmann W (1988) Primary health care: on measuring participation. Soc Sci Med 26(9):931–940
10. White AT (1982) Why community participation? A discussion of the arguments. Assignment Children
11. Manderson L, Valencia LB, Thomas B (1991) Bringing the people in: community participation and the control of tropical disease
12. Wellman B, Boase J, Chen W (2002) The networked nature of community: online and offline. It Soc 1(1):151–165
13. Putnam RD (1995) Bowling alone: America's declining social capital. J Democr 6(1):65–78
14. Coleman J (1990) Foundations of social theory. Harvard UniversityPress, Cambridge, MA
15. Adler PS, Kwon S-W (2002) Social capital: prospects for a new concept. Acad Manag Rev 27:17–40

16. Mathwick C, Wiertz C, de Ruyter K (2008) Social capital production in a virtual P3 community. J Consum Res 34:832–849
17. McCully W, Lampe C, Sarkar C, Velasquez A, Sreevinasan A (2011) Online and offline interactions in online communities. In: Proceedings of the 7th international symposium on wikis and open collaboration. ACM, pp 39–48
18. Wikimania (2017) https://wikimania2017.wikimedia.org/wiki/Wikimania. Accessed on 20 Dec 2017
19. World of warcraft. https://worldofwarcraft.com/en-us/. Accessed on 20 Dec 2017
20. Carter D (2005) Living in virtual communities: an ethnography of human relationships in cyberspace. Inf Community Soc 8(2):148–167
21. Parks MR, Floyd K (1996) Making friends in cyberspace. J Comput-Mediat Commun 1(4):0–0
22. Sessions LF (2010) How offline gatherings affect online communities: when virtual community members 'meetup'. Inf Commun Soc 13(3):375–395
23. MetaFilter. https://www.metafilter.com/. Accessed on 20 Dec 2017
24. Oldenburg R (1989) The great good place: Café, coffee shops, community centers, beauty parlors, general stores, bars, hangouts, and how they get you through the day. Paragon House Publishers
25. Steinkuehler CA, Williams D (2006) Where everybody knows your (screen) name: online games as "third places". J Comput-Mediat Commun 11(4):885–909
26. Rains SA (2005) Leveling the organizational playing field—virtually: a meta-analysis of experimental research assessing the impact of group support system use on member influence behaviors. Commun Res 32(2):193–234
27. Walther JB (2004) Language and communication technology: introduction to the special issue. J Lang Soc Psychol 23(4):384–396
28. Walther JB (1996) Computer-mediated communication: impersonal, interpersonal, and hyperpersonal interaction. Commun Res 23(1):3–43
29. Bryant SL, Forte A, Bruckman A (2005) Becoming wikipedian: transformation of participation in a collaborative online encyclopedia. In: Proceedings of the 2005 international ACM SIGGROUP conference on supporting group work. ACM, pp 1–10
30. Preece J (2000) Online communities: designing usability and supporting sociability. Wiley
31. Kim AJ (2000) Community building on the web: secret strategies for successful online communities. Addison-Wesley Longman Publishing Co., Inc
32. Preece J, Maloney-Krichmar D, Abras C (2003) History of emergence of online communities. In: Wellman B (ed) Encyclopedia of community. Berkshire Publishing Group, Sage (In press)
33. Project PIAL (2002) Getting serious online: as Americans gain experience, they use the web more at work, write e-mail with more significant content, perform more online transactions, and pursue more serious activity. Pew Internet & American Life Project. http://www.pewint ernet.org/reports/poc.asp?Report=55
34. Boyd DM, Ellison NB (2008) Social network sites: definition, history and scholarship. J Comput-Mediat Commun 13(1):210–230
35. Rheingold H (1993) The virtual community. Addison-Wesley, Reading, MA
36. Bagozzi RP, Dholakia UM (2002) Intentional social action in virtual communities. J Interact Mark 16(Spring):2–21
37. Balasubramanian S, Mahajan V (2001) The economic leverage of the virtual community. Int J Electron Commer 5(3):103–138
38. Hagel JI, Armstrong AG (1997) Net gain: expanding markets through virtual communities. Harvard Business School Press, Boston, MA
39. Ridings CM, Gefen D, Arinze B (2002) Some antecedents and effects of trust in virtual communities. J Strateg Inf Syst 11:271–295
40. Wellman B, Hampton K (1999) Living networked on and offline. Contemp Sociol 28:648–654
41. Foote J, Gergle D, Shaw A (2017) Starting online communities: motivations and goals of Wiki founders. In: Proceedings of the 2017 CHI conference on human factors in computing systems. ACM, pp 6376–6380

42. Baxter H (2002) An introduction to online communities. Community manager of Knowledge-board. com
43. Zhu H, Kraut RE, Kittur A (2014) The impact of membership overlap on the survival of online communities. In: Proceedings of the SIGCHI conference on human factors in computing systems. ACM, pp 281–290
44. Hagel JI, Armstrong AG (2006) Net gain-profit imNetz: Märkteerobernmitvirtuellen Communitys. Redline Wirtschaft, Sonderausgabe, Heidelberg
45. Chen C-J, Hung S-W (2010) To give or to receive? Factors influencing members' knowledge sharing and community promotion in professional virtual communities. Inf Manag 47:226–236
46. Dholakia U, Blazevic V, Wiertz C, Algesheimer R (2009) Communal service delivery: how customers benefit from participation in firm-hosted virtual P3 communities. J Serv Res 12(2):208–226
47. Arndt J (1967) Role of product-related conversations in the diffusion of a new product. J Mark Res 4:291–295
48. Murray K (1991) A test of services marketing theory: consumer information acquisition activities. J Mark 55:10–25
49. Ridings CM, Gefen D (2004) Virtual community attraction: why people hang out online. J Comput-Mediat Commun 10. http://jcmc.indiana.edu/vol10/issue1/ridings_gefen.html. Accessed 20 Dec 2017
50. Dholakia UM, Bagozzi RP, Pearo LK (2004) A social influence model of consumer participation in network- and small-group-based virtual communities. Int J Res Mark 21(3):241–263
51. Bagozzi RP (1994) Structural equation model in marketing research: basic principles. In: Bagozzi RP (ed) Principles of marketing research. Blackwell Publishers, Cambridge, MA, pp 317–385
52. Park N, Kee KF, Valenzuela S (2009) Being immersed in social networking environment: Facebook groups, uses and gratifications, and social outcomes. CyberPsychol Behav 12:729–733
53. Sangwan S (2005) Virtual community success: a uses and gratifications perspective. In: Proceedings of the 38th annual Hawaii international conference on system sciences—HICSS '05, 193c
54. Wasko MML, Faraj S (2000) It is what one does: why people participate and help others in electronic communities of practice. J Strateg Inf Syst 9:155–173
55. Hennig-Thurau T, Gwinner KP, Walsh G, Gremler DD (2004) Electronic word-of-mouth via consumer-opinion platforms: what motivates consumers to articulate themselves on the Internet? J Interact Mark 18:38–52
56. Wübben M, Wangenheim F (2008) Instant customer base analysis: managerial heuristics often 'Get It Right'. J Market 72:82–93
57. Ridings CM, Gefen D, Anrize B (2006) Psychological barriers: lurker and poster motivation and behavior in online communities. Commun Assoc Inf Syst 18:329–354
58. Nonnecke B, Preece J (2000) Lurker demographics: counting the silent. In: CHI '00 Proceedings of the SIGCHI conference on human factors in computing systems, New York: ACM, pp 73–80
59. deValck K, van Bruggen GH, Wierenga B (2009) Virtual communities: a marketing perspective. Decis Support Syst 47:185–203
60. Alarcon-del-Amo M-D-C, Lorenzo-Romero C, Gomez-Borja M-A (2011) Classifying and profiling social networking site users: a latent segmentation approach. Cyberpsychol Behav Soc Netw 14(9):547–553
61. Armstrong A, Hagel J III (1996) The real value of on-line communities. Harv Bus Rev 74:134–141
62. Porter CE (2004) A typology of virtual communities: a multi-disciplinary foundation for future research. J Comput-Mediat Commun 10(1), Article 3. http://jcmc.indiana.edu/vol10/issue1/porter.html. Accessed 7 Nov 2011
63. Rohrmeier P (2012) Social networks and online communities–managing user acquisition, activation and retention. Doctoral dissertation, UniversitätMünchen

64. Borden NH (1964) The concept of the marketing mix. J Advert Res 4:2–7
65. Chen Y, Xie J (2008) Online consumer review: word-of-mouth as a new element of marketing communication mix. Manag Sci 54:477–491
66. Duncan T, Moriarty SE (1998) A communication-based marketing model for managing relationships. J Mark 62:1–13
67. Herr PM, Kardes FR, Kim J (1991) Effects of word-of-mouth and product attribute information on persuasion: an accessibility–diagnosticity perspective. J Consum Res 17:454–462
68. Ling K, Beenen G, Ludford P, Wang X, Chang K, Li X, Resnick P (2005) Using social psychology to motivate contributions to online communities. J Comput-Mediat Commun 10(4):00–00
69. Nov O, Naaman M, Ye C (2009) Motivational, structural and tenure factors that impact online community photo sharing. In: ICWSM
70. Kraut RE (2012) Evidence-based social design: mining the social sciences to build online communities. MIT Press
71. Gartner Inc. (2011) Gartner survey highlights consumer fatigue with social media. http://www.gartner.com/it/page.jsp?id=1766814. Accessed 20 Dec 2017
72. Goldenberg J, Han S, Lehmann DR, Hong JW (2009) The role of hubs in the adoption process. J Mark 73:1–13
73. Shankara V, Urban GL, Sultan F (2002) Online trust: a stakeholder perspective, concepts, implications, and future directions. J Strateg Inf Syst 11(3–4):325–344
74. Moorman C, Zaltman G, Deshpandé R (1992) Relationships between providers and users of market research: the dynamics of trust within and between organizations. J Mark Res 29:314–328
75. Hsu M-H, Ju TL, Yen C-H, Chang C-M (2007) Knowledge sharing behavior in virtual communities: the relationship between trust, self-efficacy, and outcome expectations. Int J Hum Comput Stud 65(2):153–169
76. Casalo LV, Flavian C, Guinaliu M (2008) Promoting consumer's participation in virtual brand communities: a new paradigm in branding strategy. J Mark Commun 14(1):19–36
77. Masic I, Sivic S, Toromanovic S, Borojevic T, Pandza H (2012) Social networks in improvement of health care. Materia Socio-Medica 24(1):48
78. Liu H, Lim EP, Lauw HW, Le MT, Sun A, Srivastava J, Kim Y (2008) Predicting trusts among users of online communities: an epinions case study. In: Proceedings of the 9th ACM conference on electronic commerce. ACM, pp 310–319
79. Han JJ, Zheng RJ, Xu Y (2007) The effect of individual needs, trust and identification in explaining participation intentions in virtual communities. In: Proceedings of the 40th annual Hawaii international conference on system sciences (HICSS'07), 179c
80. Cutillo LA, Molva R, Strufe T (2009) Safebook: a privacy-preserving online social network leveraging on real-life trust. IEEE Commun Mag 47(12)
81. Nissenbaum H (2011) A contextual approach to privacy online. Daedalus 140(4):32–48
82. Dam WB (2009) School teacher suspended for Facebook gun photo. http://www.foxnews.com/story/2009/02/05/schoolteacher-suspended-for-facebook-gun-photo/
83. Mail D (2011) Bank worker fired for Facebook post comparing her 7-an-hour wage to Lloyds boss's 40 0 0-an-hour salary. http://dailym.ai/fjRTlC
84. Dwyer C (2011) Privacy in the age of Google and Facebook. IEEE Technol Soc Mag 30(3):58–63
85. Narayanan A, Shi E, Rubinstein BI (2011) Link prediction by de-anonymization: how we won the kaggle social network challenge. In: The 2011 international joint conference on neural networks (IJCNN). IEEE, pp 1825–1834
86. GNIP. http://support.gnip.com/apis/. Accessed on 20 Dec 2017
87. Kayes I, Iamnitchi A (2017) Privacy and security in online social networks: a survey. Online Soc Netw Media 3:1–21
88. Hwang T, Pearce I, Nanis M (2012) Socialbots: voices from the fronts. Interactions 19(2):38–45
89. Stringhini G, Wang G, Egele M, Kruegel C, Vigna G, Zheng H, Zhao BY (2013) Follow the green: growth and dynamics in twitter follower markets. In: Proceedings of the 2013 conference on internet measurement conference. ACM, pp 163–176

90. Felt A, Evans D (2008) Privacy protection for social networking APIs. In: 2008 Web 2.0 security and privacy (W2SP'08)
91. Fiesler C, Bruckman A (2014) Copyright terms in online creative communities. In: CHI'14 extended abstracts on human factors in computing systems. ACM, pp 2551–2556
92. Bonneau J, Anderson J, Danezis G (2009) Prying data out of a social network. In: IEEE international conference on advances in social network analysis and mining, 2009 (ASONAM'09), pp 249–254
93. Nissenbaum H (2004) Privacy as contextual integrity. Wash Law Rev 79(1):119–158
94. Zhao Z, Resnick P, Mei Q (2015) Enquiring minds: early detection of rumors in social media from enquiry posts. In: Proceedings of the 24th international conference on world wide web. International World Wide Web Conferences Steering Committee, pp 1395–1405
95. Farzan R, DiMicco JM, Brownholtz B (2009) Spreading the honey: a system for maintaining an online community. In: Proceedings of the ACM 2009 international conference on supporting group work. ACM, pp 31–40
96. DiMicco J, Millen DR, Geyer W, Dugan C, Brownholtz B, Muller M (2008) Motivations for social networking at work. In Proceedings of the 2008 ACM conference on computer supported cooperative work. ACM, pp 711–720
97. Thom-Santelli J, Millen DR (2009) Learning by seeing: photo viewing in the workplace. In: Proceedings of the SIGCHI conference on human factors in computing systems. ACM, pp 2081–2090
98. Geyer W, Dugan C, DiMicco J, Millen DR, Brownholtz B, Muller M (2008) Use and reuse of shared lists as a social content type. In: Proceedings of the SIGCHI conference on human factors in computing systems. ACM, pp 1545–1554
99. Sarkar C, Wohn DY, Lampe C (2012) Predicting length of membership in online community everything2 using feedback. In: Proceedings of the ACM 2012 conference on computer supported cooperative work companion. ACM, pp 207–210
100. Burke M, Marlow C, Lento T (2009) Feed me: motivating newcomer contribution in social network sites. In: Proceedings of the SIGCHI conference on human factors in computing systems. ACM, pp 945–954
101. Churchill E, Girgensohn A, Nelson L, Lee A (2004) Blending digital and physical spaces for ubiquitous community participation. Commun ACM 47(2):38–44
102. Lampe C, Johnston E (2005) Follow the (slash) dot: effects of feedback on new members in an online community. In: Proceedings of the 2005 international ACM SIGGROUP conference on supporting group work. ACM, pp 11–20
103. Fontaine MA, Millen DR (2004) Understanding the benefits and impact of communities of practice. In: Knowledge networks: innovation through communities of practice, pp 1–13
104. Kayes I, Zuo X, Wang D, Chakareski J (2014) Did you blog yesterday? Retention in community blogs. In: Proceedings of the 2014 international conference on social computing. ACM, pp 1–2
105. Zheng S (2015) Retention in MOOCs: understanding users' motivations, perceptions and activity trajectories. In: Proceedings of the 33rd annual ACM conference extended abstracts on human factors in computing systems. ACM, pp 247–250
106. Adamopoulos P (2013) What makes a great MOOC? An interdisciplinary analysis of student retention in online courses
107. Daniel J (2012) Making sense of MOOCs: musings in a maze of myth, paradox and possibility. J Interact Media Educ 2012(3)
108. Anderson A, Huttenlocher D, Kleinberg J, Leskovec J (2014) Engaging with massive online courses. In: Proceedings of the 23rd international conference on world wide web. ACM, pp 687–698
109. Baghaei N, Freyne J, Kimani S, Smith G, Berkovsky S, Bhandari D, Paris C (2009) SOFA: an online social network for engaging and motivating families to adopt a healthy lifestyle. In: Proceedings of the 21st annual conference of the Australian computer-human interaction special interest group: design: open 24/7. ACM, pp 269–272

110. Fogg BJ (2002) Persuasive technology: using computers to change what we think and do. Ubiquity 2002:5
111. Cheng R, Vassileva J (2005) User motivation and persuasion strategy for peer-to-peer communities. In: Proceedings of the 38th annual Hawaii international conference on system sciences, 2005 (HICSS'05). IEEE, pp 193a–193a
112. Lu D (2016) Engaging newcomers in hybrid communities. In: Proceedings of the 19th international conference on supporting group work. ACM, pp 491–494
113. Weinberg BD, Williams CB (2006) The 2004 US Presidential campaign: impact of hybrid offline and online 'meetup' communities. Direct Data Digit Mark Pract 8(1):46–57
114. Hampton KN (2007) Neighborhoods in the network society the e-Neighbors study. Inf Commun Soc 10(5):714–748
115. Hara N (2008) Internet use for political mobilization: voices of participants. First Monday 13(7)
116. Kraut RE, Resnick P, Kiesler S, Burke M, Chen Y, Kittur N, Riedl J (2012) Building successful online communities: evidence-based social design. MIT Press
117. Tajfel H (1974) Social identity and intergroup behaviour. Information (International Social Science Council) 13(2):65–93
118. Lampe C, Wash R, Velasquez A, Ozkaya E (2010) Motivations to participate in online communities. In: Proceedings of the SIGCHI conference on human factors in computing systems. ACM, pp 1927–1936
119. Papacharissi Z, Rubin AM (2000) Predictors of internet use. J Broadcast Electron Media 44(2):175–196
120. Ruggiero TE (2000) Uses and gratifications theory in the 21st century. Mass Commun Soc 3(1):3–37
121. Allen NJ, Meyer JP (1990) The measurement and antecedents of affective, continuance and normative commitment to the organization. J Occup Organ Psychol 63(1):1–18
122. Lakhani KR, Wolf RG (2005) Why hackers do what they do: understanding motivation and effort in free/open source software projects. In: Feller J, Fitzgerald B, Hissam SA, Lakhani KR (eds) Perspectives on free and open source software. MIT Press, Cambridge, MA
123. Wasko MM, Faraj S (2000) "It is what one does": why people participate and help others in electronic communities of practice. J Strateg Inf Syst 9:155–173
124. Constant D, Sproull L, Kiesler S (1996) The kindness of strangers: the usefulness of electronic weak ties for technical advice. Organ Sci 7(2):119–135
125. Butler B, Sproull L, Kiesler S, Kraut R (2002) Community effort in online groups: who does the work and why. Leadership at a distance: Research in technologically supported work, pp 171–194
126. Ehrlich K, Muller M, Matthews T, Guy I, Ronen I (2014) What motivates members to contribute to enterprise online communities? In: Proceedings of the companion publication of the 17th ACM conference on computer supported cooperative work & social computing. ACM, pp 149–152
127. Wu S, Das Sarma A, Fabrikant A, Lattanzi S, Tomkins A (2013) Arrival and departure dynamics in social networks. In: Proceedings of the sixth ACM international conference on web search and data mining. ACM, pp 233–242
128. Zhu H, Kraut R, Kittur A (2012) Effectiveness of shared leadership in online communities. In: Proceedings of the ACM 2012 conference on computer supported cooperative work. ACM, pp 407–416
129. Subramani MR, Rajagopalan B (2003) Knowledge-sharing and influence in online social networks via viral marketing. Commun ACM 46(12):300–307
130. Yang S, Allenby GM (2003) Modeling interdependent consumer preferences. J Mark Res 40(3):282–294
131. He J, Chu WW (2010) A social network-based recommender system (SNRS). In: Data mining for social network data. Springer US, pp 47–74
132. Su Z, Tung AK, Zhang Z (2012) Supporting top-K item exchange recommendations in large online communities. In: Proceedings of the 15th international conference on extending database technology. ACM, pp 97–108

133. Shede. http://www.shede.com. Accessed on Dec 2017
134. RecSys Challenge (2018). https://recsys.acm.org/recsys18/challenge/. Accessed on Feb 2018
135. Carroll JM, Rosson MB (1998) Network communities, community networks. In: CHI 98 conference summary on human factors in computing systems. ACM, pp. 121–122
136. Girvan M, Newman ME (2002) Community structure in social and biological networks. Proc Natl Acad Sci 99(12):7821–7826
137. Nguyen NP, Dinh TN, Nguyen DT, Thai MT (2011) Overlapping community structures and their detection on social networks. In: 2011 IEEE third international conference on privacy, security, risk and trust (PASSAT) and 2011 IEEE third international conference on social computing (SocialCom). IEEE, pp 35–40
138. Nguyen NP, Dinh TN, Xuan Y, Thai MT (2011) Adaptive algorithms for detecting community structure in dynamic social networks. In: 2011 Proceedings of IEEE INFOCOM. IEEE, pp 2282–2290
139. Nguyen NP, Xuan Y, Thai MT (2010) A novel method for worm containment on dynamic social networks. In: Military communications conference, 2010 (MILCOM 2010). IEEE, pp 2180–2185
140. Nguyen NP, Dinh TN, Tokala S, Thai MT (2011) Overlapping communities in dynamic networks: their detection and mobile applications. In: Proceedings of the 17th annual international conference on mobile computing and networking. ACM, pp 85–96
141. Palla G, Derényi I, Farkas I, Vicsek T (2005) Uncovering the overlapping community structure of complex networks in nature and society. Nature 435(7043), 814. (Tyler JR, DM)
142. Tyler JR, Wilkinson DM, Huberman BA (2003) Email as spectroscopy: automated discovery of community structure within organizations. In: Communities and technologies. Springer, Dordrecht, pp 81–96
143. Wilkinson DM, Huberman BA (2004) A method for finding communities of related genes. Proc Natl Acad Sci 101(suppl 1):5241–5248
144. Newman ME, Girvan M (2004) Finding and evaluating community structure in networks. Phys Rev E 69(2):026113
145. Clauset A, Newman ME, Moore C (2004) Finding community structure in very large networks. Phys Rev E 70(6):066111
146. Blondel VD, Guillaume JL, Lambiotte R, Lefebvre E (2008) Fast unfolding of communities in large networks. J Stat Mech Theory Exp 2008(10):P10008
147. Radicchi F, Castellano C, Cecconi F, Loreto V, Parisi D (2004) Defining and identifying communities in networks. Proc Natl Acad Sci USA 101(9):2658–2663
148. Lancichinetti A, Fortunato S, Kertész J (2009) Detecting the overlapping and hierarchical community structure in complex networks. New J Phys 11(3):033015
149. Xie J, Kelley S, Szymanski BK (2013) Overlapping community detection in networks: the state-of-the-art and comparative study. ACM Comput Surv (CSUR) 45(4):43
150. Adamcsek B, Palla G, Farkas IJ, Derényi I, Vicsek T (2006) CFinder: locating cliques and overlapping modules in biological networks. Bioinformatics 22(8):1021–1023
151. Farkas I, Ábel D, Palla G, Vicsek T (2007) Weighted network modules. New J Phys 9(6):180
152. Onnela JP, Saramäki J, Kertész J, Kaski K (2005) Intensity and coherence of motifs in weighted complex networks. Phys Rev E 71(6):065103
153. Newman ME (2002) Assortative mixing in networks. Phys Rev Lett 89(20):208701
154. Shen H, Cheng X, Cai K, Hu MB (2009) Detect overlapping and hierarchical community structure in networks. Physica A 388(8):1706–1712
155. Kumpula JM, Kivelä M, Kaski K, Saramäki J (2008) Sequential algorithm for fast clique percolation. Phys Rev E 78(2):026109
156. Ahn YY, Bagrow JP, Lehmann S (2010) Link communities reveal multiscale complexity in networks. Nature 466(7307):761
157. Rosvall M, Bergstrom CT (2008) Maps of random walks on complex networks reveal community structure. Proc Natl Acad Sci 105(4):1118–1123
158. Evans TS (2010) Clique graphs and overlapping communities. J Stat Mech Theory Exp 2010(12):P12037

159. Gregory S (2011) Fuzzy overlapping communities in networks. J Stat Mech Theory Exp 2011(02):P02017
160. Zhang S, Wang RS, Zhang XS (2007) Identification of overlapping community structure in complex networks using fuzzy c-means clustering. Physica A Stat Mech Appl 374(1):483–490
161. Nepusz T, Petróczi A, Négyessy L, Bazsó F (2008) Fuzzy communities and the concept of bridgeness in complex networks. Phys Rev E 77(1):016107
162. Wang X, Jiao L, Wu J (2009) Adjusting from disjoint to overlapping community detection of complex networks. Physica A Stat Mech Appl 388(24):5045–5056
163. Psorakis I, Roberts S, Ebden M, Sheldon B (2011) Overlapping community detection using bayesian non-negative matrix factorization. Phys Rev E 83(6):066114
164. Gregory S (2008) A fast algorithm to find overlapping communities in networks. In: Joint European conference on machine learning and knowledge discovery in databases. Springer, Berlin, Heidelberg, pp 408–423
165. Gregory S (2009) Finding overlapping communities using disjoint community detection algorithms. In: Complex networks. Springer, Berlin, Heidelberg, pp 47–61
166. Goldberg M, Kelley S, Magdon-Ismail M, Mertsalov K, Wallace A (2010) Finding overlapping communities in social networks. In: 2010 IEEE second international conference on social computing (SocialCom). IEEE, pp 104–113
167. Lancichinetti A, Radicchi F, Ramasco JJ, Fortunato S (2011) Finding statistically significant communities in networks. PLoS ONE 6(4):e18961
168. Lancichinetti A, Radicchi F, Ramasco JJ (2010) Statistical significance of communities in networks. Phys Rev E 81(4):046110
169. Raghavan UN, Albert R, Kumara S (2007) Near linear time algorithm to detect community structures in large-scale networks. Phys Rev E 76(3):036106
170. Gregory S (2010) Finding overlapping communities in networks by label propagation. New J Phys 12(10):103018
171. Chen D, Shang M, Lv Z, Fu Y (2010) Detecting overlapping communities of weighted networks via a local algorithm. Physica A Stat Mech Appl 389(19):4177–4187
172. Liu J, Zhong W, Abbass HA, Green DG (2010) Separated and overlapping community detection in complex networks using multiobjective evolutionary algorithms. In: 2010 IEEE congress on evolutionary computation (CEC). IEEE, pp 1–7
173. Rees BS, Gallagher KB (2013) Detecting overlapping communities in complex networks using swarm intelligence for multi-threaded label propagation. In: Complex networks. Springer, Berlin, Heidelberg, pp 111–119
174. Wen S, Jiang J, Xiang Y, Yu S, Zhou W, Jia W (2014) To shut them up or to clarify: restraining the spread of rumors in online social networks. IEEE Trans Parallel Distrib Syst 25(12):3306–3316
175. Lin YR, Chi Y, Zhu S, Sundaram H, Tseng BL (2009) Analyzing communities and their evolutions in dynamic social networks. ACM Trans Knowl Discov Data (TKDD) 3(2):8
176. Asur S, Parthasarathy S, Ucar D (2009) An event-based framework for characterizing the evolutionary behavior of interaction graphs. ACM Trans Knowl Discov Data (TKDD) 3(4):16
177. Kumar R, Novak J, Tomkins A (2010) Structure and evolution of online social networks. In Link mining: models, algorithms, and applications. Springer, New York, NY, pp 337–357
178. Leskovec J, Kleinberg J, Faloutsos C (2005) Graphs over time: densification laws, shrinking diameters and possible explanations. In: Proceedings of the eleventh ACM SIGKDD international conference on knowledge discovery in data mining. ACM, pp 177–187
179. Lin YR, Sundaram H, Chi Y, Tatemura J, Tseng BL (2007) Blog community discovery and evolution based on mutual awareness expansion. In: IEEE/WIC/ACM international conference on web intelligence. IEEE, pp. 48–56
180. Palla G, Barabási AL, Vicsek T (2007) Quantifying social group evolution. Nature 446(7136):664
181. Spiliopoulou M, Ntoutsi I, Theodoridis Y, Schult R (2006) Monic: modeling and monitoring cluster transitions. In: Proceedings of the 12th ACM SIGKDD international conference on knowledge discovery and data mining. ACM, pp 706–711

182. Prabavathi G, Sreeshanmugapriya T (2016) A review on community detection in dynamic social networks. Int J Innov Res Comput Commun Eng 4(7). https://doi.org/10.15680/ijircc e.2016. 0407061

183. Zhang Y, Wang J, Wang Y, Zhou L (2009) Parallel community detection on large networks with propinquity dynamics. In: Proceedings of the 15th ACM SIGKDD international conference on knowledge discovery and data mining. ACM, pp 997–1006

184. Sun J, Faloutsos C, Papadimitriou S, Yu PS (2007) Graphscope: parameter-free mining of large time-evolving graphs. In Proceedings of the 13th ACM SIGKDD international conference on knowledge discovery and data mining. ACM, pp 687–696

185. Hui P, Yoneki E, Chan SY, Crowcroft J (2007) Distributed community detection in delay tolerant networks. In: Proceedings of 2nd ACM/IEEE international workshop on mobility in the evolving internet architecture. ACM, p 7

186. Yang T, Chi Y, Zhu S, Gong Y, Jin R (2011) Detecting communities and their evolutions in dynamic social networks—a Bayesian approach. Mach Learn 82(2):157–189

187. Holland PW, Leinhardt S (1976) Local structure in social networks. Sociol Methodol 7:1–45

188. Xu KS, Kliger M, Hero III AO (2014) Adaptive evolutionary clustering. Data Min Knowl Discov 28(2):304–336

189. Palla G, Pollner P, Barabási AL, Vicsek T (2009) Social group dynamics in networks. In: Adaptive networks. Springer, Berlin, Heidelberg, pp 11–38

190. Duan D, Li Y, Jin Y, Lu Z (2009) Community mining on dynamic weighted directed graphs. In: Proceedings of the 1st ACM international workshop on complex networks meet information & knowledge management. ACM, pp 11–18

191. Kim MS, Han J (2009) A particle-and-density based evolutionary clustering method for dynamic networks. Proc VLDB Endow 2(1):622–633

192. Cazabet R, Amblard F, Hanachi C (2010) Detection of overlapping communities in dynamical social networks. In: 2010 IEEE second international conference on social computing (SocialCom). IEEE, pp 309–314

193. He-Li S, Jian-Bin H, Yong-Qiang T, Qin-Bao S, Huai-Liang L (2015) Detecting overlapping communities in networks via dominant label propagation. Chin Phys B 24(1):018703

194. Aston N, Hu W (2014) Community detection in dynamic social networks. Commun Netw 6(02):124

195. He K, Li Y, Soundarajan S, Hopcroft JE (2018) Hidden community detection in social networks. Inf Sci 425:92–106

196. Pendry LF, Salvatore J (2015) Individual and social benefits of online discussion forums. Comput Hum Behav 50:211–220

197. DiMaggio P, Hargittai E, Neuman WR, Robinson JP (2001) Social implications of the internet. Ann Rev Sociol 27:307–336. https://doi.org/10.1146/annurev.soc.27.1.307

198. Howard PEN, Rainie L, Jones S (2001) Days and nights on the internet: the impact of a diffusing technology. Am Behav Sci 45(3):383–404. https://doi.org/10.1177/000276420104 5003003

199. McKenna KYA, Bargh JA (1998) Coming out in the age of the Internet: Identity "demarginalization" through virtual group participation. J Pers Soc Psychol 75(3):681–694. https://doi.o rg/10.1037/0022-3514.75.3.681

200. Isinkaye FO, Folajimi YO, Ojokoh BA (2015) Recommendation systems: principles, methods and evaluation. Egypt Inform J 16(3):261–273

201. Fatemi M, Tokarchuk L (2013) A community based social recommender system for individuals & groups. In: 2013 international conference on social computing (SocialCom). IEEE, pp 351–356

202. Sun A, Chen X (2016) Online education and its effective practice: a research review. J Inf Technol Educ 15

The 'Verticals', 'Horizontals', and 'Diagonals' in Organisational Communication: Developing Models to Mitigate Communication Barriers Through Social Media Applications

Rima Namhata and Priyadarshi Patnaik

Abstract India's workplace culture is distinctively multilingual and multi-dialectical, where English is often considered as the lingua franca. On many occasions, the choice of one official workplace language can automatically advantage or disadvantage different members of an organisation based on their competency and confidence with it. It can also impact their interpersonal relations, one of the markers of communication, between employees, who are the social capital for an organisation, and function through social groupings. This paper aims at proposing models for overcoming the communication barriers which have a three-tier structure—vertical, horizontal and diagonal—by engaging, motivating, exploring and broadening perspectives of the organisational workforce. It will include scope to build intrapersonal, interpersonal, leadership, persuasiveness, decision-making, problem-solving, and leadership skills. This can mitigate performance anxiety and boost the morale of the workforce which can ensure effective time-use and enhance productivity. The paper will examine, analyse, and propose a model for soft skills development within the framework of a three-tier communication structure for the millennials who constitute the soul of an organisational culture. An attempt will be made through the medium of social media applications like, WhatsApp, Twitter, and LinkedIn to promote, process and shape communication strategies for the workforce to make the workplace communication professionally communicative and competent.

Keywords Organisational communication · Communication behaviour
Social media applications · Soft skills · Communication flows

R. Namhata (✉) · P. Patnaik
Indian Institute of Technology Kharagpur, Kharagpur, West Bengal, India
e-mail: rimanamhata@gmail.com

P. Patnaik
e-mail: priyadarshi1@yahoo.com

© Springer International Publishing AG, part of Springer Nature 2019
S. Patnaik et al. (eds.), *Digital Business*, Lecture Notes on Data Engineering
and Communications Technologies 21, https://doi.org/10.1007/978-3-319-93940-7_14

1   Introduction

In India, the workplace culture is invariably multilingual and multidialectal as there is diverse use of vernacular languages and English is considered the lingua franca in politics, education and business quite in league with the Western paradigm. This, however, owes to two reasons, the first being its colonial past. Social scientist, Daswani [1] study remarks that,

> The prolonged, and often acrimonious, debate between the Anglicists and the Orientalists undoubtedly reflects the deep understanding that the participants in the debate had, of the social and political ramifications of the choice of a language as a medium of education. The Anglicists won the debate and English was established as the medium of Education. Predictably, with the medium came the structure of Education prevalent in Imperial England. The total structure of Indian education was recast in the English mould (p. xvi).

Second, the colonial past was carried forward to a present where both the functional and socio-cultural significance of the language has increased. India as a colony, under the British empire, had to imbibe English for the administrative purposes, that gradually percolated into the Indian educational system. English fast gained supremacy right from the days of the British Raj which continue to the 21st century. This is highlighted by scholars like Crystal, who mention that English enjoys a privileged position in the whole world because of the geographical-historical and socio-cultural aspects. English has become the formula of 'lingua franca', a symbol of globalisation, diversification, progress, identity and change (1997) [2]. Gradually it became the language of mainstream official interactions in India, and the change became intense owing to the seeds of globalisation, and its gigantic influence. The transition is indicated by Sunil Kale, a social scientist, who writes "… four consecutive democratically elected governments in India in the 1990s were able to pass important legislation that carried India far from its historical moorings in centralised planning and towards a market-oriented system' [3, p. 210]. This market-led economy necessitated new kinds of workers with an emphasis on the linguistic skills, especially English. Khubchandani, speaking about language, remarks in *Language Demography and Language in Education* [4]:

> …everyday communication is a form of dyadic behaviour, the choice of using a particular language or a creative blending of different speech varieties is determined…by the institutional factors of identification; language here (is) serving as a label for status, prestige, and cosmetic decoration, understood under the rubrics of stylistics and rhetoric (p. 18).

In the present context, English is the oft and extensively used language in its functional role in education, banking services, hospitality sectors, and services sectors, to name a few, in India.

The choice of English over the vernacular languages prevails across the Indian sub-continent; and this is despite India's multilingual and, more so, multidalectical spread. This can put individuals in a pro-English organisational environment at an advantageous or a disadvantageous position; as this depends on the social groupings and the individual level of comfort. This visibly has its own alienation effect. However, functional English as an umbrella term can be and is extended further to the use of communication skills. Since this chapter has organisational communication at its core, and it functions through the three-tier paradigm in its communications flows—vertical, horizontal and diagonal—while understanding the communication channel and the flows, we also find differences in the communication flows of organisations.

This chapter makes an attempt to propose a symbiotic and holistic model through the use of technology, especially the social media applications like LinkedIn, WhatsApp, and Twitter handle, to mitigate the communication differences witnessed in communication flows by addressing the communication skills (inclusive of soft skills) of the workplace capital, in order to make the workplace communication professionally meaningful. It conclude by proposing a model separately on how to integrate the 'vertical', 'horizontal' and 'diagonal' flows with the aids of Social applications—LinkedIn, WhatsApp, and Twitter in a way that can educate and mobilise the workforce in an organisation to facilitate interpersonal skills, time-management, active listening, prompt responses, motivational skills, and etiquette building skills that can add meaning and effectiveness to organisational communication. An obvious rationale for the use of only these three select social tools—LinkedIn, WhatsApp, and Twitter—and not considering the others is that most other social media applications such as Facebook, Instagram, MySpace, etcetera are largely time-consuming, and add significantly less value to the organisational communication structure. They promote procrastination since the users are often obsessed with their intuitive interfaces and unrelated stories and posts. The scope for digressing from the original idea or message is low as far as Twitter is considered, due to the 280 characters cap per post (initially 140). Besides, it is considered professionally vain to indulge in non-professional trivia on LinkedIn which is the driving force for focused and productive discussions in that domain, which centre around the 'organisation' theme. The case for using WhatsApp as a tool in improving organisational communication flows is due to the fact that it is currently one of the most widely accepted and used IM platform (1.3 billion active users as of July 2017), and is also available across varied digital ecosystems.

In this conceptual paper, the introduction is followed by literature review divided into three segments. The first part reviews the dynamics of communication and the need for linguistic skills in workplace sectors, the second part examines the literature existing on the models of communication, the three-tier organisational hierarchical communication flows, and the third part presents the history of technologisation of communication, and capitalising social tools to build the organisational discourse and its productive use as a learning and training tool. At the end, based on a detailed review of the strengths and weaknesses of the social media tools indicated, a model for technology integration for effective organisational communication is proposed.

Part I

2 Review of Literature

Literature suggests the importance of oral communication skills as a vehicle for workplace communication. Since emphasis on linguistic skills for organisational culture has its origin from the West, the next sections will review the importance of oral communication skills in the corporate workplace, an essentially Western import.

2.1 History of Communication Skills

A glance at the history of language formulation and the teaching of communication skills will show us the contemporary obsession with 'Communication skills' and 'Communication problems.' Deborah Cameron, in her book, mentions of 'linguistic diversity [as] a problem, and linguistic uniformity as a desirable ideal' (2000) paved by globalisation. Umberto Eco's 'search for the perfect language' (1995), to mythically unite a diverse global population has generated debates for the last two decades. But the attempt to adapt of a single language like Volapük[1] (1880) to bring linguistic unity and thereafter the birth of Esperanto,[2] to mythically unite the world language population, also failed. The twentieth century harboured the desirability of English as the lingua franca. Dr. Judith Kuriansky an eminent American expert on communication, in conversation with Deborah Cameron does not ask for abandoning the other dialects/native languages and to communicate in one language, i.e. English, but rather encourages explorations into the norms of relating to other people. Citing examples of those norms, Dr. Kuriansky says:

> Speaking directly is better than speaking indirectly. Speaking positively is better than criticising. Negotiating is better than arguing. Sharing your feelings is better than being silent and withdrawn (2002, p. 67).

In other words, certain interactional speech norms accompany speakers in dialoguing or in communication, and 'speech styles across languages' (p. 68) to be maximally 'effective' for the purposes of 'communication'.

Having said this, the rhetorics of communication, or the effectiveness of communication, can be accomplished through books and works of communication experts, penned for professionals and the uninitiated. Customised modules can be uploaded in You-tube, or through New media's distribution channels. Rightly, here Cameron mentions that 'dissemination' of communication at the grassroots level:

> … is accomplished through instruction and training in particular linguistic practices. Forms of instruction and training which aim to develop 'communication skills' … are increasingly common in all kinds of contemporary institutions, ranging from elementary schools to multinational corporations (p. 3).

Cameron [19] elucidates that the norms used in language teaching, especially foreign or second language teaching are quite different from communication skills training directed at all where

---

[1]Volapük: in 1880, a priest created a language thinking the whole world could use. Words from French, English, and German were used and were named as Volapük. The language was hard to use as it had odd sounds and Latin case endings.

[2]Esperanto: a few years later, another language took over Volapük, and that is Esperanto. The language was lyrical and easier to master.

most recipients are either native monolingual or highly proficient bilingual speakers of the language in which (and through which) they are learning to 'communicate'. On the other hand there are forms of instruction, many of them in the category of language teaching for specific purposes or for business, which incorporate concerns like 'negotiation', 'meeting skills', 'presentation skills' etc., into programs aimed at particular groups of L2 (second language) learners such as managers in multinational companies (p. 4).

So, communication skills training if incorporated in the corpus of language teaching to privilege and promote the subtler nuances of communication with an idea to educate the occupational groups of service sectors like tourism, hospitality, and travel will be effective, as these sectors demand proficiency in a foreign language. For the past decades, communication skills have maximised employability, and often been distinguished from hard skills or ICT skills. In communication skill, oral communication is cited as an essential soft skill. In the new globalised economy, the 'New Work Order' has new demands on its workers [5] and the demands are related to the linguistic aspects [6, 7] where the new economy, a new form of capitalism, has taken prevalence. Here, this special skill of communication for creating lasting impression in the businesses and service sectors is emphasised with training programmes being made an integral part of the organisational functioning.

2.2 New Capitalism: Services Sector and Linguistic Requirements

Long before the phase of capitalism came into dominance, linguistic aspects were nevertheless imperative in labour market stratification. If the industrial economy looked for manual labourers often referred to as 'hands', the new capitalism emphasised on language use as rudimentary for every worker's functionality. The former USA labour Secretary Robert B. Reich [8] in his text *The work of Nations*, mentions of a new division of labour that supersedes the traditional manual/non-manual distinction by, 'symbolic analysts'—skilled in 'words, numbers, images, and digital bits' who will be dominant either in person or behind the scenes. Cameron further substantiates it by saying that this puts pressure on literacy skills (data inputting and record keeping), and more generally, interpersonal communication skills (being pleasant and attentive to customers and clients in face-to-face or on the telephone [9, p. 5]).

If in the advanced economies, manufacturing industries are in decline and creative and service sectors are on the rise with more emphasis on the linguistic skills, reports from IBEF (India Brand Equity Foundation) show its equivalence in India where services sector in India is not only the dominant sector, but has attracted significant foreign investment flows and includes activities in the sectors of hotels and restaurants, transport, financing, real estate, insurance, and social and personal services. IBEF report bears testimony to the statistical representation of the exponential growth of the services sector in India, having contributed to a growth of 53.8% of its gross value, added in 2016–2017 [10]. In addition, a report in *The Hindu* states, "India has the second fastest growing services sector with its compound annual growth rate at nine percent, just below China's 10.9 percent, during the last

11-year period from 2001 to 2012" [33]. So, with the Services Sector predominantly on the rise, the linguistic requirements with life skills are the dominant skill sets required in the Services Sector and workplace in general. Because of these economic developments, the rhetorics of many such organisations revolve around 'talking' that has been "promoted from taken-for-granted social accomplishment of all normal humans to a complex task requiring special effort to master" [9, p. 5].

2.3 Rhetorics of Communication and Workplace Demands

The present preoccupation with oral skills of communication is not solely a reflection of the economic norms, but more attributed to the 'self-improvement culture' so followed by the new capitalism. Author Deborah Cameron's understanding on 'self-improvement culture' is but "the informal instruction in oral communication skills, undertaken by individuals voluntarily and for personal rather than professional reasons" [9, p. 6]. The culture of self-improvement emerged in the 1960s–1970s with an emphasis on popular psychology books and a considerable emphasis on communication. Not surprisingly, the author, Cameron, suggests that being able to 'communicate' is to talk 'openly and honestly' about one's experiences, listen non-judgmentally to the talk of other people …, and indicates this (as) "the key to solving problems and improving relationships." [9, p. 7]. The author further elaborates as to some of the rudiments of self-improvement activities direct towards communication proposed are being 'honest and direct', using 'assertiveness training', using 'transactional analysis' which are quite favourable to the 'thinking and practice of new capitalism', and are now being used in workplace training [9, p. 7]. It is this 'technologisation of discourse' [6] coined by Norman Fairclough "whereby communication techniques elaborated for a particular purpose are taken out of their original context and used for a quite different purpose" that is at operation here. If this helps people practise self-improvement, to build personal relationships, then in the context of business, the same techniques would help to bond with customers, build relationships and encourage them to buy, and return to their sellers again. The scholars teaching oral communication skills in schools advocate for these skills not only in the areas of vocational courses, but refer them as 'life skills', and approve of these to be applied in other domains as well. In the words of one such advocate, Phillips, "all children benefit from learning skills that will make them better friends, better life-partners, better employees and better human beings" [11, p. 7]. This argument has been put forward by the social theorist Anthony Giddens in his book *Modernity and Self-Identity* [12], where he suggests that 'late modern' societies are a faraway call from 'traditional' or 'pre-modern ways of life'. In this context, the author suggests how an individual has become a 'reflexive project' where he has to 'work on' rather than his being taken for granted. On the other hand, in the traditional societies, people were expected to take a similar course to that of their parents. The contrary is seen in late modern societies where the pace of social change is so rapid that the experience of the older generations does not fit as a model for the next generations.

Instead of using a 'pre-existing social narrative' late modern individuals need to construct their own. Giddens rightly remarks that in today's highly individualistic and mobile culture people no longer have the same close-knit communities, but instead form relationships that pose challenges in a world of strangers; a reason why Giddens [12] emphasises on 'communication skills' and most importantly the skills of 'self-expression' (p. 12) and 'mutual disclosure' (p. 96). So, concerns on spoken interaction are underpinned by several factors and especially workplace demands. Having said this, the shift witnessed is towards skill-based or competence-led curricula in education, and training, with an amplified sense of attention to 'speaking and listening'. Put forward by Milroy and Milroy [13, p. 47], in the workplace, communication training, in a way, can reduce or even eliminate variation in people's ways of interacting, similar to the practice of wearing uniforms. And just like uniforms/appearance is treated as an aspect of 'branding', employees are required to engage and deliver standard verbal codes to make the input and output of communication channels easy, convenient and stress-free, leading to satisfaction in politeness management. When the intended meaning of the receiver and the perceived meaning of the receiver are the same, communication is achieved. However, the requirement for effective communication belies the simplicity of this definition. After hundreds of studies across organisations, Daniel Goleman in his book, *Emotional Intelligence* [14], states that for organisations to run successfully one needs individuals with high levels of 'emotional intelligence' or skills of social awareness and understanding. This typically includes motivating and influencing others through interpersonal skills, being sensitive while giving honest feedback, and the ability to empathise, and develop relationships, reflecting and correcting one's own behaviour if required, and learning the ability to handle emotions of both self and others. The fact remains that, in the contemporary business context, the communication process now includes emotional intelligence, social awareness and communication, which need to be and can be developed and honed.

Part-II

3 Communication Model and the Channels: A Brief Review

This section will briefly delineate the definitions provided by the communication scholars on the paradigm of communication per se and the communication flows operational in an organisation's context. A diagram will be proposed for the three-tier communication flows.

3.1 *The Model of Communication*

Organisational communication is non-linear. The words of Richmond and McCroskey [15, page not given] describe it as "the process by which individuals stimulate meaning in the minds of other individuals, by means of verbal and nonverbal messages in the context of a formal organisation". Another scholar, Miller [16] has to say:

> Most scholars would agree that an organisation involves a social collectivity (or a group of people) in which activities are coordinated in order to achieve both individual and collective goals. By coordinating activities, some degree of organisational structure is created to assist individuals in dealing with each other and with others in the larger organisational environment. With regard to communication, most scholars would agree that communication is a process that is transactional (i.e., it involves two or more people interacting within an environment) and symbolic (i.e., communication transactions 'stand for' other things, at various levels of abstraction) (p. 1).

Primarily, to understand communication, we need to have a source and a receiver. A source is a person or a group transmitting any information or message with an intended meaning, whereby the message is an encoded one. These messages can be verbal, written or even non-verbal (gestures, para-language). And the receiver is one for whom the message is intended, who decodes the message. The process between encoding to decoding is carried out by the commonly referred communication channels such as face-to-face interactions, e-mails, notices, reports, circulars, memorandum, telephone communications or even voice-mails, e-bulletins, personal weblogs, chat rooms, and newsletters. This model is seen in almost all business communication books across the globe. Communication channels in today's world have become more exhaustive with the interface of the communication technologies, namely Skype, Twitter, video conferencing, texting, blogs, Wikis and more of the social networking applications, which are quite often being used in informal and unofficial conversations' Much of that could be put into effective use formally to ease, hone and enhance the sharing of knowledge and employee efficacy in a given organisation.

3.2   Communication Flows

Communication Flows as a component of organisational communication is three-tiered: (a) vertical (downward and upward), (b) horizontal (lateral), and (c) diagonal (upward, downward and lateral and combinations of the three). Vertical communication is hierarchical or formal following a chain of command from the superiors to the subordinates responding to formal communication with an account of reports or meetings [34]. Most frequently used channels in this flow are memos, notices, circulars, and reports. The second is the Horizontal or lateral communication where the information process, channel, and flow is generative, easy, fast, mostly among peers of an organisation. This encourages dynamic dialoguing, instantaneous information distribution and exchange of ideas [17] adding to collaborative environments without the hassles of formal and written communication channels that make each process lengthy. Information no longer remains centralised or controlled [17]. Diagonal communication, a newer concept, identified by Wilson [18] is exchanged between the different structural forms of an organisation, especially between managers and employees in different functional departments, and primarily so in bigger organisations. A diagrammatic/(flowchart) representation below elaborates on the communication flows (Fig. 1).

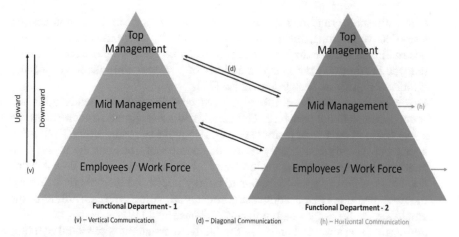

Fig. 1 Schematic of horizontal, vertical, and diagonal communication in an organisation suggested by the authors

3.3 Barriers to Communication

However, the key issue in any communication are the barriers that hinder it. A review of literature on the barriers of organisational communication suggest different categories of barriers faced in effective organisational communication. They are classified below:

A. **Physical Barriers**: A physical barrier in an environment is the natural condition that acts as a barrier to communication in sending a message from sender to receiver. Communication is generally easier over shorter distances because more communication channels are available and less technology is required. The modern technology often serves to reduce the impact of physical barriers [19]. An appropriate channel can be used to overcome the physical barriers so that the advantages and disadvantages of each communication channel should be understood. Physical barriers like doors, walls, distance, etc. do not let the communication become effective while sending the message by the sender. When the organisation's units or branches are physically scattered in various places then the communication made to them may remain ineffective due to physical distance [20].

B. **Psychological Barriers**: The psychological barrier to communication is the influence of the psychological state of the communicators. The sender and receiver create an obstacle to effective communication. Psychological distance prevents the communication , filters part of it, or causes misinterpretation. Lack of retention ability and inadequate attention to the message make communication less effective. Distrust, threat, fear, and lack of ability to communicate also cause obstruction in the free flow of the communication. For example, a receiver with reduced hearing may not grasp an entirety of a spoken conversation especially if there is significant background noise [21].

C. **Organisational Barriers**: Effective organisation largely depends upon the sound organisational structure. The classical organisational structure with the scalar chain of command restricts free and frequent communication. Each manager receives information only from one source and transmits the message to another single level. If there are too many levels in an organisation then it is difficult to pass the correct information to the right person at the right time through the right medium. Organisational structure greatly affects the capabilities of the employees as far as the communication is concerned [19].

D. **Semantic Barriers**: Semantics is the study of meaning, signs, and symbols used for communication. The word is derived from 'sema', a Greek word whose meaning signs. Semantic barriers to communication are the symbolic obstacles that distort a sent message in some other way than intended, making the message difficult to understand. The words, signs, and figures used in the communication, which is explained by the receiver in the light of his experience, it also creates doubtful situations. This happens because the information is not sent in simple language. The people interpret the same information in different ways depending on their social background, knowledge, education, and experience. Illegible handwriting is a crude example of a semantic barrier, and poor writing skills is another [19, 22].

The above review schematizes the barriers to effective organisational communication structure. From the above review, a tabular format has been drawn below to identify separately the barriers of communication processes in the three-tier communication flow (Table 1).

Table 1 Barriers to communication in a three-tier organisation

Communication flows	Description	Barriers to communication
Vertical Communication	• Vertical communication occurs between hierarchically positioned persons and can involve both downward and upward communication flows [37, 41] • Classified as downward communication (managers to employees) and upward communication (employees to managers) [35, 36] • Examples: Job delegation mails (down), employee surveys/grievance programmes/suggestion programmes/addresses (up)	• Top management generally refrains from communicating directly with the employees [38] • Constructive criticism by the management is not always perceived in a positive sense [38] • Employees are afraid to speak their minds due to fear of reprisal [38] • Employees feel that their ideas are filtered as they travel up the hierarchy [41] • Managers are reluctant to hear the employees out, citing lack of time [41]
Horizontal/ Lateral Communication	• Horizontal communication facilitates the coordination of related activities between professional peers or people at the same hierarchical level of the organisation [37, 41] • Research has shown high satisfaction level (85%) for lateral communication in organisational environment (Frank, 1984) [39, 49] • Lateral communication at the worker level is assumed to be less problematic within a functional area [38] • Examples: Informal discussions, phone calls, social media, teleconferencing, memos, meetings, etc.	• Lateral communication, if uncontrolled, can pose a problem for the top management to keep a tab on the workforce [41] • Time-consuming if vertical communication is required to ratify decisions • May create indiscipline and discontent if strict procedural rules of communication are not followed—especially true in case of mob/union mentality [38]
Diagonal Communication	• Cross-functional communication between employees at different levels of the organisational hierarchy is described as diagonal communication [39] • Increasingly common in larger organisations with matrix or project-based structures • It reduces the chances of distortion or misinterpretation by encouraging communication between the relevant parties [40] • Reduces a manager's communication workload because he/she doesn't have to act as an intermediary between his direct reports and other managers [40].	• Use of diverse jargon across functional departments may lead to dilution/distortion of information due to vague or unclear communication • Employees who receive a high level of communication over a short period of time may be susceptible to information overload and struggle to process the same • The most primary forms of diagonal communication are usually verbal, hence there is little to no accountability for the information transferred • Diagonal communication usually breaches the line of authority, and might subsequently create ego issues [40] • In some cases (diagonally downwards), it may create a case of conflicting instructions

Part-III

4 New Technologies for Communication

This section will look into the technological advancements in communication technologies and their contemporary uses in the corporate and social contexts.

"There is a revolution going on in the communications industry", [17, p. 190] reports Lucent Technologies, one of the multinational giants in American Telecommunications equipment company. This proliferation is witnessed in voice-mail messages, e-mails, internet users and the internet traffic triplicating in a quarter year. Appetite for increased information results in harnessing information technology and is used by organisations to their competitive advantage. At the soul of all this stands the integral process of communication. In this era of virtual and augmented reality, where artificial intelligence is entwined with the lives of people, with Siri[3] and Alexa,[4] where the voice-based virtual assistants of Apple and Amazon take commands and perform tasks, it can always be a daunting task to make communication, a channel of human connection, leave its impact on the lives of the individuals. More so, the case of learning is testimony to the application of online web tools. An online report on BBC states that smartphones are important aids for learning, but should be used with productive purposes. In the same report, a survey in four schools of London indicates that one of the schools, IPACA,[5] uses smartphones as a learning device. The director of IPACA, Gary Spracklen, is in support of integrating the 21st technological advancement to that of learning. He ideates, "With a smartphone we can cross-reference the textbooks—we can look at the Syrian crisis at the moment, the different population flows that are changing throughout Europe. We can't do that with a textbook." The children of the academy are still into reading books but the library has online access to facilitate cloud-based learning [23].

The late decades of the 20th century and the early years of the 21st century brought in the web tools to remotely connect people in global locations. It had its massive influence touching all sectors of health, hospitality, services, manufacturing, aviation, finance, education and learning. These 'social networking tools,' an umbrella term for web-based and mobile-based technology, allow users to create and share content, encourages interactivity and generates an online community as well. This expanse was also witnessed with the conferencing tools like Skype, Adobe Con-

---

[3] Siri is a Voice-based Artificial Intelligence device that uses voice recognition to chat with humans. An intelligent virtual assistant is capable of taking human commands and performing tasks. Initially started by apple and could be used by Apple iPhone 4S and newer. Siri can work across multiple iPhone and iPad devices and first started at 2011.

[4] Alexa is a voice-based Artificial intelligence device of Amazon, and functions similar to that of Siri. It started in November 2014, and was used in the Amazon, Echo, and Amazon Echo Dot services. It can play music, can complete to-do lists, setting alarms, providing information on news, weather, traffic, play audio books and can control smart gadgets.

[5] IPACA: School on the isle of Portland.

nect, or Blackboard Collaborate [24] to allow users to communicate using voice, video, presentations, online training materials, through web conferencing, virtual classes, and webinars, to present and collaborate on learning modules. Collaborative Knowledge forum tools like: Wikipedia, Quora, and Yahoo Answers create knowledge-based communities to disseminate information and awareness [24]. In addition, a study report by Peter McGuire in *Irishtimes.com* shows how online support systems have revolutionised learning by making it enjoyable [25]. If online study resources have gone beyond classroom teaching, smartphones are in use from Primary to Higher secondary standards, and colleges are being facilitated with webinars, virtual classrooms, and MOOC[6] (the examples of Coursera EDx,[7] NPTEL[8]). Online classroom teaching for in ESL/EFL[9] classes is aided through the refined use of LMSs[10] platforms like—*Moodle, RCampus* and *Learnopia*[11] to user-friendly applications like—Skype, Facebook, Instagram, Telegram, and WhatsApp. Consequently, online tools have penetrated deep into organisational cultures as well.

4.1 Technologisation of Communication

It is no wonder that work culture has seen revolutionary changes in artificial intelligence and machine learning, robots becoming the order of the day. Henceforth how will it affect business and work culture, are some of the probing questions. An article titled, 'Technology moves to the Head of the 21st Century Classroom' from *MIT Technology Review* [31] reflects upon a similar question of how to leverage the digital transformation of 'cloud, virtualisation, software-defined networking', interactive apps, 3D simulation to help learners with hands-on discovery, and connect and share learning experiences, something that is never possible through lectures and rote learning, (September, 2017) to improve the services and value given to the customers, employees, and partners. For instance, LinkedIn has a large database of facts and stories across organisations, training, team building, motivational speakers,

---

[6]MOOC: Massive Open Online Course aimed at unlimited open participation through web and catering distance education.

[7]edX: a MOOC from the world's best universities like Harvardx, MITx, Berkeleyx. Created by founding partners and Universities MIT and Harvard over a range of subjects, from management, finance, Liberal Arts, Philosophy, Computer Science and the like. It is a non-profit online initiative to impart education.

[8]NPTEL: A massive MOOC open online participation and certification initiated from India's IITs and IISc. These are online courses on various topics against a nominal fee for exam certificate.

[9]EFL/ESL: the first is an example where the teachers teaches English as a foreign language in a country where English is not the native language. A student learning English would fall in this category. ESL is where English is taught as a second language in a country where their primary language is English. For example, an Indian learning English in Canada.

[10]LMS: Learning and Management System is a software application where online instructors create, manage, deliver content and track student progress in e-learning courses.

[11]Moodle, RCampus, Learnopia: commercial web portals that share LMS to create and manage course content for teachers/instructors. In a word it helps to host online courses free of cost.

and edupreneurs where keeping oneself updated is easy. In a study of 4200 companies, conducted by McKinsey, 72% have reported using internal social tools to facilitate employee communication; use of stand-alone technologies such as Slack, Yammer, and Chatter or embedded applications such as Microsoft Teams and JIRA are growing at an astounding rate [26]. Even though knowledge management databases existed, they had limited success compared to their newer counterparts, the social tools which not only provide knowledge sharing but information and connection as well.

Despite its utility and vastness in its applicability to training and knowledge, social applications are considered as distracters, often eliminating the need for human interaction. This is mentioned by David Bryne in *MIT Technology Review* [27]. He observes that buying books through online merchants like Amazon help us in getting books quickly, but often eliminating the need to interact with human beings in the process. So, technology development may not aim at eliminating human connection but can become an outcome. The more are the algorithmic recommendations through AI principles, the lesser are the social interactions (pp. 7–8). Social media tools that often appear to connect people are actually a 'simulation of real connection'. The same article cites a study from UC San Diego and Yale suggest that the more people do Facebook, the worse they feel in their lives, often making people seem sad and envious (pp. 7–8). An article *Social Media Messages are Becoming More Complex and Nobody Knows Why?* in *MIT Technology Review* (July 2017) also suggests negative health implications of longer screen time [32].

However, the given commonalities through shared interests can bring the employees of an organisation together and consequently connect them to different domains as well. But a report from *Harvard Business Review* suggests, for internal social tools to deliver on their promise, it takes both the employees and the organisations to be comfortable with personal and professional interactions online. Without going into the controversy, the tools can always be harnessed for utilitarian purposes. Social tools, instead of being used by 'lurkers' and 'observers,' can be used by "content producers for writing posts, sharing information, creating documents and videos. People can acquire at least two types of knowledge this way: *direct knowledge* and *metaknowledge*" [26]. For instance, through direct knowledge when an employee posts a domain-specific problem, and when there is someone who adds a solution to it, the other employees can gain insights by looking at that message and the solution. And meta-knowledge in a way shows the insiders (employees) the technical or domain-specific expertise of employees, which becomes so visible through social messages. Just by way of reference it is also possible that by observing and picking up bits of unsaid things in communications, 'many employees form a picture of who knows what and whom.' In the context of this, a rich metaphor is drawn by Clive Thompson from *nytimes.com* comparing the use of social tools similar to a pointillist painting where no single dot by itself makes sense as much as putting the dots together [28].

4.2 Capitalising Social Tools as an Organisational Discourse

Investor, philanthropist, and founder of Virgin Records, Sir Richard Charles Nicholas Branson's words come in handy to talk of business and more so of organisational culture, where he writes, "A business has to be involving, has to be fun, and it has to exercise your creative instincts" [29, p.12]. So, in order to chart out a course to capitalise social tools in an organisational discourse, and facilitating communication, just analysing the past behaviour of employees, and their way of conduct will be insular in shaping the communication process. So, an inclusive approach by combining the factors of defining the purpose, creating a global connectedness, increase in innovation, honing 'ambient awareness', and spelling out the rules of conduct while implementing social tools can in a way change the dynamics of the functioning of organisational communication [47]. An extended and effective functionality can be to develop and implement training modules, lifeskills or even behavioural lessons to the stakeholders at large. A look at the existing literature on social media applications as effective learning tools is nothing new. They are indeed effective if used wisely. The emergence of new channels of communication such as the internet has added to the complexity and flexibility of communication processes. It helps since speakers can now communicate without foregrounding their cultural identities, or can hide them on purpose [30, p. 8]. Hence, tailoring etiquette modules to suit the need of the millennials in an organisation will be a building process to mitigate the hierarchical, paternalistic system and make it convenient and less resistive for the employees. The essentialist views of organisational culture held assessments of vertical channel of communication as supreme and inflexible, which has nowadays been mitigated by diagonal communication. So, in this world of socio-digital interaction, organisational communication discourse revolves around the participants engaged in the act, who are less conscious of self-images, but talk about topics they are interested in (see Sect. 4.2). A brief glance at the above factors and the literature studied brings out the prowess of these social tools if used wisely:

- **Defining the purpose**: Executive can make clear at the beginning the need of the organisation to the use social tools for improving employee relationships as a medium and not to stalk their personal lives. It takes critical foresight to use new technologies that can provide value and training to the organisational population.
- **Creating a global connectedness**: Many companies have operational offices onsite and off-site locations across the globe. So, often, to create a shared identity and build relationships with employees across offices, social tools facilitate professional connections.

- **Increase in innovation**: Social tools can be an effective medium to trigger new ideas for a project, can provide solutions, and can also act as tools for a newer form of online or in-house training.
- **'Ambient awareness'**: This term implies awareness of communication and behaviour around one's environment, where one may or may not be connected. This adds value to friendly interaction and builds camaraderie even if it has nothing to do with work. One who is socially striking and friendly helps to share bond and warmth with coworkers, and short one or two bonding (personal) lines beyond work make people feel easy at workplace.
- **Spell out the rules of conduct**: Since social tools are informal channels of communication, people can inadvertently share personal and confidential information. So, this necessitates keeping the rules of conduct clear by the management. Or else, social tools can ensnare communication channel.
- **Social tools to implement training modules**: These tools can be effectively put to use in designing and disseminating training nuggets to look into the gap areas of communication, develop lifeskills, and effectively train the workforce.

5   Social Tools

We have identified the features for three social tools, viz. WhatsApp, Twitter, and LinkedIn, and shall discuss their functionality with respect to some of their prominent features. These tools possess the potential to overcome the barriers faced in communication in a three-tier organisation. They do have some drawbacks, too, but we have also discussed how they may be mitigated. This can, in a way, facilitate soft skills development, and improve the efficacy of the communication process of the organisation's work force.

In the following pages, we shall discuss the features of WhatsApp, Twitter, LinkedIn (in that order), the role they play in mitigation of the barriers faced in organisational communication, and their potential in developing the employees soft skills, along with a diagrammatic representation of each (Table 2 and Fig. 2).

Table 2 WhatsApp as a social tool to address communication barriers in the three-tier communication flows

Features of the social channel	Communication barriers addressed	Probable misuse of the feature	Remedial measure of the misuse	Scope for soft skills development	Supporting screenshot[a]
Group Chats	• Uncoordinated horizontal communication is checked • Speedy responses ensure fast transfer of information • Ensures participation of a majority of the members of the communication channel • If top management is involved, vertical communication can be speedy, and less intimidating	• Too many participants can make the communication more chaotic and less helpful • Procrastination is inevitable, owing to number of participants	• Limiting number of participants in the groups • Having at least one group admin from the top management • Tighter control from human resources	• Employees afraid to pitch in ideas might get a morale boost • Solution-oriented discussions are encouraged [43] • Active participation ensures prompt clarification of ideas, right away [43]	
Restricted Groups	• A new feature in WhatsApp, wherein only the admin of a chat group can post messages [42] • Ensures efficient vertical (downward) communication, without pointless discussions			• Using the informal channel for downward communication can potentially strengthen the employee-employer rapport • Meaningfully precise and time-effective communication is ensured	
Broadcast Messages	• Urgent news and notices can be shared within lateral and/or vertical circles instantly with this feature • Since the message is received individually, there is little scope for pointless discussions (as in group chats)	• Unimportant information and trivia are not supposed to be broadcasted	• An event code may be established by the top management/human resources as to when a piece of information may be broadcasted	• Enhances time management	

(continued)

Table 2 (continued)

Features of the social channel	Communication barriers addressed	Probable misuse of the feature	Remedial measure of the misuse	Scope for soft skills development	Supporting screenshot[a]
Picture Status	• A feature that allows the user to set a picture/video status (for 24 h) with user-controlled viewing privileges • Employees who are otherwise shy of sharing their merits/achievements directly may use this feature to reminisce the workforce • Alternately, top management may use it to introduce the workforce to landmark events in the organisation	• Since statuses are set at a personal level, the management/human resources have little control • Multiple statuses can be set, hence the important ones may get lost		• Employees may be encouraged to share their achievements • The ignorant crowd within the organisation may be kept up-to-date with the organisational affairs • May act as a morale booster and in turn, help productivity • Provides motivation to employees	
IP-based Voice/Video Calling	• WhatsApp natively supports internet-based voice and video calling facilities, so they are practically free of cost (except for internet data consumption) • In situations where second/third party ratification is required, this feature may come in handy • It helps alleviate physical communication barriers to some extent • Diagonal communication barriers are better handled face-to-face using this feature	• Excessive use of video calls may lead to internet data and bandwidth consumption, which is a cost to the organisation	• The Internet Service Provider (ISP) for the organisation may place a cap on data consumption by video calls	• Employees are encouraged to engage in one-to-one correspondences with colleagues from within and outside the functional department	

(continued)

Table 2 (continued)

Features of the social channel	Communication barriers addressed	Probable misuse of the feature	Remedial measure of the misuse	Scope for soft skills development	Supporting screenshot[a]
Pinned Chats	• WhatsApp offers an option to pin a particular chat to the top of the chat list • This helps if there's a chance of missing important notification from the members of the organisation in a slew of other conversations			• Short-circuiting and loss of information is avoided • Since the information reaches all intended parties in the channel, productivity is increased • Acts as a catalyst for efficient information transfer	
WhatsApp Web Client	• WhatsApp offers a desktop-based web client on various platforms, that helps users operate the application from their respective browsers/desktops • This does not address any communication barrier, in particular, but helps reduce the time spent by users on their phones [44] • Desktop notifications can be enabled, so that users check WhatsApp only when required [44] • The features available in the web client are limited to personal and group chats, and file sharing			• By cutting down the time users spend on their phones, and making less features available, this feature helps boost productivity among the employees	

[a]Explanatory screenshots for WhatsApp have been taken from the images available in public domain

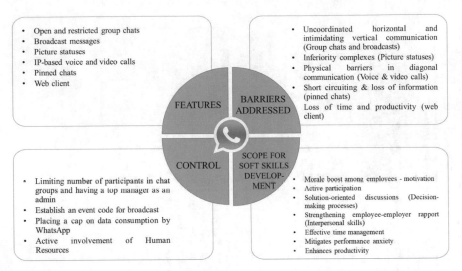

- Open and restricted group chats
- Broadcast messages
- Picture statuses
- IP-based voice and video calls
- Pinned chats
- Web client

- Uncoordinated horizontal and intimidating vertical communication (Group chats and broadcasts)
- Inferiority complexes (Picture statuses)
- Physical barriers in diagonal communication (Voice & video calls)
- Short circuiting & loss of information (pinned chats)
- Loss of time and productivity (web client)

FEATURES **BARRIERS ADDRESSED**

CONTROL **SCOPE FOR SOFT SKILLS DEVELOP-MENT**

- Limiting number of participants in chat groups and having a top manager as an admin
- Establish an event code for broadcast
- Placing a cap on data consumption by WhatsApp
- Active involvement of Human Resources

- Morale boost among employees - motivation
- Active participation
- Solution-oriented discussions (Decision-making processes)
- Strengthening employee-employer rapport (Interpersonal skills)
- Effective time management
- Mitigates performance anxiety
- Enhances productivity

Fig. 2 WhatsApp as a social tool to address communication barriers in the three-tier communication flows

There are certain limitations, when using the tool of WhatsApp. When it comes to WhatsApp, the limitation is being connected only to the people you know/work for/works with/has their contact numbers. But when the need arises to reach out for help/expertise/opinion on a global scale, the communication flows must traverse the boundaries of the three-tier organisational model. This is where social tools media like twitter (Table 3 and Fig. 3) and LinkedIn (Table 4 and Fig. 4) come in handy [48].

Table 3 Twitter as a social tool to address communication barriers in the three-tier communication flows

Features of the social channel	Communication barriers addressed	Probable misuse of the feature	Remedial measure to the misuse	Scope for soft skills development	Supporting screenshot[a]
Tweets	• Any post on Twitter, known as a Tweet, can include text (up to 280 characters), photos, web links, polls, etc. • Due to the character cap, brevity and precision is maintained in communication • Helps the user put out thoughts/opinions/queries on the social media to incite relevant and helpful responses (comments) from a global audience; thus defeating physical barriers to communication on a global scale • A tweet may be designed to be audience-specific, that facilitates participants from a particular functional department may respond (helps enhance diagonal communication) [45] • In the absence of an internet connection, a user may tweet using SMS from their mobile number linked to the Twitter account	• Needlessly long posts require a user to break it into a series of tweets, which is counter-intuitive for the reader • Users may resort to deliberate instiga-tion/trolling/cyber controversies • Employees may spend too much time online • The broadcast nature of communica-tion may incite irrelevant, or at times, negative comments	• The Human Resources may actively monitor the online activities of the employees • Internet data capping is advisable to limit the procrastination	• Promotes solution-centric thought process • Enables a wider perspective owing to the enormous scope of communication • Ensures succinctness of the problem statement to acquire the most relevant and helpful solutions • Taking a poll in a tweet facilitates quick conveyance of the opinion of one's target audience (work group, in this case) • Promotes connectedness, since users get an idea of what their colleagues are working on	

(continued)

Table 3 (continued)

Features of the social channel	Communication barriers addressed	Probable misuse of the feature	Remedial measure to the misuse	Scope for soft skills development	Supporting screenshot[a]
Retweet (RT)	• A retweet is in effect, sharing a tweet that we feel needs more audience, with a click • Helps users convey the exact words of a person to their social circles, without any distortion • Further extends communicative scope of the tweet by diminishing physical barriers			• Helps inculcate an attitude of respect and acknowledgment for the originality of ideas	
Trends/Hashtag	• Twitter facilitates stacking of several tweets around the world about a particular event or phenomenon. This is known as 'Trends', and may be initiated by using a hashtag (#) • Tweets in response to a particular trend may be both informative and opinionated (depending on the nature of the trend). This helps users understand the perspectives and thought processes of their work group [45] • This is particularly helpful, since asking for an opinion over a controversial trend may often lead to biased and prejudiced arguments	• Users may indulge in controversies arising from the aforementioned arguments and thus waste time	• Employees may be advised to stay away from such actions, and only to focus on the tweets concerning the organisational domain	• Inculcates all-round awareness about the issues around the globe • Keeps the employees about the latest trends in their respective work domains • Helps the user stay on the pulse of their respective industry/sector	
Direct Messaging (DM)	• Twitter also enables a user to send a direct message (DM) another user • If a person does not wish to reply to a tweet or engage in a conversation, he/she may send a DM to incite a more personalised, relevant, and secure response [45] • This boosts one-to-one communication by practically eliminating all barriers • Alternatively, DM may be used as an instant messaging platform	• Controversial tweets may lead to verbal attacks on DM • Users are likely to procrastinate while using DM as an instant messenger	• Employees may be advised not to use DM unless they are out of ways to contact the concerned person	• Encourages employees to seek advice/opinions from the stalwarts of their respective domains	

(continued)

Table 3 (continued)

Features of the social channel	Communication barriers addressed	Probable misuse of the feature	Remedial measure to the misuse	Scope for soft skills development	Supporting screenshot[a]
Selective Following	• The Twitter feed of a user shall only display updates of people/organisations/events that they are following (in certain cases, of global importance) [45] • This helps reduce the clutter on their respective homepages, which may help improve productivity, since the time required to find the relevant content is reduced • The algorithms of Twitter 'learn' what content is relevant/useful to the user, and brings up filtered and personalised results accordingly • This also enables a user to create a person-based, event-based, or organisation-based RSS fees			• Helps save the employee's time • Indirectly trains the employees to customise Twitter for an efficient and productive usage	
Moments	• Events of landmark importance across the globe are now enlisted in a separate tab on Twitter, called 'Moments' • It is a digital repository of phenomena that are demographically chosen to have a significant impact on the world • Since these kinds of tweets have the most active respondents (generally), it is better to keep them sorted separately from the main feed so that it is not spammed with irrelevant content			• Enhances productivity and saves time by organizing the user's feed	

[a]Explanatory screenshots for Twitter have been taken from the images available in public domain

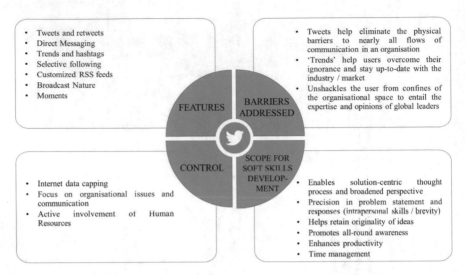

- Tweets and retweets
- Direct Messaging
- Trends and hashtags
- Selective following
- Customized RSS feeds
- Broadcast Nature
- Moments

- Tweets help eliminate the physical barriers to nearly all flows of communication in an organisation
- 'Trends' help users overcome their ignorance and stay up-to-date with the industry / market
- Unshackles the user from confines of the organisational space to entail the expertise and opinions of global leaders

- Internet data capping
- Focus on organisational issues and communication
- Active involvement of Human Resources

- Enables solution-centric thought process and broadened perspective
- Precision in problem statement and responses (intrapersonal skills / brevity)
- Helps retain originality of ideas
- Promotes all-round awareness
- Enhances productivity
- Time management

FEATURES BARRIERS ADDRESSED CONTROL SCOPE FOR SOFT SKILLS DEVELOP-MENT

Fig. 3 Twitter as a social tool to address communication barriers in the three-tier communication flows

Taking into account the scope of this paper, we shall limit our discussion to the features offered under the domain www.linkedin.com. As an organisation, LinkedIn offers various products and services like LinkedIn Learning Solutions, LinkedIn Marketing Solutions, LinkedIn Talent Solutions, LinkedIn Profinder, Slideshare, etc., that address very specific requirements of the various strata of the corporate world. LinkedIn also offers certain paid services/features that we will not be covering in the following table, in favour of better accessibility.

Table 4 LinkedIn as a social tool to address communication barriers in the three-tier communication flows

Features of the social channel	Communication barriers addressed	Probable misuse of the feature	Remedial measure to the misuse	Scope for soft skills development	Supporting screenshot[a]
Posts and Shares	• LinkedIn enables users to share articles, posts, and images of varied nature—A motivational, educational, inspirational, subject matter, industry trends, technology, life skills and experiences, best practices, etc. [46] • These posts go a long way in establishing LinkedIn as an effective professionalism enhancing tool • This feature may also be used by individuals to explore new professional arenas, based on their skill sets • The physical and organisational barriers to communication are largely overcome due to this feature	• Certain market/industry trends may be overwhelming for employees, and may affect their morale negatively	• Users must limit themselves to constructive posting, and making the best out of this feature • The Human Resources may actively monitor the online activities of the employees • Internet data capping is advisable to limit the procrastination	• Employees may refine their professional literary skills by following the posts of influential individuals on LinkedIn	
Talent Acquisition Portal	• One of the most important feature of LinkedIn is its talent acquisition portal—organisations sort the vacancies available at their disposal by location, hierarchical level, skill sets, and other relevant filters [46] • This provides the user a unified access to multiple job portals, and a chance of leveraging their skill sets to climb the corporate ladder • The stigma of unemployability is successfully mitigated with this feature, thus combating psychological barriers to communication	• It is in extremely poor taste to use the talent portal service with the current organisation's internet resources	• The Human Resources may block the talent portal service of LinkedIn selectively	• In a bid to enhance their digital footprint on the corporate channel, the employees have a chance to refine their profiles	

(continued)

Table 4 (continued)

Features of the social channel	Communication barriers addressed	Probable misuse of the feature	Remedial measure to the misuse	Scope for soft skills development	Supporting screenshot[a]
Connections and Followers	• The crux of LinkedIn is to expand one's professional contact database through primary, secondary, and tertiary connections, based on industry, proximity, or influence[46] • A user can entail immense help and open up new arenas of professional development by connecting with like-minded veterans around the globe • Following certified Influencers on LinkedIn helps a user track their updates and posts, which offers insights into their experiences and life lessons [46] • One may send direct messages to one's connections in order to mitigate the physical barriers in communication			• Having a rich and diverse connections database aids all-round development of the employee's professional persona • In case of entrepreneurship, these connections act as a solid resource pool for talent	
Endorsements and recommendations	• LinkedIn offers users a choice to endorse their connections for the skill set they possess • Endorsements can be given to individual skills—a large number of endorsements generally signifies the proficiency of a person with that particular skill [46] • Similarly, a personalised recommendation by the one or more connections goes a long way in boosting the credibility of one's professional profile [46] • Users can both give and ask for endorsements and recommendations • These features essentially help mitigate issues with interpersonal communication			• Receiving skill endorsements and recommendations from the work circle may provide a huge morale boost to the employees • Enhances visibility and credibility of one's corporate profile in the social channel	

(continued)

Table 4 (continued)

Features of the social channel	Communication barriers addressed	Probable misuse of the feature	Remedial measure to the misuse	Scope for soft skills development	Supporting screenshot[a]
Knowledge-sharing groups	• LinkedIn hosts a huge collection of knowledge-sharing groups that are formed based on industry, sector-wise domain, trends, services, skill sets, demographics, etc. [46] • One may request to join a group, and once the request is sanctioned, one may learn from/contribute to the real-time discussions occurring in these groups • An organisation may have a group of its own, consisting of its work force to facilitate real-time problem sharing and solving with the help of this platform • This feature helps alleviate almost all barriers to efficient organisational communication	• The discussion portal may turn into a hotbed for controversies if the group is not effectively moderated	• A top management representative must play the role of a moderator for such groups (in an organisational context) • Users must refrain from commenting in a group discussion unless they are completely sure about their contribution	• Enables a richer know-how of the latest trends in one's industry/domain • Solution-centric thinking is promoted through active participation in the discussions occurring in this group	
Network Analytics	• LinkedIn keeps a track of one's profile activity (both incoming and outgoing) • The user is notified of the number of views to his/her profile, professional background of the viewers, searches that presented their profile, popularity among connections, organisations looking for similar candidates, etc.			• Such statistics enable the users to streamline their communication in order to keep their profiles among the top-viewed ones on the platform	

[a]Explanatory Screenshots for LinkedIn have been taken from personal profile

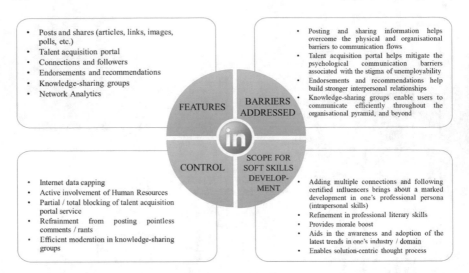

- Posts and shares (articles, links, images, polls, etc.)
- Talent acquisition portal
- Connections and followers
- Endorsements and recommendations
- Knowledge-sharing groups
- Network Analytics

FEATURES

BARRIERS ADDRESSED

- Posting and sharing information helps overcome the physical and organisational barriers to communication flows
- Talent acquisition portal helps mitigate the psychological communication barriers associated with the stigma of unemployability
- Endorsements and recommendations help build stronger interpersonal relationships
- Knowledge-sharing groups enable users to communicate efficiently throughout the organisational pyramid, and beyond

CONTROL

SCOPE FOR SOFT SKILLS DEVELOPMENT

- Internet data capping
- Active involvement of Human Resources
- Partial / total blocking of talent acquisition portal service
- Refrainment from posting pointless comments / rants
- Efficient moderation in knowledge-sharing groups

- Adding multiple connections and following certified influencers brings about a marked development in one's professional persona (intrapersonal skills)
- Refinement in professional literary skills
- Provides morale boost
- Aids in the awareness and adoption of the latest trends in one's industry / domain
- Enables solution-centric thought process

Fig. 4 LinkedIn as a social tool to address communication barriers in the three-tier communication flows

6 Conclusion: Scope, Limitations, and Future Directions

In conclusion, the applicability of the social tools—WhatsApp, Twitter, LinkedIn—in trying to mitigate the barriers of three-tier communication process have been proposed here as three models separately. This is because cross-platform communication is something that these tools do not permit. However, in the future, it is always possible to explore how they can interact. The current models proposed attempt to answer the two basic questions—first, who will implement it and the second, why should it be implemented? We suggest that the proposed model(s) with three separate tools need to be implemented by the top management. However, being a part of the social media, they will evolve their own rules, thus giving rise to a certain empowerment and challenging hierarchy; but once in use, it will have the flexibility of incorporating social norms, and would not follow a hierarchical system. The process of organisational communication has its respective scopes and limitations. The empowerment of the workforce capital will help to build on refined professional skills or has the scope to build on the ability of soft skills. Certain traits of soft skills like time-management and increased productivity, brevity in responses owing to intrapersonal skills, adoption to and enhanced awareness of the latest trends in one's domain, and solution-oriented thought process—will be developed. Besides, innovative leadership skills get displayed by coming up with solution-centric responses, and increases active participation because of the lack of physical and psychological barriers. A morale boost mitigates performance anxiety and enhances productivity, improves on employee-employer rapport based on shared interests. The social tools in the organisational context will be evolving, changing, and because of their evolutionary nature, successful practices will grow and unsuccessful ones will automatically die out. The regular internal official communication of memo, agenda, proposals reports, circulars, and notices have to follow the rule-book, but the internal communication on meetings, presentations, and other brief-follow ups when communicating in the three-tier structure can be addressed through social tools. True, external correspondence and formal correspondence can never be replaced by social tools, but this is a limitation of the role of organisational communication. Second, since, like all social tools, it might have a life of its own; it may go out of control, and this will have to be addressed cautiously. Finally, a technology based and rule-based model is expected to evolve spontaneously from such interactions in the future.

References

1. Daswani CJ (ed) (2001) Language education in multilingual India. Introduction. Veerendra Printers and Paris, UNESCO, New Delhi, pp x–xviii
2. Crystal D (2003) English as a global language. Cambridge University Press, Cambridge, U.K.
3. Kale S (2003) The political economy of India's second-generation reforms. In: Ganguly Sumit (ed) India as an emerging power. Frank Cass publishers, London, pp 207–225
4. Khubchandani LM (2001) Language demography and language in education. In: Daswani CJ (ed) Language education in multilingual India. United Nations Educational Scientific and Cultural Organisation, New Delhi, pp 3–47
5. Gee JP, Hull G et al (1996) The New Work Order: behind the language of the new capitalism. Allen and Unwin, St Leonards, NSW
6. Fairclough N (1992) Discourse and social change. Polity Press, Cambridge
7. Cope B, Kalantzis M (eds) (2000) Multiliteracies: literacy learning and the design of social futures. Routledge, London
8. Reich R (1991) The work of nations: preparing ourselves for 21st century capitalism. Simon and Schuster, London
9. Block D, Cameron D (2002) Globalisation and the teaching of 'communication skills. In: Cameron Deborah (ed) Globalisation and language teaching. Routledge, London, pp 67–82
10. IBEF database (2010) Indian economy overview. https://www.ibef.org/economy/indian-econ omy-overview. Accessed 15 Dec 2017
11. Phillips (1998) Communication: a key skill for education. Report published in the BT forum
12. Giddens A (1991) Modernity and self-identity: self and society in the late modern age. Stanford University Press, USA
13. Milroy J, Milroy L (1999) Authority in language: investing Standard English. Routledge, New York
14. Goleman D (1996) Emotional intelligence. Bloomsbury, London
15. Richmond VP, McCroskey JC (2009) Organisational communication for survival: making work, work. Pearson/Allyn & Bacon, Boston
16. Miller K (2015) Organisational communication: approaches and processes. Cengage/Wadsworth, Belmont, CA
17. Schermerhorn JR Jr, Osborn JG, Hunt JG (2002) Organisational behavior. Wiley, University of Phoenix
18. Wilson DO (1992) Diagonal communication links within organisations. Int J Bus Commun 29(2):129–143. https://doi.org/10.1177/002194369202900202
19. Barriers to Effective Communication (2013). https://www.skillsyouneed.com/ips/barriers-com munication.html
20. Physical Barriers to Communication (2016). https://www.businesstopia.net/communication/p hysical-barriers

21. Psychological Barriers to Communication (2016). https://www.businesstopia.net/communicat ion/psychological-barriers
22. Semantic Barriers of Communication (2016) Businesstopia. https://www.businesstopia.net/co mmunication/semantic-barriers-communication
23. Jeffreys B (2015) Can a smartphone be a tool for learning? Resource document. BBC. http:// www.bbc.com/news/education-34389063. Accessed 26 Dec 2017
24. Bechingham S (2017) Conferencing. Social media for learning. https://socialmediaforlearnin g.com/collaboration-tools/conferencing. Accessed 29 Dec 2017
25. McGuire P (2016) The best online study tools: students give their verdict. https://www.irisht imes.com/news/education/the-best-online-study-tools-students-give-their-verdict-1.2810540. Accessed 26 Dec 2017
26. Leonardi P, Neeley T (2017) What managers need to know about social tools. Harv Bus Rev 118–126. https://hbr.org/2017/11/what managers-need-to-know-about-social-tools. Accessed 12 Dec 2017
27. Bryne D (2017) Eliminating the human. MIT Technol Rev. Accessed 8 Sept 2017
28. Thompson C (2008) Brave new world of digital intimacy. http://www.nytimes.com/2008/09/0 7/magazine/07awareness-t.html. Accessed 20 Dec 2017
29. Ugwuoke CA (2011) Time to decide: a career choice manual. AuthorHouse publisher, UK
30. Sharifian F, Jamarani M (eds) (2013) Language and intercultural communication in the new era. Background. Routledge, New York, USA, pp 1–19
31. Amino Labs (2017) Technology moves to the Head of the 21st Century Classroom. MIT Technol Rev. https://www.technologyreview.com/s/608774-technology-moves-to-the-head-o f-the21st-century-classroom/. Accessed 8 Sept 2017
32. ArXiv labs (2017) Social media messages are becoming more complex and nobody knows why. MIT Technol Rev. https://www.technologyreview.com/s/608345-Social-Media-Message s-are-Becoming-more-Complex-and-Nobody-Knows-Why/. Accessed 8 Sept 2017
33. Bhargava Y (2014) India has fastest growing services sector. The Hindu. http://www.thehind u.com/business/budget/india-has-second-fastest-growing-services-sector/article6193500.ece
34. Grimsley S (2015) Internal communication in an organization: definition, strategies and examples. https://study.com/academy/lesson/internal-communication-in-an-organisation-defi nition-strategies-examples.html. Accessed 4 Jan 2017
35. Grimsley S (2015) Downward communication: definition, advantages and disadvan-tages. https://study.com/academy/lesson/downward-communication-definition-advantages-di sadvantages.html. Accessed 4 Jan 2017
36. Manker AD (2015) Upward communication: definition, advantages, disadvantages and exam-ples. https://study.com/academy/lesson/upward-communication-definition-advantages-disadv antages-examples.html. Accessed 4 Jan 2017
37. Grimsley S (2015) Horizontal communication: definition, advantages, disadvantages and exam-ples. https://study.com/academy/lesson/horizontal-communication-definition-advantages-disa dvantages-examples.html. Accessed 4 Jan 2017
38. Nanda (2007) Organizational communication. http://orgcommunication-nanda.blogspot.in/20 07/12/vertical-horizontal-and-diagonal.html. Accessed 4 Jan 2017
39. Nordmeyer B (n.d.) The difference in diagonal and horizontal flow of communi-cation. https://yourbusiness.azcentral.com/difference-diagonal-horizontal-flow-communicati on-17341.html. Accessed 4 Jan 2017
40. Uhlig DK (2012) Importance of diagonal communication routes. http://smallbusiness.chron.c om/importance-diagonal-communication-routes-35496.html. Accessed 4 Jan 2017
41. Joseph C (2011) Advantages and disadvantages of a vertical and horizontal organiza-tion. http://smallbusiness.chron.com/advantages-disadvantages-vertical-horizontal-organisati on-24212.html. Accessed 4 Jan 2017
42. Sulleyman A (2017) WhatsApp: new feature lets you force everyone else in your group to stop chatting. http://www.independent.co.uk/life-style/gadgets-and-tech/news/whatsapp-res tricted-groups-admin-stop-messages-silence-everyone-else-mute-a8086546.html. Accessed 7 Jan 2017

43. Medium G (2015) Lead nurturing using WhatsApp. https://genymedium.com/lead-nurturing-using-whatsapp/. Accessed 7 Jan 2017
44. WhatsApp desktop app to eliminate communication barriers (2016). http://www.arabnews.co m/news/whatsapp-desktop-app-eliminate-communication-barriers. Accessed 7 Jan 2017
45. Zhao D, Rosson MB (2009) How and why people twitter: the role that micro-blogging plays in informal communication at work. Soc Behav Sci 4:243–252. https://doi.org/10.1145/15316 74.1531710
46. Brinkman A (2014) 11 Incredibly useful LinkedIn features you might not be using. https://blo g.hubspot.com/insiders/linkedin-features. Accessed 7 Jan 2017
47. Senapati A (2016) Importance of effective communication in organisation. LinkedIn. https:// www.linkedin.com/pulse/importance-effective-communication-organisation-avinash-senapat i/. Accessed 7 Jan 2017
48. Barve V (2015) Social media for organizational communication. LinkedIn. https://www.linked in.com/pulse/social-media-organisational-communication-vikas-barve/. Accessed 7 Jan 2017
49. Frank AD (1984) Trends in Communication: Who Talks to Whom? Personnel December, 41–47

Subjective Interestingness in Association Rule Mining: A Theoretical Analysis

Rupal Sethi and B. Shekar

Abstract The main aim of "Knowledge Discovery in Databases" is to extract and interpret interesting patterns present in real-world datasets. Measures to identify interesting patterns (called Interestingness Measures) may be categorized on the basis of statistical significance (*Objective Measures*), or on the basis of data and subjectivity of the user (*Subjective Measures*) which includes user's domain knowledge. Three major steps in dealing with subjective measures are (1) *knowledge acquisition* from the user in terms of his beliefs, (2) the *matching methodology* for comparing generated association rules and user's belief, and (3) generation of *interesting rules* that may be unexpected, novel or actionable. We propose and construct a theoretical framework for studying subjective interestingness in association rule mining, which takes care of these steps. We attempt to fit prior work done on subjective interestingness into this framework, thus identifying relevant research gaps. The notion of subjective interestingness confines to knowledge discovery by managers in a supermarket focusing on their expectations based on the data available. Perceptions behind customer purchases are not explicitly considered. We pose a major research question in subjective interestingness: *What is the nature of subjective interestingness among associations of items, in terms of manager's expectations and customers' purchase patterns?*

1 Introduction

Knowledge Discovery in Databases (KDD) is a non-trivial process for identifying valid, new, potentially useful and ultimately understandable patterns in data [20]. The main aim behind this process is to extract and interpret interesting patterns present in

R. Sethi · B. Shekar (✉)
Decision Sciences and Information Systems Area,
Indian Institute of Management Bangalore, Bangalore 560076, India
e-mail: shek@iimb.ac.in

R. Sethi
e-mail: rupal.sethi14@iimb.ac.in

© Springer International Publishing AG, part of Springer Nature 2019
S. Patnaik et al. (eds.), *Digital Business*, Lecture Notes on Data Engineering
and Communications Technologies 21, https://doi.org/10.1007/978-3-319-93940-7_15

large real-life datasets [70]. Thus, KDD is categorised as a *decision support technology* that facilitates managerial decision making [61]. A pattern (or association rule) is interesting if it is valid, new and comprehensive [19], and accurate and comprehensible [13]. Measures to determine whether an association rule is interesting, more known as Interestingness measures, may be categorized into two broad classes—one on the basis of statistical significance and these are termed as *Objective Measures*, and the other on the basis of subjectivity of the user and these are termed as *Subjective Measures*. Many researchers have argued that objective measures are not sufficient to select interesting rules since they lack domain knowledge from the user's point of view [42]. In order to include user's domain knowledge into the KDD process, subjective measures of interestingness are an important part of research.

The notion of subjective interestingness has been largely limited to knowledge discovery by managers in a supermarket typically focusing on their expectations, background knowledge about the data, and sometimes about their goals [28]. Customer-perceptions about product purchase are not explicitly considered while studying subjective interestingness in Association Rule (AR) Mining. It is known that some customers who buy beer also buy diapers. This is known through objective measures such as support and confidence [46]. However the reason for this unexpected purchase does not fall under the purview of any subjective interestingness measure. This chapter would take us one step closer to addressing such issues by enhancing our understanding of subjective interestingness.

Our approach to subjective interestingness compares manager's expectations and user-perceptions through the lens of *affordance*. The concept of affordance originates in ecological psychology. *Affordances* are viewed as relational action possibilities that emerge from interaction between an object and its user [24]. This interaction is contingent on the features of an object and the abilities of a goal-seeking user [54]. In the absence of either of these, affordance may not exist. The current formalization of affordances pertains to defining it as a function of object-features and user-properties [15]. Description of an object is a fairly direct concept that includes its features such as appearance, functionality and design elements. However user characteristics have been viewed from two angles: *property* and *ability*. *User property* [22, 58] points to extrinsic features of an actor such as height, weight and arm length. For example, a flight of steps *affords* climbing to an adult but not to a child. This is due to the ratio between height of the steps and height of an adult [63]. On the other hand, *user ability* [15, 44] points to a user's capabilities shaped through her knowledge, past experience and culture. For example, a social network site such as Facebook *affords* commenting only to someone who is trained in the usage of a computer or a mobile handset. If affordances are relative to user characteristics, an object cannot appear to have the same set of affordances to all actors [33]. For instance, a remote control may be used by someone to control an air conditioner while someone else may use it as a paper weight.

We expand on the current formalization [15] of affordances by introducing a third angle—*the observer's perspective*. Addition of the third angle in our affordance model is important as it includes managerial insights from an interestingness point-of-view. A manager should be able to assess affordances that are actualized by user-object interaction. In a market basket scenario, user-object interaction pertains to decisions made by customers while buying products as a single lot. We include observer's (manager's) *expectations* of user-object interaction into Chemero's formalization of affordance. This is because she might not be aware of the intentions behind buying certain products together. For example, a manager may expect customers buying a carbonated, sweet cola drink to buy salted chips for complementary taste. Instead, they may buy sweet chocolates along with it. This deviation in customer behaviour may provide interesting information to the manager resulting in a greater utility of discovered knowledge.

2 Knowledge Discovery in Databases

In this information age we are exposed to a vast amount of data that gets generated from every transaction that occurs in our routine lives. Be it purchasing groceries at the supermarket, booking movie tickets online, planning a vacation or searching a keyword on Google, everything is a source for huge amount of data. With the advent of the Internet and modern technology, huge chunks of data may be easily generated, stored and retrieved from databases at the click of a button. However there still remains a huge gap between data generation and data comprehension [17]. Research on data comprehension is categorised under the umbrella term *Knowledge Discovery in Databases* (KDD). The KDD process pertains to the discovery and extraction of useful patterns from large amounts of data through various data mining algorithms. Thus knowledge discovery is the transformation of raw data into actionable and useful information (or knowledge). This knowledge is in the form of patterns that stand out from the large pool of data, and provides some useful insights for the stakeholders of the databases. Consider Google Flu Trends[1] (GFT). Analysis of relevant search words and subsequent prediction of the US states that are likely to suffer from Swine Flu was a part of the KDD process. Extraction of keywords used by the US population on Google was done through web mining with the help of various data mining techniques.

2.1 Association Rule Mining

Association Rule Mining (ARM) is one of the important techniques of data mining used in the KDD process. ARM generates patterns from large transactional databases in the form of associations between products. The concept of ARM was initiated by Agrawal and Srikant [4, 5] when they implemented Apriori algorithm to mine association rules from large datasets. The basic problem was to look for frequently co-occurring items and the associations between them in transaction-based databases.

---

[1]Here we comment on the methodology adopted by GFT and not on the failure of the project.

Table 1 Sample purchase transactions

1	Fruits, bread, jam, butter
2	Butter, jam, cereals, milk
3	Toothpaste, jam, fruits
4	Coffee, cereals
5	Milk, bread, butter, jam
6	Coffee, cereals, butter, bread
7	Bread, coffee, jam, butter, milk, cereals
8	Milk, fruits, bread, toothpaste

An association rule is of the form of $A \rightarrow B$ which says that if items in set A are present in the transaction, then items in B are also likely to be present in the same transaction. The most prevalent application of ARM is market basket analysis [3]. Market basket data comprises huge number of transactions corresponding to the purchases made by customers. In a market basket scenario, a typical AR would look like *Bread, Butter $\rightarrow$ Milk*. This says that customers buying bread and butter would buy milk along with it. The antecedent of the rule is Bread and Butter and the consequent is Milk. Two threshold measures used for generation of ARs are support and confidence. *Support* is the probability that both antecedent and consequent are found together in the same itemset, while *confidence* is the probability of an itemset containing the consequent given it contains the antecedent. The statistical definition for both the measures is as follows:

$$support = P(A \cup B) \tag{1}$$

$$confidence = \frac{P(A \cup B)}{P(A)} \tag{2}$$

Consider a snapshot of a database of market transactions in the form of purchase itemsets (Table 1).

Consider the association rule *Bread, Butter $\rightarrow$ Milk*.

$$support = \frac{Number\ of\ transaction\ having\ bread,\ butter\ and\ milk}{Total\ number\ of\ transactions} = \frac{2}{8} = 0.25$$

$$confidence = \frac{Number\ of\ transaction\ having\ bread,\ butter\ and\ milk\ together}{Total\ number\ of\ transactions\ having\ bread\ and\ butter} = 0.5$$

Generation of ARs are done with the help of thresholds specified by the user for both support and confidence. These measures are the basic de facto interestingness measures for ranking association rules. Cardinality of rules increases with decrease in the thresholds.

3 Interestingness

The main aim behind the KDD process is to extract and interpret *interesting* patterns of data from large real-world datasets. *Interestingness* has varied connotations in the field of pattern recognition and data mining. It is quantified through measures that select and rank patterns according to their potential interest to the user [23]. According to Merriam Webster, *interest* is defined as a "feeling that accompanies or causes special attention to an object or class of objects".[2] Interesting patterns evoke strong attention from the user examining the data. The facets of interestingness may be dependent on the domain or on the user. Thus the concept of interestingness is difficult to operationalize and capture. Researchers in the field of knowledge discovery and pattern recognition typically concentrate on facets they consider relevant to the domain of study. Some have conceptualised interesting patterns as accurate and comprehensible [13], while some see it as valid, new and comprehensive [19]. Once interestingness measures have been defined, the goal then is to develop KDD tools that help discover and rank patterns that score high on these measures.

Since applications of ARM are spread across various fields, the notion of *interestingness* also varies with them. What may be interesting for a manager of a supermarket may be entirely useless for a doctor treating chronic diseases. For instance, in market basket analysis, managers tend to look for patterns that are unexpected with respect to their prior knowledge and hence may be acted upon to increase profits for their firms. The standard *beer and diaper* example is an excellent illustration of the notion of interestingness in market application. By mining the purchase data of Walmart it was found that when young fathers went to supermarkets on Fridays, they bought beer along with diapers [46]. This purchase pattern was surprising for the managers of Walmart as a result of which they placed a tray of beer cans along with diapers to increase the sales of both the products collectively. On the other hand, in the field of medicine, along with rare occurrences of patterns (under-expressed genes), frequent associations (over-expressed genes) also matter for doctors as interesting [9]. Thus it would not be appropriate to limit the interestingness measure to a certain threshold since a rare or frequent pattern might be of interest to the user.

Algorithms such as Apriori generate a large number of rules that are often redundant or irrelevant for the domain under study [64]. Standard thresholds for *support* and *confidence* are not sufficient to generate fewer relevant rules. Hence researchers have suggested multiple and relative *support* and *confidence* thresholds for such algorithms [21, 38, 69]. On the other hand, another branch of pattern recognition literature [1, 2, 11] claims that support and confidence are *not statistically robust* measures of interestingness since a low threshold encompasses large number of rules having redundant information, and a high threshold limits the rules to too small a number that only contains required obvious information that may not be interesting. As a result of this gap interestingness measures mainly occur in two broad categories—one on the basis of statistical significance which are the *objective measures*; and the other

---

[2]https://www.merriam-webster.com/dictionary/interest, Accessed on 20th March 2017

on the basis of subjectivity of the user like unexpectedness, novelty and actionability, which are termed as *subjective measures*.

3.1 Objective Interestingness

In KDD and pattern recognition literature the stress is more on how to measure interestingness rather than what is interesting. There are numerous objective interestingness measures that are based only on data in the form of transactions. The rationale behind these objective measures is to statistically validate the ranking of association rules in order to provide most interesting rules [23]. Rules that exceed the threshold value of the objective measure specified by the user are deemed interesting. If two objective measures provide the same ranking to a rule they are fundamentally considered the same [57]. Thus interestingness provided by an objective measure is blurred among the rankings given by it.

There has been extensive work done on **defining** [10, 11, 31, 32, 41, 48, 56], **comparing** [16, 18, 26, 30, 67], **clustering** [25, 29, 32, 45, 57, 60, 67] and **analysing** [23, 27, 67] various objective interestingness measures. Several problems pertaining to objective measures have been identified and dealt with. They span large redundant rule generation, over reliance on support and confidence measures as defacto interestingness measures, and clustering of measures that result in same ranking of rules.

One of the early works on objective interestingness measures was by Piatetsky-Shapiro [48]. They proposed a framework having three properties that characterize "good" association rules. Shapiro calculated the interestingness of deviations in a healthcare information system called KEFIR which may be actionable by the manager to avoid defective output. The first property stated that the interestingness is zero when both the antecedent and the consequent are independent of each other. The second property stated that interestingness increases when occurrences of both the antecedent and consequent together increases. The third property stated that if individual occurrences of antecedent and consequent increase, then the interestingness of the rule will decrease. Other properties were also introduced by various researchers based on association rules [31, 32, 41, 56]. Blanchard et al. [10] proposed that interestingness can also be measured by the amount of information provided by the antecedent about the consequent. They also claimed that among other objective measures, information-theoretic measures are more intelligible and useful.

There are a number of objective measures defined in the literature and recently there has been a rigorous attempt by Tew et al. [57] to minimize some of these measures based on clustering of ranking behaviour. However there is no attempt to conceptually interpret objective measures other than statistically improve them. Many researchers have argued that objective measures are not sufficient to select interesting rules since they lack domain knowledge from the user's point of view [38, 72].

Users play an important role in the interpretation and application of interesting patterns. Thus, it is important that interestingness measures incorporate not only data related aspects but also the views of the user examining the patterns. The importance of subjective interestingness measures is furthered increased since a pattern that is of interest to one user may not be of any interest to another user [52].

3.2 Subjective Interestingness

A subjective interestingness measure takes into account both data and user's knowledge. Such a measure is appropriate when: (1) Background knowledge of users varies, (2) Interests of the users vary and (3) Background knowledge of users evolve [23]. Subjective measures have been studied as three major concepts—unexpectedness, novelty and actionability. A pattern is *unexpected* if it contradicts user's existing beliefs or knowledge [23, 37, 38, 46, 52]. Unexpectedness is a measure of subjective interestingness because it enables a user to rethink about her previous beliefs. A pattern is *novel* if it was not known a priori to the user [23]. There has not been significant work in this direction since novelty is difficult to estimate. Data mining techniques cannot trace novelty through user's knowledge or ignorance due to the limitation of completely knowing the user's prior knowledge. A pattern is *actionable* when it enables decision making about future actions in a particular domain [23, 35, 62]. Since actionability is entirely based on user's discretion of either acting upon the generated rule or not, researchers have not looked at this facet of subjective interestingness through data mining. Geng and Hamilton [23] define semantic measures as the third category of interestingness measures that includes utility and actionability. These measures take into account the additional utility function of the user which reflects her goals for mining the rules. Essentially utility measures are a subset of actionability measures both pointing to user objectives and resultant decision-making from interesting rules. A user might specify a profit function that assigns makes weights to attributes (horizontal weights) or transactions (vertical weights). This leads to the concept of weighted association rule mining based on user's utility function [40, 12].

The three major steps in the study of subjective measures are (1) *knowledge acquisition* of the user in terms of her beliefs, (2) *matching methodology* for comparing generated association rules and user's belief and (3) generation of *interesting rules* that may be unexpected, novel or actionable. Knowledge acquisition of the belief system of the user results in hard beliefs and soft beliefs [52]. A belief is categorised as hard or soft based on a confidence measure assigned to it. This measure is based on the probability that the belief holds given an *evidence*[3] for it. A *hard belief* is one that cannot be changed with new evidence. If an evidence contradicts a hard belief, then the evidence is said to be inaccurate. On the other hand, a *soft belief* is fairly

---

[3]The word "evidence" has its roots in epistemology and philosophy of science. Jaegwon Kim in his book 'What is "Naturalized Epistemology"? defines evidence as justification of a belief.

flexible and may be changed according to new evidence. Silberschatz and Tuzhilin argue that if an evidence contradicts a hard belief of the user, then it is surely a very interesting evidence or an association rule. They consider data that keeps changing with time, and if additional new data results in modification of the belief system, this new data is said to consist of interesting association rules.

Another categorization of belief system is based on the degree of *preciseness* of knowledge of the user [35–38]. Liu et al. [38] separate beliefs into user's vague feeling and precise knowledge. They define three specifications of knowledge—general impressions, reasonably precise concepts and precise knowledge. One may distinguish these based on the keywords used: impression, concept and knowledge. They formally define these categories in terms of taxonomy structure of the data. *General impressions* represent user's vague feelings about associations between classes of items with the direction of association being unknown to the user. *Reasonably precise concepts* indicate associations between classes along with directions. *Precise knowledge* represents exact association between items (not classes) in the proper form of an association rule. Thus, borrowing from both categorizations of beliefs, we divide beliefs into hard or accurate beliefs and soft or general impressions.

Having classified the beliefs based on the granularities of human knowledge, we focus on **matching methodologies** used in the literature to compare the belief system with existing data patterns. We arrive at two categories of methodologies: *constraint-based* and *taxonomy-based*. Constraint-based matching methodologies include statistical matching between a rule and a belief. Taxonomy-based matching methodologies include distance matching between a rule and a belief represented in a tree structure.

Silberschatz and Tuzhilin [52] use Bayesian-based *probabilistic matching* between belief and evidence. Ng et al. [43] use *constraint-based matching* between user-specified constraints about associations between itemsets, and the actual transaction dataset. A similar approach was adopted by Srikant et al. [53] where rules that satisfy user specified constraints were the only rules to be generated.

Another form of constraint-based matching methodology adopted by Padmanabhan and Tuzhilin [46] is *logical contradiction* between a belief and a rule. Rule A → B is unexpected with respect to belief X → Y on the dataset D if (1) B and Y logically contradict each other (2) A and X hold on a large subset of tuples in D. They represent itemsets as discrete variables comprising profiles of customers, demographic details, purchase details, etc. Unexpectedness is brought out only between values of different discrete variables. A sample unexpected rule generated by Padmanabhan and Tuzhilin [46] is "*occupation=professional, household=large → day=weekday*" with respect to manager's belief "*occupation=professional → day=weekend*" (p. 315). Consider an illustration of the logical contradiction approach in terms of market basket data. Let the manager's belief be Bread → Butter, then the algorithm presented by Padmanabhan and Tuzhilin will search for rules where Bread → ~Butter. As a result, a rule Bread → Beer may be termed as unexpected to the belief and hence interesting. There are two major disadvantages with this approach. First, they run the algorithm on a set of predefined beliefs of the user. These beliefs are subject to change and may not be accurate enough for matching. Here the general impressions of the user are not taken into consideration as done in Liu et al. [38]. Next the logical contradiction approach may result in an overestimated set of interesting rules with consequents not matching with the consequent of the belief. In this example, all rules that have bread as antecedent and any consequent apart from butter will be generated as interesting.

Srikant and Agrawal [53] introduced the concept of generalized ARs which initiated researchers defining interestingness measures using the *taxonomy approach*. Savasere et al. [51] use the taxonomy of dataset to mine unexpected ARs. According to them, a rule is interesting if it deviates from the manager's expectation that is based on prior belief. The major assumption based on the taxonomy of items, is called the *uniformity assumption* [51]. It states that items that belong to the same parent in a taxonomy are expected to have similar types of associations. In other words, siblings in a taxonomy are substitutable. For example, if Lays Chips are bought with Pepsi, one expects Lays Chips to be bought with Coke as well. If the actual support of Lays Chips and Coke deviates from the support of Lays Chips and Pepsi, then Lays Chips and Coke generate an unexpected AR.

A similar approach to uniformity assumption [51] has been adopted by Yuan et al. [68]. They also use the concept of *locality of similarity* in defining sibling rules from the taxonomy. Sibling rules are a pair of positive association rules where both the siblings are expected to be related to the same consequent. For example, if Pepsi → Lays is an AR that is generated through apriori algorithm [4], then Coke → Lays should also be generated. Unexpectedness is captured through the difference in confidence values of the rules Pepsi → Lays and Coke → Lays.

Liu et al. [38] on the other hand address the notion of varied granularities of user's knowledge. The matching methodology they used was based on *syntactic distance measure* between a rule and a belief. Resulting rules will be conforming or unexpected with respect to a belief on the basis of the distance measure calculated from the taxonomy. If the distance between the antecedent and the consequent of a rule and that with respect to a belief is less than a threshold, then the rule is said to conform to the belief; otherwise it is deemed unexpected from the antecedent, consequent or both sides.

4 Classification Framework for Subjective Interestingness

We propose a two-dimensional classification framework for studying subjective interestingness in association rule mining, presented in Fig. 1. The horizontal axis represents the *matching methodology*—constrained-based and taxonomy-based. The vertical axis highlights the *granularity* of user's knowledge or belief system as hard or accurate and soft or general impressions. We fit the previous work done on subjective interestingness in the proposed classification framework. This is given in Fig. 2. Constraint-based methodology mainly focussed on hard beliefs. We review three major categories of work that were based on constraint-based methodology

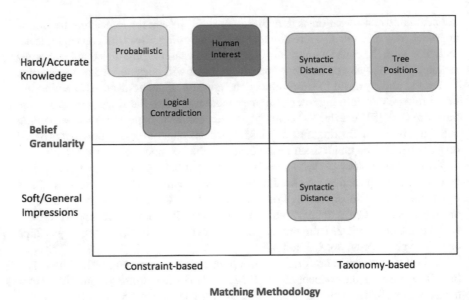

Fig. 1 Two-dimensional classification framework for subjective interestingness

and hard beliefs: probabilistic approach, logical contradiction and human interest. All of them are in the form of constraints. The probabilistic matching approach comprise Bayesian analysis, correlation and support based pruning of association rules [31, 50, 52, 56]. The approach based on Human interest is essentially user-specified constraints using weights [40, 39], optimization functions [7, 43, 47] and utility functions [14, 53, 66]. Research on logical contradiction [34, 46, 59, 71] is also based on matching constraints. Here hard beliefs are used to generate interesting rules. Iterative algorithms are used for the refinement of belief systems. Hence the user may be able to use accurate and reasonably precise concepts for matching.

Under taxonomy-based matching, we further classify the work on the basis of *syntactic distance* and *tree positions*. Syntactic distance approach comprises research on Lexical Analysis [8] and Natural Language Processing [49] and these are based on hard beliefs only. It is interesting to note that the methodology adopted by Liu et al. [38] and Raghavan and Mooney [49] spans across both kinds of beliefs, hard and soft. Work based on matching tree positions in the taxonomy with hard beliefs is placed at the top right corner of the framework [6, 51, 53, 55, 68, 65]. It considers sibling substitution behaviour for matching a rule and a belief.

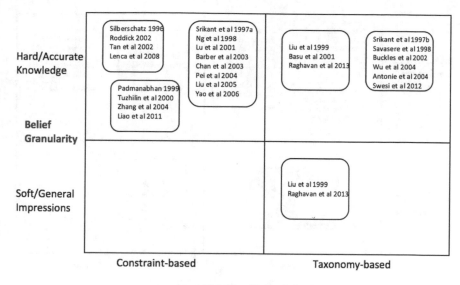

Fig. 2 Positioning of previous literature in the classification framework

5 Discussion and Conclusions

The notion of subjective interestingness has been limited to knowledge discovery by managers in a supermarket, typically focusing on their expectations or background knowledge about the data [28]. Customer perceptions about purchasing a product are not explicitly considered while studying subjective interestingness in AR Mining. For example, we know that some users who buy beer also buy diapers. This discovery was a result of objective measures such as support and confidence [46]. However, apart from being unexpected with respect to a manager's prior belief, the reason for this unexpected purchase will still be unknown and does not fall in the purview of any subjective interestingness measure.

The theoretical framework of literature on subjective interestingness presented in Fig. 2 highlights that there has not been sufficient work towards amalgamating constraint-based matching methodology with the general impressions of a manager. Therefore we pose a major research question that needs to be investigated in the area of AR mining and subjective interestingness: *What is the nature of subjective interestingness among associations of items, in terms of manager's expectations and customers' purchase patterns?*

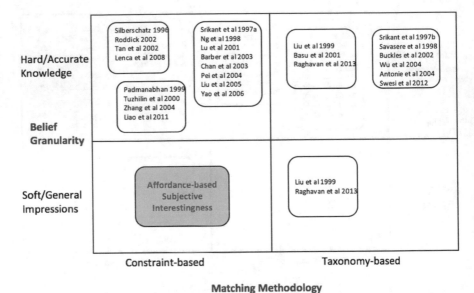

Fig. 3 Positioning the research question in the context of previous literature

This question may be addressed by enhancing the knowledge of subjective interestingness. We propose that the approach to subjective interestingness should compare manager's expectations and users' perceptions. Manager's expectations are tapped in terms of her beliefs in the form of general impressions (soft beliefs). Customer perceptions in the form of constraint-based matching are added to the analysis of subjective interestingness. This is presented in Fig. 3.

The motivation behind addressing this research gap comes from the differential nature of interestingness. What may be interesting for a manager, may not be interesting for a customer. A manager might view interestingness in accordance to expectations based on her prior knowledge since she is unaware of the user's immediate goals. For example, a manager might expect a customer to buy butter along with bread. However, some might buy butter along with mechanical hardware. Manager's beliefs point to butter's features of oral consumption with bread. However customers' perceptions may point towards using butter as a substitute to grease when the latter may be expensive or unavailable in small quantity. Thus butter because of its oily nature is bought along with mechanical hardware. Thus there is a contradiction between manager's expectations, and user intentions behind the purchase.

Hence we open new venues for future researchers to look at this facet of subjective interestingness and possibly help mine reasons along with interesting patterns of data.

References

1. Adamo JM (2001) Data mining for association rules and sequential patterns: sequential and parallel algorithms. Springer, New York
2. Aggarwal CC, Yu PS (1998) Mining large itemsets for association rules. IEEE Data Eng. Bull 21(1):23–31
3. Aggarwal CC (2015) Data mining: the textbook. Springer
4. Agrawal R, Srikant R (1994) Fast algorithms for mining association rules. In Proceedings 20th international conference very large data bases, VLDB, 1215, pp 487–499
5. Agrawal R, Imieliński T, Swami A (1993) Mining association rules between sets of items in large databases. ACM SIGMOD Rec 22(2):207–216
6. Antonie ML, Zaïane OR (2004) Mining positive and negative association rules: an approach for confined rules. European conference on principles of data mining and knowledge discovery. Springer, Berlin, Heidelberg, pp 27–38
7. Barber MJ, Clark JW (2009) Detecting network communities by propagating labels under constraints. Phys Rev E 80(2):026129
8. Basu S, Mooney RJ, Pasupuleti KV, Ghosh J (2001) Evaluating the novelty of text-mined rules using lexical knowledge. In: Proceedings of the seventh ACM SIGKDD international conference on knowledge discovery and data mining. ACM, pp 233–238
9. Becquet C, Blachon S, Jeudy B, Boulicaut JF, Gandrillon O (2002) Strong-association-rule mining for large-scale gene-expression data analysis: a case study on human SAGE data. Genome Biol 3(12), research0067-1
10. Blanchard J, Guillet F, Gras R, Briand H (2005) Using information-theoretic measures to assess association rule interestingness. In: Fifth IEEE international conference on data mining, 8
11. Brin S, Motwani R, Silverstein C (1997) Beyond market baskets: generalizing association rules to correlations. AcmSigmod Rec 26(2):265–276 ACM
12. Cai CH, Fu AWC, Cheng CH, Kwong WW (1998) Mining association rules with weighted items. In: Database engineering and applications symposium, 1998. Proceedings. IDEAS'98. International. IEEE, pp 68–77
13. Carvalho DR, Freitas AA, Ebecken N (2005) Evaluating the correlation between objective rule interestingness measures and real human interest. European conference on principles of data mining and knowledge discovery. Springer, Berlin, Heidelberg, pp 453–461
14. Chan R, Yang Q, Shen YD (2003) Mining high utility itemsets. In: Third IEEE international conference on data mining, 19–26
15. Chemero A (2003) An outline of a theory of affordances. Ecol Psychol 15(2):181–195
16. Chen MS, Han J, Yu PS (1996) Data mining: an overview from a database perspective. IEEE Trans Knowl Data Eng 8(6):866–883
17. Cios KJ, Pedrycz W, Swiniarski RW (1998) Data mining and knowledge discovery. Data mining methods for knowledge discovery. Springer, US, pp 1–26
18. Duda R, Gaschnig J, Hart P (1979) Model design in the PROSPECTOR consultant system for mineral exploration. In: Expert systems in the microelectronic age, vol 1234, pp 153–167
19. Fayyad U, Piatetsky-Shapiro G, Smyth P (1996) From data mining to knowledge discovery in databases. AI Mag 17(3):37
20. Frawley WJ, Piatetsky-Shapiro G, Matheus CJ (1992) Knowledge discovery in databases: an overview. AI Mag 13(3):57
21. Freitas AA (1999) On rule interestingness measures. Knowl-Based Syst 12(5):309–315
22. Galvao AB, Sato K (2005) Affordances in product architecture: linking technical functions and users' tasks. In: ASME 2005 international design engineering technical conferences and computers and information in engineering conference. American Society of Mechanical Engineers, pp 143–153
23. Geng L, Hamilton HJ (2006) Interestingness measures for data mining: a survey. ACM Comput Surv (CSUR) 38(3):9
24. Gibson JJ (1977) Perceiving, acting, and knowing: toward an ecological psychology. In: The theory of affordances, pp 67–82

25. Huynh XH, Guillet F, Briand H (2005) A data analysis approach for evaluating the behavior of interestingness measures. Discovery science. Springer, Berlin, Heidelberg, pp 330–337
26. Kamber M, Shinghal R (1996) Evaluating the Interestingness of characteristic rules. In: KDD, pp 263–266
27. Kannan S, Bhaskaran R (2009) Association rule pruning based on interestingness measures with clustering. arXiv:0912.1822
28. Kontonasios KN, Spyropoulou E, De Bie T (2012) Knowledge discovery interestingness measures based on unexpectedness. Wiley Interdiscip Rev: Data Min Knowl Discov 2(5):386–399
29. Lallich S, Teytaud O, Prudhomme E (2007) Association rule interestingness: measure and statistical validation. In: Quality measures in data mining. Springer, Berlin, Heidelberg, pp 251–275
30. Lavrač N, Flach P, Zupan B (1999) Rule evaluation measures: a unifying view. Springer, Berlin, Heidelberg, pp 174–185
31. Lenca P, Meyer P, Vaillant B, Lallich S (2008) On selecting interestingness measures for association rules: user oriented description and multiple criteria decision aid. Eur J Oper Res 184(2):610–626
32. Lenca P, Vaillant B, Meyer P, Lallich S (2007) Association rule interestingness measures: experimental and theoretical studies. Quality measures in data mining. Springer, Berlin, Heidelberg, pp 51–76
33. Leonardi PM (2013) When does technology use enable network change in organizations? A comparative study of feature use and shared affordances. Manag Inf Syst Q 37(3):749–775
34. Liao SH, Chen YJ, Lin YT (2011) Mining customer knowledge to implement online shopping and home delivery for hypermarkets. Expert Syst Appl 38(4):3982–3991
35. Ling CX, Chen T, Yang Q, Cheng J (2002) Mining optimal actions for profitable CRM. In: IEEE international conference on data mining, pp 767–770
36. Liu B, Hsu W (1996) Post-analysis of learned rules. AAAI/IAAI 1:828–834
37. Liu B, Hsu W, Chen S (1997) Using general impressions to analyze discovered classification rules. In: KDD, pp 31–36
38. Liu B, Hsu W, Ma Y (1999) Mining association rules with multiple minimum supports. In: Proceedings of the fifth ACM SIGKDD international conference on knowledge discovery and data mining, pp 337–341
39. Liu DR, Shih YY (2005) Integrating AHP and data mining for product recommendation based on customer lifetime value. Info Manag, 42(3):387–400
40. Lu S, Hu H, Li F (2001) Mining weighted association rules. Intell Data Anal 5(3):211–225
41. Major JA, Mangano JJ (1995) Selecting among rules induced from a hurricane database. J Intell Inf Syst 4(1):39–52
42. McGarry K (2005) A survey of interestingness measures for knowledge discovery. Knowl Eng Rev 20(01):39–61
43. Ng RT, Lakshmanan LV, Han J, Pang A (1998) Exploratory mining and pruning optimizations of constrained associations rules. ACM SIGMOD Rec 27(2):13–24
44. Norman DA (2013) The design of everyday things: revised and expanded edition. Basic Books
45. Ohsaki M, Kitaguchi S, Yokoi H, Yamaguchi T (2005) Investigation of rule interestingness in medical data mining. Active mining. Springer, Berlin, Heidelberg, pp 174–189
46. Padmanabhan B, Tuzhilin A (1999) Unexpectedness as a measure of interestingness in knowledge discovery. Decis Support Syst 27(3):303–318
47. Pei J, Han J, Lakshmanan LV (2004) Pushing convertible constraints in frequent itemset mining. Data Min Knowl Disc 8(3):227–252
48. Piatetsky-Shapiro G, Matheus CJ (1994) The interestingness of deviations. In: Proceedings of AAAI workshop on knowledge discovery in databases
49. Raghavan S, Mooney RJ (2013) Online inference-rule learning from natural-language extractions. In: AAAI workshop: statistical relational artificial intelligence
50. Roddick JF, Spiliopoulou M (2002) A survey of temporal knowledge discovery paradigms and methods. IEEE Trans Knowl Data Eng 14(4):750–767

51. Savasere A, Omiecinski E, Navathe S (1998) Mining for strong negative associations in a large database of customer transactions. In: 14th IEEE international conference on data engineering, pp 494–502
52. Silberschatz A, Tuzhilin A (1996) What makes patterns interesting in knowledge discovery systems. IEEE Trans Knowl Data Eng 8(6):970–974
53. Srikant R, Vu Q, Agrawal R (1997) Mining association rules with item constraints. KDD 97:67–73
54. Stoffregen TA (2003) Affordances as properties of the animal-environment system. Ecol Psychol 15(2):115–134
55. Swesi IMAO, Bakar AA, Kadir ASA (2012) Mining positive and negative association rules from interesting frequent and infrequent itemsets. In: 9th IEEE international conference on fuzzy systems and knowledge discovery (FSKD), pp 650–655
56. Tan PN, Kumar V, Srivastava J (2002) Selecting the right interestingness measure for association patterns. In: Proceedings of the eighth ACM SIGKDD international conference on Knowledge discovery and data mining, pp 32–41
57. Tew C, Giraud-Carrier C, Tanner K, Burton S (2014) Behavior-based clustering and analysis of interestingness measures for association rule mining. Data Min Knowl Disc 28(4):1004–1045
58. Turvey MT (1992) Affordances and prospective control: an outline of the ontology. Ecol Psychol 4(3):173–187
59. Tuzhilin A, Adomavicius G (2002) Handling very large numbers of association rules in the analysis of microarray data. In: Proceedings of the eighth ACM SIGKDD international conference on Knowledge discovery and data mining, pp 396–404
60. Vaillant B, Lenca P, Lallich S (2004) A clustering of interestingness measures. Discovery science. Springer, Berlin, Heidelberg, pp 290–297
61. Wang H (1997) Intelligent agent-assisted decision support systems: integration of knowledge discovery, knowledge analysis, and group decision support. Expert Syst Appl 12(3):323–335
62. Wang K, Tang L, Han J, Liu J (2002) Top down fp-growth for association rule mining. In: Pacific-Asia conference on knowledge discovery and data mining. Springer, Berlin, Heidelberg, pp 334–340
63. Warren WH (1984) Perceiving affordances: visual guidance of stair climbing. J Exp Psychol Hum Percept Perform 10(5):683
64. Wei JM, Yi WG, Wang MY (2006) Novel measurement for mining effective association rules. Knowl-Based Syst 19(8):739–743
65. Wu ST, Li Y, Xu Y, Pham B, Chen P (2004) Automatic pattern-taxonomy extraction for web mining. In: Proceedings of the 2004 IEEE/WIC/ACM international conference on web intelligence. IEEE Computer Society, pp 242–248
66. Yao H, Hamilton HJ (2006) Mining itemset utilities from transaction databases. Data Knowl Eng 59(3):603–626
67. Yao YY, Zhong N (1999) An analysis of quantitative measures associated with rules. Methodologies for knowledge discovery and data mining. Springer, Berlin, Heidelberg, pp 479–488
68. Yuan X, Buckles BP, Yuan Z, Zhang J (2002) Mining negative association rules. In: Proceedings of Seventh International Symposium on Computers and Communications, pp 623–628
69. Yun H, Ha D, Hwang B, Ryu KH (2003) Mining association rules on significant rare data using relative support. J Sys Soft, 67(3):181–191
70. Zhang C, Zhang S (2002) Association rule mining: models and algorithms. Springer
71. Zhang H, Padmanabhan B, Tuzhilin A (2004) On the discovery of significant statistical quantitative rules. In: Proceedings of the tenth ACM SIGKDD international conference on knowledge discovery and data mining, pp 374–383
72. Zhong N, Yao YY, Ohishima M (2003) Peculiarity oriented multidatabase mining. IEEE Trans Knowl Data Eng 15(4):952–960

Identifying Sentiment of Malayalam Tweets Using Deep Learning

S. Sachin Kumar, M. Anand Kumar and K. P. Soman

Abstract The current chapter focus on providing a comparative study for identifying sentiment of Malayalam tweets using deep learning methods such as convolutional neural net (CNN), long short-term memory units (LSTM). The baseline methods used to compare are support vector machines (SVM), regularized least square classification with random kitchen sink mapping (RKS-RLSC). Malayalam is a low resource language spoken in Kerala state, India. Due to the unavailability of data, tweets were collected and labeled manually based on its polarity as neutral, negative and positive. RKS mapping is a well explored approach in which data are nonlinearly mapped to higher dimension where linear classifier can be used. The evaluation measure chosen for the experiments are F1-score, recall, accuracy and precision. The experiments also provide a comparison with classical methods such as logistic regression (LR), adaboost (Ab), random forest (RF), decision tree (DT), k-nearest neighbor (KNN) on the basis of accuracy as the measure. For the experiments using CNN and LSTM, we report the effectiveness of activation functions such as rectified linear units (ReLU), exponential linear units (ELU) and scaled exponential linear units (SELU) for the sentiment identification of Malayalam tweets over SVM and RKS-RLSC.

1 Introduction

Sentiment analysis is a key text analytics task. The sentiment expressed in text data form contains information related to happiness, sadness, anger, love etc., carries opinions regarding events related to politics, sports, movies, products and services etc. In other words, text data become the source of several opinions. The internet is everyday flooded with new data in billions [1] in the form of text, images, audio/video etc. Not just data, meta data's are also generated. The text data exist as (1) structured (organized manner), (2) unstructured. These text data are favorites

S. Sachin Kumar (✉) · M. Anand Kumar · K. P. Soman
Centre for Computational Engineering and Networking, Amrita Vishwa Vidyapeetham,
Coimbatore, India
e-mail: sachinnme@gmail.com

© Springer International Publishing AG, part of Springer Nature 2019
S. Patnaik et al. (eds.), *Digital Business*, Lecture Notes on Data Engineering
and Communications Technologies 21, https://doi.org/10.1007/978-3-319-93940-7_16

for companies, researchers, organizations, government as it will have vital sug-
gestions/recommendations related to the products/movies/events etc. The advent
of social media sites (Facebook, google+, twitter etc.) gave different platforms to
share opinions in web or to initiate discussions. The users of these sites express
their opinions in their mother tongue (native language) or English. Such data will
contain reviews and opinions regarding products, politics, sports, entertainment etc.
With the choice of language the opinions can be conveyed. Massive users are there
in the present scenario who uses social media. This mode communication brought
a change in the conventional simplex way of exchange of information. This mode
reaches much faster to masses. This made it favorite to many researchers, companies,
governments etc. This will help the data scientist to develop insight or extract information about any trend which can acts as aid towards, making any decisions. Even
though the data accommodates such useful information, it also poses challenges for
the analysis like missing data, breaks or incomplete text sequences, detection of temporal patterns, incorrect grammars, mistakes in spellings etc. Twitter is one among
the popular social media service commonly in use. Mining the tweets can provide
general views or opinions on how people respond to any event, crisis, change in
trends, present topic of discussions etc.

Extensive work is going on in the sentiment analysis of text in English language
[1–4]. The analysis becomes more challenging when the tweets are in Non-English
languages. This motivates the importance of carrying out research in the sentiment
analysis of text existing in native languages. In this work, we took twitter data in
Malayalam language (native language of Kerala State, India), a corpus containing
13,000 tweets and manually labeled each based on its sentiment/polarity. The tweets
collected contains opinions related to cuisine, places, political events, movie reviews
etc. Some examples are shown below, written in phonetic form.

- kEraLaththile ONasadya… lOkam aRiyappe-Tunna vibhavamANu
- kANan aTipoLi sthalangngnaL mUnARiluNTu
- kOyikkOTan biriyANI…supparrrr
- chAya kuTikkunnathu SarIraththinu hAnIkaramANu
- ayyO… ivan lApuTOpu mOshTiCCatu kaNTO…?? kaLLan ANEngkilum samma
  tiCCu mOne!
- viTTile kASu aTiCCumARRi nATu viTTallO
- ishTappeTTAl…oru…laikku…I…paTaththinum…I…pEjinum

The current work provides a comparison on polarity identification using machine
learning methods such as support vector machine (SVM), regularized least method
via random mapping, and deep learning methods such as convolutional neural nets
(CNN), long short-term memory units (LSTM).

2 Related Work

Few research works based on lexicon, senti-wordnets, corpora of emotions, polarity, subjectivity word-list, Taboada's adjective list, WordNet-Affect list are proposed for languages like English, Hindi, Chinese, Bengali, Japanese, Telugu, Tamil etc. [5–14]. In [15–17] proposed different methods to use dictionaries relating sentiment words. Authors used machine learning approaches like maximum entropy (ME), support vector machines (SVM), Naive Bayes (NB). In their experiments, different feature such as unigram, bigrams, unigram frequency, combinations, unigram with POS tags adjectives etc are used.

Extensive work are undergoing for sentiment analysis (SA) in English, Japanese, Chinese, Spanish, German etc. Similar to this, several researches are undergoing for SA in native Indian languages like Marathi, Bengali, Hindi, Manipuri, Punjabi, Malayalam, Tamil etc. Due to the difficulty and complexity involved in processing text in Indian languages, very few system only exists for SA. The research work in this direction demands lexicon dictionaries, taggers, labeled corpus, parsers, senti-wordnets etc. The problem of sentiment analysis in semantically disambiguation is discussed in [18]. The authors proposed a framework for SA in Hindi language using senti-wordnet in Hindi. The authors improves the senti-wordnet by adding more senti-words. The missing words undergoes translation to English first, and then looked in English-senti-wordnet list to retrieve its polarity. In [19], the authors used annotated corpus of movie reviews in Hindi. The paper discusses on updating the senti-wordnet for Hindi language by adding opinion words and put-forth new procedures to handle relations, negation in discourse. A strategy for SA was proposed in [20]. The authors proposed three approaches; (1) a machine learning model created with labeled corpora of movie reviews in Hindi. (2) translating the Hindi text content to English and classify the sentiments. (3) a score based strategy along with Hindi senti-wordnet for classifying sentiment. In [21], proposed ways to create senti-wordnets for Indian Languages with WordNet, corpus, dictionary. For validation, an interactive game was created and made online. In [22–24], a survey is provided on SA and opinion mining for text in Hindi. SA of movie review for text in Tamil is discussed in [25]. The feature used was frequency count. The data collected was hand-tagged into positive, negative. The paper compares different machine learning algorithms such as logistic regression (LR) Naive Bayes, random kitchen sink (RKS), support vector machines to classify movie reviews in Tamil. Mohandas et al. [26] is an initial work in SA of text in Malayalam at sentence level. The authors proposed a method to get the positive and negative emotion response. A word list which contains the positive, negative emotion responses were prepared and used for unsupervised classification.

3 Materials and Methods

3.1 Support Vector Machines

SVM is a supervised machine learning based method which learns a linear classifier (hyper plane) from the data. For a given dataset for training where is the number of data points, denotes the class labels, the well known objective function of linear svm in primal form is defined as

$$\min \frac{1}{n}\|w\|^2 + \lambda \sum_{i=1}^{n} e_i$$
$$st : y_i(w \cdot x_i - b) \geq 1 - e_i$$
$$e_i \geq 0 \qquad (1)$$

w denotes weight vector, e denotes error and λ denotes control parameter.

3.2 Random Mapping: Random Kitchen Sink Method

Random kitchen sink algorithm (RKS) is a new machine learning algorithm for classification of non-linearly separated data set. The advantage over conventional non-linear kernel method is that the method is suitable for learning and classifying large data sets. When large data set is involved in training non-linear SVM, usually large proportion of data points become support vectors and hence it must be stored for classifying new data points. This not only demands more space but also more time for classification tasks. In this context, random kitchen sink algorithm is a suitable alternative. The space and time requirement is independent of the number of data points. It depends only on feature size, which in most practical cases is less than hundred. Ultimately, we obtain an explicit feature mapping $\phi(x)$ corresponding to radial basis function (RBF) kernel. This in combination with regularized least square algorithm for regression allows us to obtain a simple classifier that can be used for real time applications. A nonlinear mapping which maps the input feature space into a compact random Fourier feature space using a randomized map $z : \Re^d \to \Re^D$ is discussed in [27]. The kernel for mapping is defined as

$$ke(x, y) = \langle \phi(x), \phi(y) \rangle \approx z(x)^T z(y) \qquad (2)$$

ϕ denotes the implicit mapping as in SVM, whereas z denotes the explicit mapping. However, explicit mapping suffers from high computation cost when dealing with large sized matrices. A solution to this is advised in [27] and used for sentiment analysis in [28]. z is defined as a nonlinear mapping using Gaussian kernel. For a given data pair x and y, the kernel is defined as,

$$ke(x, y) = e^{-\gamma \|x-y\|^2} = e^{-\gamma (x-y)^T (x-y)} \tag{3}$$

Here $\gamma = 1/2\sigma^2$. As the Gaussian kernel is translation invariant, the Fourier transform of the kernel can be interpreted as expectation of $e^{j(x-y)^T \Omega}$. For a given data pair (x_1, x_2), let $x_1 - x_2 = z$, then the kernel function can be written as, $f(z) = e^{\frac{-1}{2} z^T \Sigma^{-1} z}$. Let $F(\Omega)$ denotes Fourier transform of $f(z)$.

$$F(\Omega) = \frac{1}{2\pi} \int\limits_{-\infty}^{\infty} f(z) e^{-jz^T \Omega} dz \tag{4}$$

$f(z)$ is Gaussian, hence $F(\Omega)$ is again Gaussian. To be precise, it is a multivariate Gaussian function with variance. We interpret $F(\Omega)$ as a Gaussian (multivariate) density function. Now,

$$F^{-1}(\Omega) = \langle \phi(x_1), \phi(x_2) \rangle = \int\limits_{-\infty}^{\infty} F(\Omega) e^{jz^T \Omega} dz \tag{5}$$

This can be interpreted as expected value of $e^{jz^T \Omega}$. That is,

$$E(e^{jz^T \Omega}) = \int\limits_{-\infty}^{\infty} F(\Omega_i) e^{jz^T \Omega} dz \tag{6}$$

The expected value of any function of random variable (here Ω) can be obtained by taking several independent samples from the associated probability density function and find its average. Since z is n-tuple, Ω is n-tuple and one particular vector Ω_i can be easily generated. For example, in Matlab by randn(n, 1) command, as many Ω_i can be generated and compute $e^{jz^T \Omega}$ for given z, then take average to obtain the required expected value. However, our aim is to obtain a generic expression for $\phi(x)$. So we proceed further to achieve the task as,

$$E(e^{jz^T \Omega}) = \frac{1}{k} \sum_{i=1}^{k} e^{jz^T \Omega_i} = \frac{1}{k} \sum_{i=1}^{k} e^{j(x-y)^T \Omega_i}$$

$$= \frac{1}{k} \sum_{i=1}^{k} e^{jx^T \Omega_i} \overline{e^{jy^T \Omega_i}}$$

$$= \langle \phi(x_1), \phi(x_2) \rangle = Ke(x_1, x_2) \tag{7}$$

$$Ke(x_1, x_2) = \frac{1}{k} \begin{bmatrix} e^{j(x_1-x_2)^T \Omega_1} \\ e^{j(x_1-x_2)^T \Omega_2} \\ . \\ e^{j(x_1-x_2)^T \Omega_k} \end{bmatrix}$$

$$= \frac{1}{k} \left\langle \begin{bmatrix} e^{jx_1^T \Omega_1} \\ e^{jx_1^T \Omega_2} \\ . \\ e^{jx_1^T \Omega_k} \end{bmatrix}, \begin{bmatrix} e^{jx_2^T \Omega_1} \\ e^{jx_2^T \Omega_2} \\ . \\ e^{jx_2^T \Omega_k} \end{bmatrix} \right\rangle$$

$$= \left\langle \begin{bmatrix} \sqrt{1/k} e^{jx_1^T \Omega_1} \\ \sqrt{1/k} e^{jx_1^T \Omega_2} \\ . \\ \sqrt{1/k} e^{jx_1^T \Omega_k} \end{bmatrix}, \begin{bmatrix} \sqrt{1/k} e^{jx_2^T \Omega_1} \\ \sqrt{1/k} e^{jx_2^T \Omega_2} \\ . \\ \sqrt{1/k} e^{jx_2^T \Omega_k} \end{bmatrix} \right\rangle \tag{8}$$

In brief,

$$FT(ke(x, y)) = FT(ke(x, y))$$
$$= FT(ke(x, y))$$
$$= E\left(e^{j(x-y)^T \Omega} \right)$$
$$= \begin{bmatrix} \frac{1}{\sqrt{D}} e^{jx^T \Omega_1} \\ \frac{1}{\sqrt{D}} e^{jx^T \Omega_2} \\ . \\ \frac{1}{\sqrt{D}} e^{jx^T \Omega_D} \end{bmatrix} \begin{bmatrix} \frac{1}{\sqrt{D}} e^{jy^T \Omega_1} \\ \frac{1}{\sqrt{D}} e^{jy^T \Omega_2} \\ . \\ \frac{1}{\sqrt{D}} e^{jy^T \Omega_D} \end{bmatrix}$$
$$= (z(x)^T z(y))$$
$$= ke(x, y) \tag{9}$$

Ω denotes the random variable for Gaussian distribution, denotes the dimension of random vector. The authors [27] proves that the vectors obtained using this approach can be linearly separated.

3.3 Regularized Least Square Method for Classification (RLSC)

This method finds the labels of the test data using a weight matrix obtained using the objective function

$$\min_{W \in \mathbb{R}^{n \times T}} \left\{ \frac{1}{n} \|Y - WX\|_F^2 + \lambda \|W\|_F^2 \right\} \qquad (10)$$

where Y is the class label matrix, X is the data matrix and W is the weight matrix. The objective function acts as the trade-off between minimizing the sum of errors and an approximate weight matrix. The minimum of the objective function is obtained by using the property of trace as follows,

$$\frac{1}{n} Tr \|Y - WX\|_F^2 + \lambda Tr \|W\|_F^2$$

$$= Tr \begin{pmatrix} Y^T Y - Y^T X W - W^T X^T Y \\ + W^T X^T X W + \lambda W^T W \end{pmatrix} \qquad (11)$$

Now, differentiating with respect to W and equating it to 0 we get,

$$\frac{\partial}{\partial W} (Tr(Y^T Y - Y^T X W - W^T X^T Y + W^T X^T X W + \lambda W^T W)) = 0$$

$$\Rightarrow -X^T Y - X^T Y + 2X^T X W + 2\lambda W = 0$$

$$\Rightarrow 2(X^T X + \lambda I)W = 2X^T Y \qquad (12)$$

$$W = (X^T X + \lambda I)X^T Y \qquad (13)$$

In the testing phase, the test signal is projected onto the weight matrix obtained in the training phase. The difference between the predicted labels and true labels are to be made minimum for accurate classification.

3.4 Deep Learning Methods—CNN, LSTM

Deep learning has become a vital approach in machine learning. It has the advantage of learning rich feature representation from data which avoids the feature selection process in traditional machine learning methods. Deep learning methods such as CNN, LSTM has shown promising results in speech and image processing, and tasks in NLP. In [29], the authors showed that language modelling using recurrent neural nets (RNN), outperformed feed-forward method as RNN captured the temporal dynamics in tasks of sequential form. However, training RNN via back-propagation is hard as during the process to capture the long-dependencies, the gradients can decays/vanish or explode [30]. In other words, RNN can keep representations of recent events (or information) as short-term memories obtained via feedback connections. This short-coming is resolved by the variants of RNN such as LSTM, GRU (gated recurrent unit), BLSTM (bidirectional LSTM). The current work only makes use of LSTM method for experiment.

3.4.1   Convolutional Neural Net-1D

CNN method is a popular deep learning approach in computer vision. The idea of CNN is adopted from the convolution operation in time-domain. Consider a discrete function of the form $f(x) \in R^l$ and another discrete function or kernel function $g(x) \in R^d$, then the convolution operation is defined as

$$h_c(y) = \sum_1^d f(x) \cdot g(y \cdot d - x + c) \tag{14}$$

where $c = d - s + 1$ is an offset and s is the stride. This operation is adopted in deep learning and the kernel function is learned from the data. The kernel function or the filter is known as weights and is estimated. It has obtained competing results in text processing tasks such as text classification, entity recognition, semantic role labeling, etc. In the current work, 1-dimensional CNN is used. Let $D = \{c_1, c_2, \ldots, c_l\}$ denote the sequence of tokens and l denotes the number of words in the sequence. V denotes the vocabulary of words, d is the dimensionality of embedding for words, $V_D \in R^{d \times l}$. The objective is to perform 1-dimensional convolution and learn filters (or kernel functions) specific to the data $f \in R^{dxp}$, where p is the window of characters to form feature map f_m over the window of tokens $V[*, k : k + p]$. The feature map thus obtained is operated by activation function known as rectified linear units (ReLU) $f_m(x) = \max(0, x)$. This is an element-wise operation. The non-linear function allows to learn a decision boundary with non-linear nature. After obtaining feature maps, pooling operation is performed to down sample the size of feature map. The most commonly used pooling approaches are average and max pooling. In the current work max pooling is used which returns the maximum value by operating on the feature maps. The window size used for pooling is 2. CNN approach learns the features by itself, eliminating the difficult task of finding the right features. CNN efficiently exploits the locality information via convolution operation applied on to the input vector (one-hot, vector indices etc.). In this paper, the tokens are mapped to their corresponding indices in the V. The output of the pooling operation is given to a fully connected softmax layer. This layer calculates the probability distribution which gives information about the labels.

$$soft \max(x^T w + b) = \frac{\exp(x^T w_i + b_i)}{\sum_{j=1}^J \exp(x^T w_j + b_j)} \tag{15}$$

where w_i, b_i corresponds to the weight vector and bias of jth class respectively.

3.4.2   Long Short-Term Memory Units (LSTM)

LSTM is also a variant of RNN with gating mechanism proposed in 1997 [31] to capture the long dependencies or to keep the information from longer time-steps.

It can be viewed as a single block known as a memory cell with two inputs to it—previous hidden state and current input. Internally in the memory cell, it decides on what information needs to persist or not. LSTM has gating mechanism to deal with vanishing gradients and it is another technique to compute hidden state vectors. LSTM also provides a different way to compute hidden state (will be vector). The basic equation to compute the hidden state at tth time-step is

$$h_t = f(Ax_t + Wh_{t-1}) \tag{16}$$

where f is generally a function with non-linear nature (activation function) like ReLU (rectified linear units), tanh() etc. The input to the function is current input x_t and previous hidden state h_{t-1} at time-steps t and $t-1$. LSTM is computed in a different way using three different gates, namely input gate, forget gate and output gate. The input gate i_t decides how much of the information needs to be passed.

$$i_t = \sigma(A^i x_t + W^i h_{t-1}) \tag{17}$$

$$fg_t = \sigma(A^{fg} x_t + W^{fg} h_{t-1}) \tag{18}$$

$$o_t = \sigma(A^o x_t + W^o h_{t-1}) \tag{19}$$

$$\hat{c}_t = \tanh(A^c x_t + W^c h_{t-1}) \tag{20}$$

$$c_t = \hat{c}_t^o i_t + c_{t-1}^o fg_t \tag{21}$$

$$h_t = o_t^o \tanh(c_t) \tag{22}$$

The forget gate fg_t decides how much of the currently computed hidden state need to be passed. The output gate o_t is useful when there are multiple layers as it decides how much of the information in internal state need to be passed. The state of the cell at time-step t is denoted as c_t. In order to compute it an intermediate representation of cell state is computed. Hence, the new internal memory state at time-step t is computed as in Eq. 21. Hence, the hidden state at time-step t is computed using memory cell state and output gate as shown in Eq. 19. Here, i, f_g, o, c, input, forget, output gates, and cell state are all vectors which has dimension same as that of the hidden state vector.

4 Experimental Setup

The focus of the current work is to provide a comparison on sentiment identification of tweets in Malayalam language using classical machine learning and deep learning methods such as SVM, LR, KNN, Ab, DT, RF, RKS-RLSC, CNN and LSTM. The experiments are prepared using MATLAB and python code. To realize the deep learning experiments, we used tensorflow [32, 33] and Keras package [34]. The current work is an extension of our previous work providing comparisons [35].

Table 1 Dataset

Polarity	Count	Perc. (%)
Neutral	6680	51.38
Negative	3137	24.13
Positive	3183	24.48
Total	**13,000**	

4.1 Dataset Description

For the experiments in the current work, due to unavailability of dataset, we collected tweets in Malayalam language. The tweets can be extracted using 'tweepy' [36]. The tweets are manually annotated according to the sentiment/polarity as neutral, negative and positive. Table 1 shows the dataset distribution. Few datasets related to sentiment analysis which can be used for experiment and studies are [37–39]. In the Indian language context of social media data analysis, the data in the shared tasks [40, 41] gives a head-start to work in this direction.

4.2 Preprocessing

Initially, a preprocessing task is performed to tokenize the tweet. The training data contains symbols like "…", $"/////"$, $"???+++++++++++"$, $"****"$, $"||"$, $"…\gg"$, $"..-.^-.."$, $"?????"$, $"?"$, $"??"$, $"?"$, $"?"$, $"—"$, $"?"$, $"?"$ etc. These symbols are removed during pre-processing operations. All the URL's, hash tags, numbers, are replaced with <URL>, <#tag>, <@name>. After pre-processing is performed, a vocabulary is created with unique tokens. Using this vocabulary lookup, each token in the tweet is replaced by its corresponding index number in the vocabulary. This creates a tweet-vector for each tweet with different vector length. In order to make all the vectors of the same length, zeros are padded. In the current work, the maximum vector length of each tweet is fixed at 50. Hence, after the preprocessing stage the dimension of the data matrix is 13000×50.

Deep learning models are data-driven and in order to learn the parameters or weights required for each models, cost functions are defined to minimize the error in this process via back-propagation using gradient descent algorithm. The weights are updated as the gradient of error of the defined cost function decreases. The cost function (or loss function) defined for the current work is cross-entropy as it is popular for NLP tasks.

$$H_{y'}(y) := - \sum_i y'_i \log(y_i) \qquad (23)$$

where y_i denotes the probability distribution of predicted ith class and y_i' denotes the true probability distribution.

4.3 Models

SVM: For the sentiment identifying task using SVM, in order to find the suitable control parameter λ, it is varied from 1 to 10 and performed a k-fold cross-validation (k = 5) with RBF kernel for each. To train using SVM, libsvm library is used [42]. Along with SVM, other classical methods used for comparison are logistic regression (LR), k-nearest neighbor (KNN), adaboost (Ab), decision tree (DT), and random forest (RF). The realization of these algorithms are taken from scikit-learn [43].

RKS-RLSC: For the experiment, each tweet-vector of length 50 is mapped to 500 dimension using RKS method. To avoid the complex number computation in Eq. 9, used an equivalent form as in Eq. 24.

$$z(x) = \sqrt{1/k} \begin{bmatrix} \cos(x^T \Omega_1) \\ \vdots \\ \cos(x^T \Omega_k) \\ \sin(x^T \Omega_1) \\ \vdots \\ \sin(x^T \Omega_k) \end{bmatrix} \tag{24}$$

Hence, each 500 dimension tweet-vector becomes 1000 dimension as cosine and sine operations are performed on each tweet-vector and appended. Hence, the dimension of the mapped input matrix is 13000 × 1000.

CNN, LSTM: For the experiments, four different models are created for each of the deep learning methods. This is performed to find out the best model for each. The tweet-vector is fed to the CNN and LSTM units. In the training using CNN-1D, the number of filters plays an important role. In order to find a proper filter size, CNN models are generated at four different filter sized namely 32, 64, 128, 256. The increase in model size comes with heavy computation and a good high-performance machine is required. Hence, 256 is chosen as the limit for the experiment. The CNN net contains one convolution layer with stride as 1 and global max pooling. The architecture is similar to the one used in [44]. Similar to CNN, for LSTM, the number of cells as the parameter needs to be fixed which gives competing result. Four different models are generated for each method by varying the number of cell parameter as 32, 64, 128, 256. In the training process, to handle over-fitting issue, dropout parameter is fixed. All the deep learning models are realized for the experiment using Keras [34]. Each of the models is trained using 'Adam' optimizer [45]. To deal with gradient explosion issue, the current work utilizes three promising and popular activation functions such rectified linear units (ReLU), exponential linear units (ELU) [46], scaled exponential linear units (SELU) [47].

1. ReLU:

$$f_a(x) = \max(0, x) \tag{25}$$

2. ELU:

$$f_a(x) = \begin{cases} x, & x > 0 \\ \alpha(\exp(x) - 1), & x \leq 0 \end{cases} \tag{26}$$

3. SELU:

$$f_a(x) = \lambda \begin{cases} x, & x > 0 \\ \alpha(\exp(x) - 1), & x \leq 0 \end{cases} \tag{27}$$

For experiment with ELU, the parameter $\alpha = 0.1$. Whereas, for the experiment with SELU, the parameter is chosen as discussed in [47].

4.4 Evaluation

For evaluation, we used the popular metrics such as precision, recall, F1-score and accuracy during evaluation.

$$\text{Precision} = \frac{TrP}{TrP + FaP} \tag{28}$$

$$\text{Recall} = \frac{TrP}{TrP + FaN} \tag{29}$$

$$F1 = \frac{2 \times (\text{Precision} \times \text{Recall})}{(\text{Precision} + \text{Recall})} \tag{30}$$

$$\text{Accuracy} = \frac{TrP + TrN}{TrP + FaN + FaP + TaN} \tag{31}$$

where TrP, FaP, FaN,TaN denotes true-positive, false-positive, false-negative and true-negative. The metrics give an option to compare the models in and among the methods based on different parameter changes.

Table 2 Accuracy obtained for classical and RKS-RLSC method

No	Algorithm	Kernel function	Accuracy (%)
1	RKS	RBF	86.5
2		Randfeats	89.3
3	Classical	LR	82.9
4		KNN	79.1
5		RF	83.3
6		DT	78.3

5 Result and Analysis

This section shows the experimental results obtained on identifying the sentiment of Malayalam tweets using SVM, RLSC, CNN and LSTM. The SVM and RLSC results are currently kept as the baseline. In order to find the suitable control parameter, λ, for SVM, the experiment with SVM was run for 5-fold cross-validation for each $\lambda = \{1, 2, \ldots, 10\}$ with 10% data considered in the test case. The kernel selected for the experiment is RBF kernel $Ke(x_i, x_j) = e^{-\sigma \|x_i - x_j\|^2}$. From Fig. 1, it can observe that for $\lambda = 6$, SVM obtained a maximum accuracy of 85.2%. However, with the experiment using RKS-RLSC, compared to SVM has shown an improved score. Table 2 shows the performance of the classical machine learning and RKS-RLSC methods. For all the classical methods, the tweet-vector representation for each tweet is given as input. Where as in the RKS method, the tweet-vector is mapped to 500 dimensional vector using random kitchen sink method [27]. The randfeats kernel obtained an improved result. The RKS-RLSC is performed using GURLS package [48]. The accuracy obtained for each kernel is an average score from a 5-fold cross-validation.

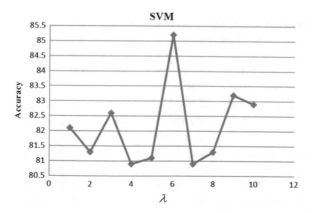

Fig. 1 Demonstrates the accuracy obtained for each control parameter λ in SVM

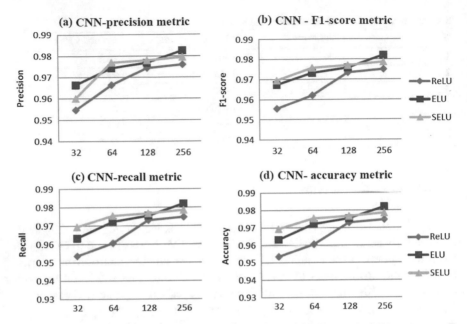

Fig. 2 Demonstrates CNN based evaluation metric results **a** precision metric, **b** F1-score, **c** recall, **d** accuracy

Figures 2 and 3 shows the metric curve obtained during evaluations for sentiment analysis of tweets in Malayalam using CNN and LSTM models on three differ-ent activation functions such as ReLU, ELU and SELU. The activation function introduces nonlinearity in the network as the data itself is nonlinear in behaviour. Using a linear activation effectively makes a complex, deep network to act as a single layer neural net which does linear transformation. In the current paper, three differ-ent nonlinear functions are used, namely ReLU, ELU and SELU. Another advantage of these nonlinear activation functions is its capability to handle gradient exploding issues and it's discussed in detail in [46, 47]. It can be observed from Fig. 2 that the behaviour of the curve looks similar, however there are small variations in the values obtained for each model. The x-axis in Fig. 2 denotes the filter size, whereas in Fig. 3 it denotes the number of hidden states.

For CNN based experiment, the filter length or kernel size is fixed at 3. Therefore, for choosing 32 as filter size, it means there will be 32 different filters (or 32 different feature maps) with each having 3 elements. During the back-propagation process, the suitable filter values are calculated. In case of CNN, these filter values act as the weight values. From Fig. 2 it can be found that the increase in filter size shows better evaluation measures. Table 3 shows the evaluation measures obtained for filter size 256 as it obtained high scores during evaluations for CNN-1D model formed using ELU activation function. This difference in evaluation score can be observed from Table 3.

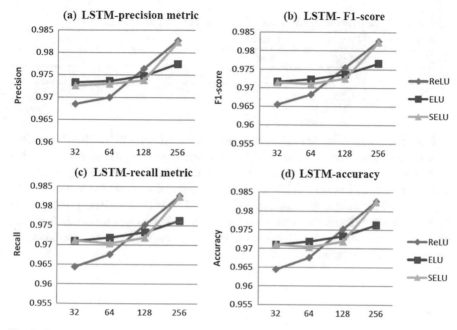

Fig. 3 Demonstrates LSTM based evaluation metric results **a** precision metric, **b** F1-score, **c** recall, **d** accuracy

Table 3 Evaluation metrics for filter size taken as 256 in CNN-1D

Function	Precision	Recall	F1-score	Accuracy
ReLU	0.9759	0.9746	0.9750	0.9746
ELU	0.9825	0.9819	0.9821	0.9819
SELU	0.9794	0.978	0.9787	0.9784

Similar to CNN, experiments with LSTM are performed choosing four different number of hidden states such as 32, 64, 128, 256. The evaluation scores obtained after 10-fold cross-validation are shown in Fig. 3. Even though the evaluation scores curve behaviour is similar, there are differences in values. It's evident from Fig. 3, the LSTM with 256 hidden states scored highest. Also, it can be seen that the LSTM models created using ReLU and SELU activation function has competing evaluation scores when compared to the LSTM models formed using ELU activation function. Table 4 shows the evaluation measures obtained for LSTM models with number of hidden state as 256.

The results obtained clearly indicate the advantage of using LSTM and CNN for creating sentiment analysis models for tweets in native language, such as Malayalam, for applications related to big data. The experiment and evaluation performed in the current paper demonstrate the effectiveness of nonlinear functions as it will help the deep network to learn better and is evident from the results obtained. More conclusive results can be obtained by increasing the corpus size along with parameter tuning.

Table 4 Evaluation metrics for LSTM model with number of hidden state as 256

Function	Precision	Recall	F1-score	Accuracy
ReLU	0.9827	0.9826	0.9828	0.9826
ELU	0.9775	0.9763	0.9766	0.9763
SELU	0.9823	0.9824	0.9823	0.9824

From the experiments, it can be observed that the RKS-RLSC method has shown improvement when compared to SVM. The change in mapping dimension to a higher value can provide convincing results, however, it need to be experimentally verified and we keep it as a future work. Even though deep learning methods are data-hungry approaches, the current experiment with CNN and LSTM for identifying sentiment of Malayalam tweets forms the initial work. As another future work we intend to collect more tweets related mostly on reviews in different domains such as entertainment, sports, food & restaurant, places etc.

6 Conclusion

The chapter presents a comparison on sentiment identification of Malayalam tweets using SVM, RKS-RLSC (keeping both as the baseline), CNN-1D and LSTM along with other classical methods such as logistic regression, adaboost, decision tree, random forest, k-nearest neighbour. For the experiments we took tweets in Malayalam language. Due to the unavailability of data, we collected it and manually labeled its sentiment as neutral, negative or positive. The SVM and RKS-RLSC approaches are considered as the baseline method to compare with the results obtained from CNN and LSTM as the deep learning methods have outperformed. The work also provides a comparison with other classical methods. The experiments using CNN and LSTM are performed with three different activation functions such as ReLU, ELU, SELU. In the experiment with CNN-1D, we report that the convolutional neural net with ELU activation function outperformed. Whereas with LSTM, the experiments with ReLU and SELU comparatively outperformed.

References

1. Zikopoulos P, Eaton C, DeRoos D, Deutch T, Lapis G (2011) Understanding big data: analytics for enterprise class Hadoop and streaming data. McGraw-Hill Osborne Media
2. Agarwal A, Xie B, Vovsha I, Rambow O, Passonneau R (2011) Sentiment analysis of twitter data. In: Proceedings of ACL 2011 workshop on languages in social media, pp 30–38
3. Saif H, He Y, Alani H (2011) Semantic smoothing for twitter sentiment analysis. In: Proceeding of the 10th international semantic web conference (ISWC)
4. Kiritchenko S, Zhu X, Mohammad SM (2014) Sentiment analysis of short informal texts. J Artif Intell Res 723–762

5. Das A, Bandyopadhyay S (2010) SentiWordNet for Indian languages. In: Proceedings of the 8th workshop on Asian Language Resources (ALR), August, pp 56–63
6. Maite T, Julian B, Milan T, Kimberly V, Manfred S (2011) Lexiconbased methods for sentiment analysis. Comput Linguist 37(2):267–307
7. Willyan DA., de Leandro NC (2014) A keyword extraction method from twitter messages represented as graphs
8. Saima A, Stan S (2007) Identifying expressions of emotion in text. In: Text, speech and dialogue. Springer, Berlin, Heidelberg, pp 196–205
9. Changhua Y, Lin KH-Y, Chen H-H (2007) Building emotion lexicon from weblog corpora. In: Proceedings of the 45th annual meeting of the ACL on interactive poster and demonstration sessions. Association for Computational Linguistics, pp 133–136
10. Hiroya T, Takashi I, Manabu O (2005) Extracting semantic orientations of words using spin model. In: Proceedings of the 43rd annual meeting of the association for computational linguistics (ACL'05), pp 133–140
11. Stefano B, Andrea E, Fabrizio S (2010) Sentiwordnet 3.0: An enhanced lexical resource for sentiment analysis and opinion mining. In: Proceedings of the 7th conference on international language resources and evaluation (LREC'10), Valletta, Malta, May
12. Theresa W, Janyce W, Paul H (2005) Recognizing contextual polarity in phrase-level sentiment analysis. In: Proceedings of the HLT/EMNLP, Vancouver, Canada
13. Maite T, Anthony C, Voll K (2006) Creating semantic orientation dictionaries. In: Proceedings of the 5th international conference on language resources and evaluation (LREC), Genoa, pp 427–432
14. Yoshimitsu T, Dipankar D, Sivaji B, Manabu O (2011) Proceedings of 2nd workshop on computational approaches to subjectivity and sentimental analysis, ACL-HLT, pp 80–86
15. Pang B, Lee L, Vaithyanathan S (2002) Thumbs up? Sentiment classification using machine learning techniques. In: Proceedings of EMNLP, pp 79–86
16. Baccianella S, Esuli A, Sebastiani F (2010) SENTIWORDNET 3.0: an enhanced lexical resource for sentiment analysis and opinion mining. In: Proceedings of LREC-10
17. Wiebe J, Mihalcea R (2006) Word sense and subjectivity. In: Proceedings of COLING/ACL-06, pp 1065–1072
18. Pooja P, Sharvari G (2015) A framework for sentiment analysis in Hindi using HSWN. IJCA 119
19. Mittal N, Agarwal B, Chouhan G, Bania N, Prateek P (2013) Sentiment analysis of Hindi review based on negation and discourse relation. In: International joint conference on natural language processing, Nagoya, Japan, Oct 2013
20. Joshi A, Balamurali AR, Bhattacharyya P (2010) A fall-back strategy for sentiment analysis in Hindi: a case study. In: Proceedings of the 8th ICON
21. Das A, Bandyopadhyay S (2010) SentiWordNet for Indian languages. Asian Federation for Natural Language Processing (COLING), China, pp 56–63
22. Sumit KG, Gunjan A (2014) Sentiment analysis in Hindi language: a survey. IJMTER
23. Sharma R, Nigam S, Jain R (2014) Opinion mining in Hindi language: a survey. IJFCST 4(2)
24. Pooja P, Sharvari G (2015) A survey of sentiment classification techniques used for Indian regional languages. IJCSA 5(2)
25. Selvan A, Anand Kumar M, Soman KP (2015) Sentiment analysis of Tamil movie reviews via feature frequency count. In: International conference on innovations in information, embedded and communication systems (ICIIECS 15). IEEE
26. Mohandas N, Nair JPS, Govindaru V (2012) Domain specific sentence level mood extraction from Malayalam text. In: Advances in Computing and Communications (ICACC)
27. Rahimi A, Recht B (2007) Random features for large-scale kernel machines. In: Advances in neural information processing systems
28. Kumar SS, Premjith B, Kumar MA, Soman KP (2015) AMRITA_CEN-NLP@ SAIL2015: Sentiment analysis in Indian language using regularized least square approach with randomized feature learning. In: Mining intelligence and knowledge exploration, Dec, pp 671–683

29. Bengio Y, Simard P, Frasconi P (1994) Learning long-term dependencies with gradient descent is difficult. IEEE Trans Neural Netw 5:157–166
30. Bengio Y, Boulanger Lewandowski N, Pascanu R (2013) Advances in optimizing recurrent networks. ICASSP
31. Hochreiter S, Schmidhuber J (1997) Long short-term memory. Neural Comput 9(8):1735–1780
32. Tensorflow. https://www.tensorflow.org/get_started/get_started
33. Theano. http://deeplearning.net/software/theano/
34. http://keras.io
35. Sachin Kumar S, Anand Kumar M, Soman KP (2017) Sentiment analysis of tweets in Malayalam language using long sort-term memory units and convolutional neural nets. In: MIKE 2017, Springer, pp 320–334
36. Tweepy. https://github.com/tweepy/tweepy
37. Kaggle dataset. https://www.kaggle.com/c/si650winter11/data
38. Arabic sentiment tweet dataset. https://archive.ics.uci.edu/ml/datasets/Twitter+Data+set+for+Arabic+Sentiment+Analysis
39. Twitter sentiment corpus. http://www.sananalytics.com/lab/twitter-sentiment/
40. Sentiment analysis in Indian languages (SAIL). http://amitavadas.com/SAIL/
41. Sentiment analysis for Indian languages (Code Mixed). http://www.dasdipankar.com/SAILCodeMixed.html
42. Chang C-C, Lin C-J (2001) LIBSVM: a library for support vector machines. http://www.csie.ntu.edu.tw/~cjlin/libsvm
43. Scikit learn library. http://scikit-learn.org/stable/supervisedlearning.html
44. Yoon K (2014) Convolutional neural networks for sentence classification. arXiv:1408.5882
45. Kingma DP, Ba J (2014) Adam. A method for stochastic optimization. arXiv:1412.6980
46. Clevert DA, Unterthiner T, Hochreiter S (2016) Fast and accurate deep network learning by exponential linear units (ELUs). In: ICLR
47. Klambauer G, Unterthiner T, Mayr A (2017) Self-normalizing neural networks. arXiv:1706.02515
48. Tacchetti A, Mallapragada PK, Santoro M, Rosasco L (2013) Gurls: a least squares library for supervised learning. J Mach Learn Res 14:3201–3205

Twitter Based Sentiment Analysis of GST Implementation by Indian Government

Prabhsimran Singh, Ravinder Singh Sawhney and Karanjeet Singh Kahlon

Abstract Bringing major changes in existing tax structure is always a monotonous task to implement, especially when it affects one and all of the business world of one of the fastest growing economy. There are numerous hidden taxes, which remain inherently correlated with the goods reaching out to the general public. Implementation of Goods and Services Tax (GST) has been the biggest reform and a bold action performed by the Government of India recently. This paper takes into consideration the overall impact of GST implementation and the opinion of the Indian public about GST. Using our mathematically improvised modeling approach, we have done the sentiment analysis of the Twitter data collected over a period consisting of Pre-GST, In-GST and Post-GST period from all the regions and states of India. Multiple datasets are adopted to bring a rationalized outlook of this economic reform in Indian corporate scenario.

Keywords Goods and Services Tax · Machine learning · Twitter Sentiment analysis

1 Introduction

Indian tax system consists of multiple taxes levied by both the central and the state governments. Things become more complicated when a commodity is being sold at different rates in different states. The actual reason behind this variation in selling price of the commodity is on account of variable tax structures being imposed by the

P. Singh (✉) · K. S. Kahlon
Department of Computer Science, Guru Nanak Dev University, Amritsar, India
e-mail: prabh_singh32@yahoo.com

K. S. Kahlon
e-mail: karankahlon@gndu.ac.in

R. S. Sawhney
Department of Electronics Technology, Guru Nanak Dev University, Amritsar, India
e-mail: sawhney.ece@gndu.ac.in

© Springer International Publishing AG, part of Springer Nature 2019   409
S. Patnaik et al. (eds.), *Digital Business*, Lecture Notes on Data Engineering
and Communications Technologies 21, https://doi.org/10.1007/978-3-319-93940-7_17

state governments. So to overcome this problem, various proposals were suggested for a single uniform tax system and eventually in 1999 Goods and Services Tax (GST) was proposed under the leadership of Prime Minister of India Mr. Atal Bihari Vajpayee. However it took another 18 long years for GST to become a reality. GST was introduced as The Constitution (One Hundred and First Amendment) Act 2017 [1], following the passage of Constitution 122nd Amendment Act Bill. A GST council was formed under the Chairmanship of Finance Minister of India [2], which will regulate the GST from time to time. Finally, on July 1, 2017 GST was implemented in India.

Like every government policy, it the common people who face the initial hardships. Most of the times newspapers and electronic media, try to cash on and show such information in an exaggerated manner. This creates a faulty image about the policy among the common people, although it may have many benefits in store for them. This paper presents the sentiment analysis of the opinion mined data on GST implementation by Indian Government using Twitter as a medium. The ultimate aim to do this opinion mining was to find out the reaction of the public to the event and confirm whether the people of India were actually happy or sad with this decision. The main highlight of our analysis is that the analysis is carried out on regional basis and state basis to give a better insight of the results.

2   Methodology Followed

The aim of this paper is to do region wise and state wise sentiment analysis towards GST implementation. To accomplish this task, we follow the methodology as given in Algorithm 1.

Algorithm 1: (Methodology)

Start
    1: Collecting the tweets from Twitter based on specific hashtag (#).
    2: Preprocessing the data (tweets).
        2.1: Removing multiple tweets from same person.
        2.2: Removing web links.
        2.3: Removing special symbols.
        2.4: Removing English stop words.
    3: Performing sentiment analysis using machine learning algorithms.
    4: Separating the tweets on region wise and state wise basis.
    5: Calculating the sentiment score for each region and state.
    6: Representing the analysis using heat maps.
  End

The algorithm starts with collection of tweets from popular social networking website (SNW) Twitter. Once tweets were collected, it was important to preprocess them before applying further operations, hence task of data preprocessing was performed. Once tweets were preprocessed, we performed sentiment analysis using

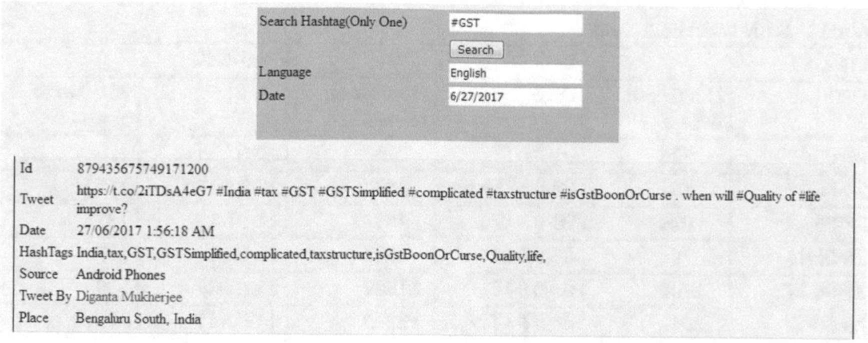

Fig. 1 System developed for tweet collection

machine learning algorithms. Based upon sentiment analysis, we calculate sentiment score for each region and state. Finally, we plot heat maps that give us better description of calculated results. The detail working of our methodology is explained in upcoming sections.

3 Tweet Collection

The very base of our analysis is the tweets collected from Twitter, a famous SNW. To do this tweet collection task in a reliable manner we developed a system using ASP.Net [3] and integrated Tweetinvi API [4] for accomplishing this task. Tweets were collected based upon the Specific hashtag (#), i.e. #GST. The system developed by us returned us the tweet ID, tweet, Date time, various hashtags in the tweet, source from where tweet was tweeted, person who tweeted the tweet and finally the location from where the tweet was tweeted. Due to privacy reasons we do not get location of every tweet. Figure 1 shows the system developed by us for tweet collection.

Since our aim was to do sentiment analysis of Indian states and region therefore tweets were strictly collected from India only. A total of 41,823 tweets were collected from June 23, 2017 to July 16, 2017. Tweets were collected in three phases (Pre-GST, In-GST and Post-GST). The aim of fetching tweets in three different phases was to do proper analysis, so that we can conclude how states and regions felt prior to the GST, during the GST and finally post GST was implemented. The details of daily tweet collection during all the three phases are show in Table 1.

Table 1 Daily tweet collection

Pre-GST		In-GST		Post-GST	
Date	Number of tweets	Date	Number of tweets	Date	Number of tweets
23-06-17	526	01-07-17	2160	09-07-17	1109
24-06-17	553	02-07-17	2522	10-07-17	1552
25-06-17	628	03-07-17	4622	11-07-17	1357
26-06-17	781	04-07-17	2555	12-07-17	1336
27-06-17	2308	05-07-17	1907	13-07-17	1393
28-06-17	2444	06-07-17	2326	14-07-17	1297
29-06-17	2776	07-07-17	2453	15-07-17	1081
30-06-17	3026	08-07-17	440	16-07-17	671
Total	13,042	Total	18,985	Total	9796

Table 2 Example of data preprocessing

Tweet before preprocessing	Tweet after preprocessing
#GST, the Road Ahead: Uninterrupted movement of trucks across states is helping save huge amount of money and time… https://t.co/dbbq0jgbOK	GST the Road Ahead Uninterrupted movement of trucks across states is helping save huge amount of money and time

4 Data Preprocessing

Data preprocessing is very crucial step in data mining, as it has a huge impact on the results. An unprocessed dataset can lead to wrong results and can ruin the analysis; hence it is necessary to preprocess the data before applying any data mining operation [5]. In data preprocessing we removed the unwanted HTML tags, web links and special symbols (@ # ^ * " / : ; > , < \ | ?) which could have led to wrong results. The entire task of data preprocessing is done in an automated fashion using the same system developed for tweet collection. Table 2 shows an example of data preprocessing.

Further, since a person can send multiple tweets on Twitter, if these multiple tweets are not removed then it could lead to biased analysis. So in order to make an unbiased analysis we have removed all the multiple tweets from same person by adopting the same technique as used by Singh et al. [6]. From the total 41,823 tweets, 6423 such multiple tweets were removed.

Table 3 Training dataset statistics

Source	Positive instances	Negative instances
IMDb	1500	1500
Demonetization	500	500
Manually labeled tweets	500	500
Total	2500	2500

Table 4 Training dataset examples

Source	Positive	Negative
IMDb	Still I do like this movie for its empowerment of women; theres not enough movies out there like this one	This is a bad film with bad writing and good actors an ugly cartoon crafted by Paul Haggis for people who can't handle anything but the bold strokes in storytelling a picture painted with crayons
Demonetization	One of the greatest and dynamic decision taken in last few decades respect ModiRulesIndianHearts DeMonetisation	U knw its real shit wen ur debit crd gts expired n u gt new crd bt cnt gt into atm to activate it bt u need 2 withdraw cash too BlackMoney
Manually labeled tweets	GST impact Colgate-Palmolive cuts prices of toothpaste toothbrush by 9%	Post GST higher rate for detergents and shampoos at 28% the highest tax bracket GST impact on FMCG sector

5 Training Datasets

For training we have used IMDb [7] and Demonetization [8] datasets consisting of 3000 and 1000 sentences respectively and corresponding sentiment of each sentence in form of "neg" (negative) and "pos" (positive). Further, to achieve the better efficiency we manually labeled 1000 tweets (500 Positive and 500 Negative) collected by us (See Sect. 3). This helped us to add domain specific training data to our training dataset. Table 3 shows the statistics of our training dataset, while Table 4 shows example of the training dataset.

6 Sentiment Analysis

Sentiment analysis is the study of analyzing people's opinion (sentiment) towards an entity. It is often called opinion mining [9]. For performing sentiment analysis we have used WEKA 3.8. WEKA is an open source software which consists of a collection of machine learning algorithms for data mining tasks [10]. Using WEKA

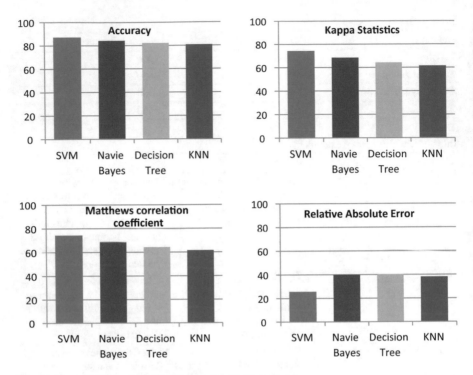

Fig. 2 Comparative analysis of different classification models

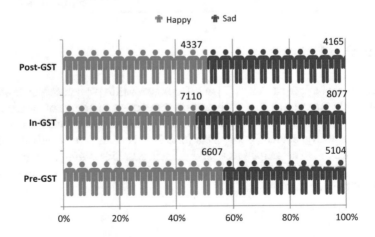

Fig. 3 Results of sentiment analysis

we have built four different classification models Support Vector Machine (SVM) [11], Naive Bayes (NB) [12], K-nearest neighbor (KNN) [13] and Decision Tree (DT) [14]. The aim of choosing multiple models were to check which model gives us best result for our training data.

The training dataset was passed through four classification models (SVM, Naive Bayes, KNN and DT) using 10-fold cross validation. The purpose of using 10-fold cross validation is to analyze how a predictive model would perform on an unknown dataset. Figure 2 shows the comparative analysis of different classification models for our training dataset. Results clearly indicate that SVM gave the best results for our training dataset; hence it will be used for performing sentiment analysis with the testing dataset.

For testing dataset we have used tweets collected from Twitter (See Sect. 3). From a total of 35,400 tweets, 18,454 (52.12%) tweets were classified as positive, while 16,946 (47.88%) tweets were classified as negative tweets. The results of sentiment analysis for all three phases (Pre-GST, In-GST and Post-GST) are show in Fig. 3.

From 11,711 tweet collected during pre-GST period, 6607 (56.41%) were positive while 5104 (43.59%) tweets were negative. Similarly, from 15,187 tweets collected during In-GST period, 7110 (46.81%) tweets were positive while 8077 (53.19%) tweets were negative. Finally, from 8502 tweets collected during post-GST period, 4337 (51.01%) tweets were positive while 4165 (48.99%) tweets were negative. The above results clearly indicate that during pre-GST period people were happy with GST implementation as overall sentiment of nation was positive. However it was during in-GST period when people started facing the hardship the overall sentiment of the nation was on the negative side, as sellers has to update their billing mechanisms according to GST and manufacturers had to do repackaging of their items with new GST inclusions. But soon Government introduced various webinars and step by step guides to GST process with the sole purpose to make people familiar with GST. As a result, things started moving back to normal, the overall sentiment of the nation during post-GST period was again on positive side.

7   Location Based Analysis

In the previous section we discussed the sentiment of the nation in all the three phases of GST. However, this is an overall analysis of nation, which does not depict the true picture of analysis and hence the need of location based analysis comes into play [8]. As discussed earlier that tweets also contained the location (place) from where the tweet was tweeted. Hence we will utilize this location parameter to do region wise and state wise analysis, which will help us to depict how the region and states felt during all the three phases of GST. The reason of doing this location based analysis is the diversity of India. We can find a 94% literacy rate in a state like Kerala which is 32.20% higher than the state Bihar. Similarly, the population of Uttar Pradesh (UP) is 199,812,341 which accounts for 16.5% of the total Indian population, on other hand Sikkim has a population of 610,577, almost 327 times smaller than Uttar Pradesh [15]. To do state wise analysis we have the concept of sentiment score as shown in Eq. 1.

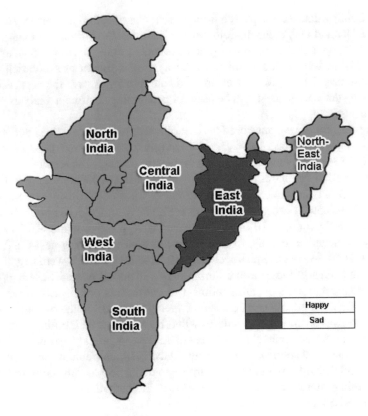

Fig. 4 Region wise analysis of Pre-GST period

$$Sentiment\ Score\ (State\ or\ Region) = \sum Positive\ Tweets - \sum Negative\ Tweets \tag{1}$$

The sentiment score gave us the description of how the people of state felt during three phases of GST. If the sentiment score is positive, then this indicates the people of that particular state are happy during that phase of GST. If the sentiment score is negative, then this indicate that people of that particular state are not happy during that phase of GST. Similarly, if the net sentiment score is "0" (zero), and then this indicates that the mood of the people is neutral during that phase of GST. A similar approach is used to find sentiment of the entire region. Below we have shown phase wise results of regional analysis and state analysis.

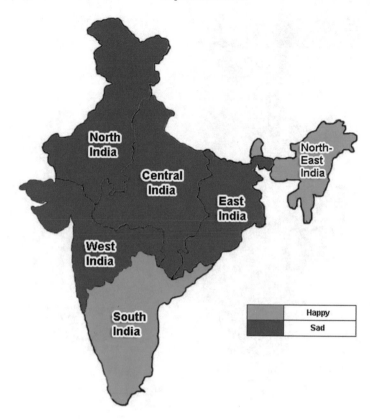

Fig. 5 Region wise analysis of In-GST period

7.1 Region Wise Analysis

In this section we will show the mood of the country for three phases of GST by dividing the entire nation into six regions.

(a) **Pre-GST**: During pre-GST period, a part from east India which comprises of Bihar, Jharkhand, Odisha and West Bengal states, all other regions where happy. Figure 4 shows the result of region wise analysis of pre-GST period.

(b) **In-GST**: During in-GST period, a part from north-east India and south India, all other regions where sad. The possible reason for this is the hardship faced by the people, which caused them inconvenience and showed unhappiness during this period. Figure 5 shows the result of region wise analysis of in-GST period.

(c) **Post-GST**: During post-GST period, a part from north India and west India, all other regions where happy. This shows once people got familiar with this new policy the overall sentiment of the sates again turned positive. Figure 6 shows the result of region wise analysis of post-GST period.

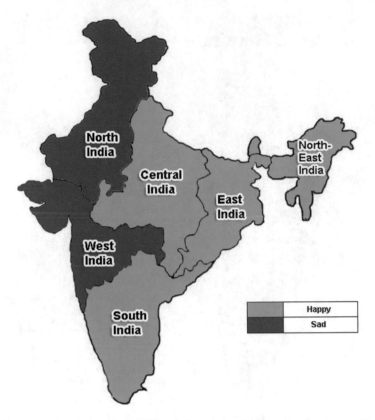

Fig. 6 Region wise analysis of Post-GST period

7.2 State Wise Analysis

In this section we will show the mood of the country for three phases of GST by dividing the entire nation into states and Union Territories (UT). Note that no data was available for Arunachal Pradesh, Meghalaya, Nagaland, Manipur, Mizoram and Tripura states, hence these state are not included for analysis in any of the three phases.

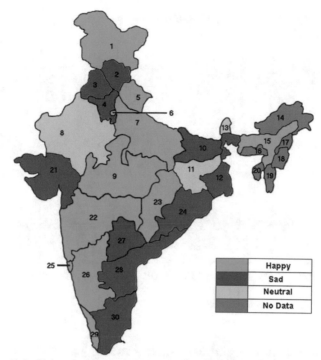

Note: [1-Jammu & Kashmir, 2-Himachal Pradesh, 3-Punjab, 4-Haryana,
5-Uttarakhand, 6-New Delhi, 7-Uttar Pradesh, 8-Rajasthan, 9-Madhya Pradesh,
10-Bihar, 11- Jharkhand, 12-West Bengal, 13-Sikkim, 14-Arunachal Pradesh,
15-Assam, 16-Meghalaya, 17- Nagaland, 18-Manipur, 19-Mizoram, 20-Tripura,
21-Gujarat, 22-Maharashtra, 23- Chhattisgarh, 24-Odisha, 25-Goa,
26-Karnataka, 27-Telangana, 28-Andhra Pradesh, 29-Kerala, 30-Tamil Nadu]

Fig. 7 State wise analysis of Pre-GST period

(a) **Pre-GST**: During pre-GST analysis, 10 states were unhappy, 13 (10 states & 3 UT) were happy and 4 (3 states & 1 UT) were neutral. So in total 62.96% of states & UT were in favor or neutral. Figure 7 shows state wise analysis of pre-GST period. The analysis of three UT is shown in Fig. 8.

(b) **In-GST**: During pre-GST analysis, 15 states were unhappy, 9 (6 states & 3 UT) were happy and 3 (2 states & 1 UT) were neutral. So in total 55.55% of states & UT were unhappy with GST. Figure 9 shows state wise analysis of In-GST period. The analysis of three UT is shown in Fig. 10.

(c) **Post-GST**: During post-GST analysis, 4 states were unhappy, 18 (6 states & 4 UT) were happy and 5 states were neutral. So in total 85.18% of states & UT were either happy or neutral with GST. Figure 11 shows state wise analysis of In-GST period. The analysis of three UT is shown in Fig. 12.

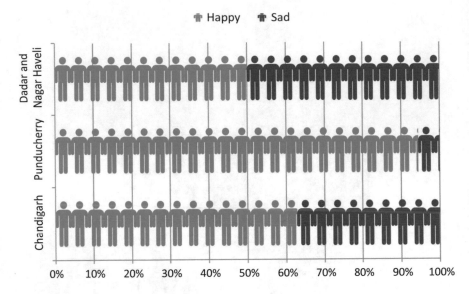

Fig. 8 Analysis of 3 UT during Pre-GST period

8 Possible Reason of Unhappiness Among Various States

As observed from the available data, there is clear indication that earlier there was displeasure among the public from few of the states with the implementation of GST. This number gradually increased to as many as 15 states when GST was getting actually implemented. But once the government provided various online webinars and tutorials to make people familiar with GST, things stated to get normal and as a result during post-GST period, only four states (Haryana, Maharashtra, Rajasthan and Uttar Pradesh) were observed to be unhappy with GST, while all other states eventually started responding positively to this government initiative.

If we minutely observe the details of these states on a day by day basis during the post-GST period, we analyzed that the overall sentiment among people throughout the nation started to drift towards positive perception during the last days of our analysis. Since the overall sentiment during the initial days of post-GST period was highly negative, the accumulated overall sentiment score for post-GST period still remained negative.

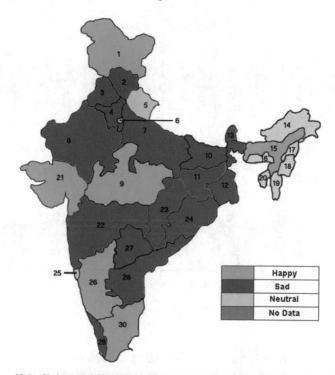

Note: [1-Jammu & Kashmir, 2-Himachal Pradesh, 3-Punjab, 4-Haryana,
5-Uttarakhand, 6-New Delhi, 7-Uttar Pradesh, 8-Rajasthan, 9-Madhya Pradesh,
10-Bihar, 11- Jharkhand, 12-West Bengal, 13-Sikkim, 14-Arunachal Pradesh,
15-Assam, 16-Meghalaya, 17- Nagaland, 18-Manipur, 19-Mizoram, 20-Tripura,
21-Gujarat, 22-Maharashtra, 23- Chhattisgarh, 24-Odisha, 25-Goa,
26-Karnataka, 27-Telangana, 28-Andhra Pradesh, 29-Kerala, 30-Tamil Nadu]

Fig. 9 State wise analysis of In-GST period

Figure 13 elaborates the day by day analysis for the state of Haryana during post-GST period. On July 9 and July 10, the overall sentiment was mildly negative, but on July 11, the sentiment was highly negative and not even a single tweet was having positive sentiment. Things started to change thereafter as the central government shared remedial measures to overcome the doubts in the minds of various traders. July 12 and July 13 were neutral in the sense that net positive tweets were equal to net negative tweets. From July 14 to July 16, the net sentiment of the state was observed to be on positive side, which indicates that probably the people started getting convinced by the governmental assurances. Since a large number of negative tweets were received in starting days of post-GST period, the net sentiment score of the Haryana state was observed to be negative. The vital observation during last days of our analysis is that the overall sentiment of the state was shifting towards

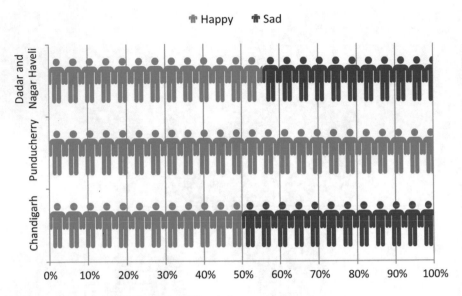

Fig. 10 Analysis of 3 UT during In-GST period

the positive side. Hence if the analysis would have been extended for a few more days, then the overall sentiment would have been become largely positive instead of current negative.

Figure 14 shows the day by day analysis for the state of Maharashtra during post-GST period. For initial 5 days i.e. July 9 to July 13, the overall sentiment of people was negative. But during the final 3 days of post-GST analysis i.e. July 14 to July 16, the overall sentiment of the state stated to drift towards positive side. This again shows that the overall sentiment of even Maharashtra state was positive during final days of our analysis.

Figure 15 shows the day by day analysis for the state of Rajasthan during post-GST period. Sentiment of the people of Rajasthan was on negative side during initial 3 days i.e. July 9 to July 11. But next 3 days were neutral, and finally from July 15 onwards the sentiment of Rajasthan was on positive side.

Figure 16 shows the day by day analysis for the state of Uttar Pradesh during post-GST period. Sentiment of the people of Uttar Pradesh was on negative for the initial four days i.e. July 9 to July 12. But from July 13 onwards the overall sentiment of the state was either neutral or positive.

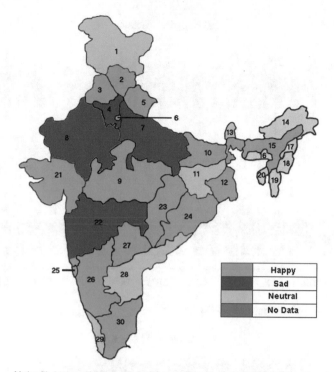

Note: [1-Jammu & Kashmir, 2-Himachal Pradesh, 3-Punjab, 4-Haryana,
5-Uttarakhand, 6-New Delhi, 7-Uttar Pradesh, 8-Rajasthan, 9-Madhya Pradesh,
10-Bihar, 11- Jharkhand, 12-West Bengal, 13-Sikkim, 14-Arunachal Pradesh,
15-Assam, 16-Meghalaya, 17- Nagaland, 18-Manipur, 19-Mizoram, 20-Tripura,
21-Gujarat, 22-Maharashtra, 23- Chhattisgarh, 24-Odisha, 25-Goa,
26-Karnataka, 27-Telangana, 28-Andhra Pradesh, 29-Kerala, 30-Tamil Nadu]

Fig. 11 State wise analysis of Post-GST period

The day by day analysis of all four states shows that these states were having negative sentiment during initial days but as we approached the final days this sentiment changed towards negative side. Since the number of negative tweets gathered during initial days were quite high as compared to number of positive tweets during final days of analysis and hence these states had an overall negative sentiment i.e. sad during post-GST period.

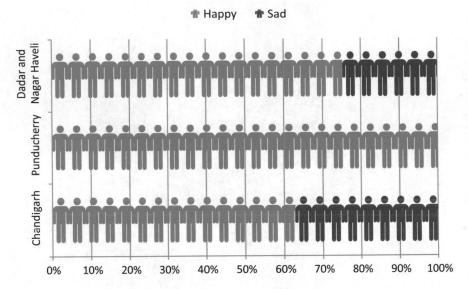

Fig. 12 Analysis of 3 UT during Post-GST period

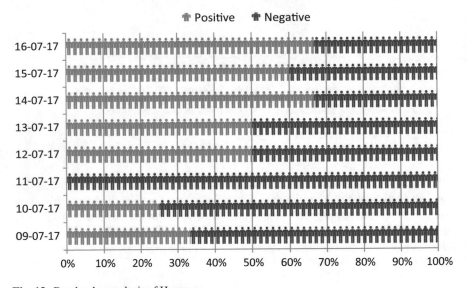

Fig. 13 Day by day analysis of Haryana

9   Conclusions

Using the mathematical modeling based testing of the datasets collected from the general public about the economic policy of a union government, especially the fastest growing economy of the world is a novel effort and the results have been

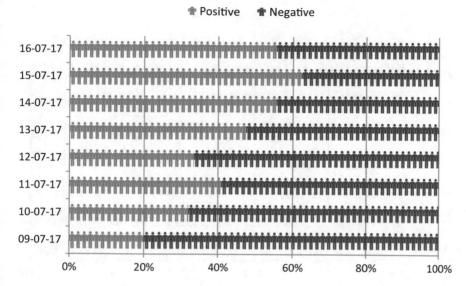

Fig. 14 Day by day analysis of Maharashtra

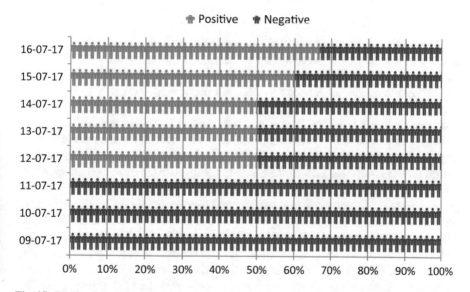

Fig. 15 Day by day analysis Rajasthan

put forward for the analysis of the economic gurus. The sentiment opinion of all the directly as well indirectly affected parties has been considered to bring out a truthful conclusion.

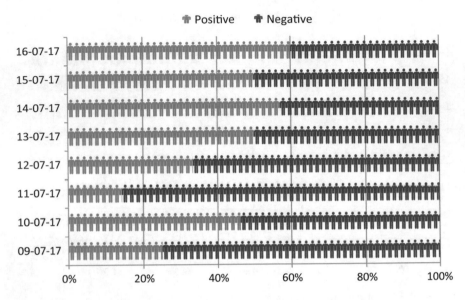

Fig. 16 Day by day analysis of Uttar Pradesh

The opinion analysis towards GST was completed three phases. Pre-GST period showed that among 27 states and UT's under consideration 10 states (37.03%) were unhappy suggesting that rest of the 62.97% of the states and UT's were in favor of GST implementation. As soon as GST got implemented on July 1, 2017 things started to change. The group of people mainly belonging to businessmen, traders and manufacturers faced the hardship as they needed to update their billing system in accordance to the GST guidelines. Further all the inventories and the stocks were required to get repackaged with new packaging which must have GST charges mentioned on these entities. This further added an extra burden of unwanted spending. As a result of this, 15 states (55.55%) became unhappy with GST during In-GST period. Soon the government realized the problems encountered by this group and subsequence hardship faced by its people and it introduced various webinars and online tutorials to make these people familiar with GST. Further the rates of some entities especially fast moving consumer goods (FMCG) got reduced due to implementation of GST. This lead to an overall positive swing in the mood of people towards GST, and hence 85.15% of the state supported the GST during post-GST period. Further our day by day analysis showed the even for the remaining unhappy states (Haryana, Maharashtra, Rajasthan and Uttar Pradesh) the sentiment started to swing towards the positive direction during the final days of the post-GS period. So we can conclude from our overall detailed analysis that majority of the people supported the GST implementation decision of the Indian government.

References

1. The Constitution (One Hundred and First Amendment) Act (2017). http://lawmin.nic.in/ld/T he%20Constitution%20(One%20Hundred%20and%20First%20Amendment)%20Act,%2020 16.pdf
2. GST Council. http://gstcouncil.gov.in/gst-council-members
3. Visual Studio (2012). https://www.visualstudio.com/en-us/downloads/download-visual-studi o-vs.aspx
4. Tweetinvi API. https://www.nuget.org/packages/TweetinviAPI/
5. Salvador G, Luengo J, Francisco H (2015) Data preprocessing in data mining. Springer, New York
6. Singh P, Sawhney RS, Kahlon KS (2017) Forecasting the 2016 US presidential elections using sentiment analysis. In: Conference on e-Business, e-Services and e-Society. Springer, Cham, pp 412–423. https://doi.org/10.1007/978-3-319-68557-1_36
7. Maas AL, Daly RE, Pham TP, Haung, Ng AY, Potts C (2011) Learning word vectors for sentiment analysis. In: Proceedings of the 49th annual meeting of the association for computational linguistics: human language technologies, vol 1. Association for Computational Linguistics, pp 142–150
8. Singh P, Sawhney RS, Kahlon KS (2017) Sentiment analysis of demonetization of 500 and 1000 rupee banknotes by Indian government. ICT Express http://dx.doi.org/10.1016/j.icte.20 17.03.001
9. Liu B (2012) Sentiment analysis and opinion mining. In: Synthesis lectures on human language technologies, vol 5, no 1, pp 1–167
10. Witten IH, Frank E, Hall, MA Pal CJ (2016) Data mining: practical machine learning tools and techniques. Morgan Kaufmann
11. Bernhard S, Christopher JCB, Alexander JS (1999) Advances in kernel methods: support vector learning. MIT Press
12. McCallum A, Nigam K (1998) A comparison of event models for Naive Bayes text classification. In: AAAI-98 workshop on learning for text categorization, vol 752, pp 41–48
13. Aha DW, Kibler D, Albert MK (1991) Instance-based learning algorithms. Mach Learn 6(1):37–66
14. Quinlan JR (1993) C4.5: programming for machine learning, vol 38. Morgan Kauffmann
15. India Census 2011 Population Report. http://www.dataforall.org/dashboard/censusinfo/. Accessed 10 Dec 2016

Event Detection Using Twitter Platform

Anuradha Goswami and Ajey Kumar

Abstract Online Social Network (OSN) has evolved through a radical transformation in the way user communicate with each other in the Web 2.0 environment. User communicate over OSN through a connected network structure, forming a group of individuals who interacts among themselves. Interaction among users, within a community or inter-community, facilitates in the formation and exchange of huge User-Generated Content (UGC) across the OSN platforms. UGC is an important source for researchers to extract relevant insights related to events of significance e.g. earthquake, product review, emerging topics, etc. In this chapter, a comprehensive survey of event detection techniques for OSN is done. First, the types of OSN based on information flow (service oriented, sharing services, Social Network Sharing News, Location Based Social Network and community building Social networks) and then the various categories of events (natural or manmade disaster events, public opinion events & emerging events) are studied. Second, events were categorized based on four dimensions—thematic, temporal, spatial and network structure. An extensive survey of dimension-wise event detection techniques is carried out and the research gaps are identified. Third, Twitter platform was taken as a case study due to its popularity among users as well as researchers. An in-depth survey of event detection techniques with respect to different dimensions applicable to Twitter data for disaster event management, detection of emerging events and prediction of emerging events is performed and respective research challenges are enlisted. Finally, an exclusive study is conducted for Twitter platform based data collection and event detection & analysis tools. The suggested open challenges will give researchers/readers ample scope to work upon.

A. Goswami · A. Kumar (✉)
Symbiosis Centre for Information Technology (SCIT), Symbiosis International (Deemed University) (SIU), Rajiv Gandhi Infotech Park, Hinjewadi, Pune 411057, Maharashtra, India
e-mail: ajeykumar@scit.edu

A. Goswami
e-mail: anuradha@scit.edu

© Springer International Publishing AG, part of Springer Nature 2019
S. Patnaik et al. (eds.), *Digital Business*, Lecture Notes on Data Engineering and Communications Technologies 21, https://doi.org/10.1007/978-3-319-93940-7_18

Keywords Web 2.0 · Online social networks · Twitter · Event detection · Disaster events · Emerging trends · Public opinion events · Data collection tools · Event detection tools

1 Introduction

The invent of Web 2.0 technology has brought up a radical transformation in the way people communicate with each other throughout the world over Internet. The foundation of Internet in the early 1990s has encouraged users to interact and communicate information over the web in different forms. The era of Web 2.0 further added to this volume of information in the form of World Wide Web content [1]. According to Cisco VNI (Visual Networking Index), the global IP traffic in 2016 stood at 96 Exabyte per month and will nearly triple by 2021, to reach 278 Exabyte per month, which implies a growth at a Compound Annual Growth Rate (CAGR) of 24% from 2016 to 2021 as shown in Fig. 1 [2]. Most of these traffics are generated by the Online Social Networks (OSN) platforms such as Youtube [3], Facebook [4], Twitter [5], etc. in the form of either text, videos, images or photos [6]. Table 1, shows the data explosion statistics from different OSN platforms in every minute of the day.

Online Social Network (OSN) has emerged as a part of this Web 2.0 technology. It is defined as web-based services that provide a platform to create public/semi-public profile within a specific domain by the users, which further initiate the communication to other users within the network [7, 8]. These users communicate over OSN through a connected network structure, forming a group of individuals who interacts among themselves following a certain pattern [9].

Fig. 1 Cisco VNI global internet traffic prediction

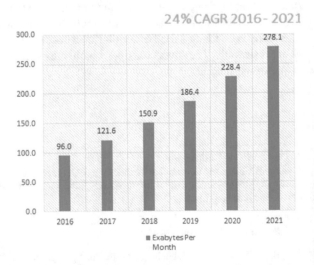

Table 1 Data generation rate for OSN

OSN	Data volume generated/minute of the day
Facebook	2,460,000 pieces of content shared
Twitter	Tweets 277,000 times amounting to 400 million tweets per day
Youtube	Users upload 72 h of new videos
Pinterest	Users pin 3472 images
Instagram	Users posts 216,000 new photos

This OSN network structure pattern is diagrammatically represented as collection of nodes (A1, B1, C1, …etc.) connected with links/edges ([A1, A5], [B1, B4] etc.), where nodes signify individuals/entities/actors and existence of an edge between nodes reveals that the nodes are related to each other socially [10], as shown in Fig. 2. The pattern could be casual connection, intimate bonding of friendship or formal business relationship. The communication pattern generates cohesive groups or communities (A, B or C as in Fig. 2.) in an OSN, where users are tightly coupled to each other with strong ties/edges through similar interests and characteristics within a community [11–15]. The inter-community ties/weak links ([A5, C5], [B4, C4], etc.) encourage sharing of information across groups/communities. For example, a group or community of technology gadgets form a strongly tied group of members/friends, where all follow the same news stream and updated with the same information related to technology world. A weak tie, on the otherhand, are loose acquaintances of the member of a community with some other community, with whom a person connects very often, and specially used to new information flow. But, in order to get a job reference, the said community members can get to use the inter-community ties, which sometimes proves to be more effective than the strong ties in a community in receiving any novel information from the rest of the social communities. Both the interaction among users, within a community or inter-community, facilitates in the formation and exchange of huge User-Generated Content (UGC) across the OSN platform [16].

UGC is defined as *"the work that is published in some context, be it on a publicly accessible website or on a page on a social networking site only accessible to a select group of people"*, or a work such that *"a certain amount of creative effort was put into creating the work or adapting existing works to construct a new one"* and something *"generally created outside of professional routines and practices"* [17]. The UGC is expressed in the form of text, blogs, micro-blogs, news, discussions or some other kind of documents [18]. Any interpersonal relationship between individuals are maintained by frequent exchange/flow of information in the form of UGC. The type and timing of information flow can have an effect on the intensity of the relationship between individuals, and is often called 'tie-strength' [19]. The UGC thus generated could be an important source of learning of opinions and sentiments [20], subjectivity [21], influences [22], observations [23], or feelings [16].

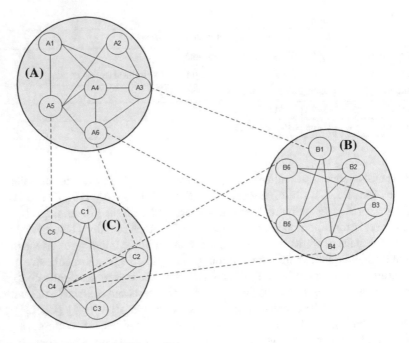

Fig. 2 Network structure of OSN

In order to derive important insights from this huge UGC of OSN during events of significance like earthquake, epidemics, election polls, etc., appropriate data collection tools are required by the researchers. The effective collection of this meaningful data in the form of UGC, imposes major computational challenges. These challenges are mainly due to the huge volume (in terabytes or petabytes), noise, and diversity of UGC data [24, 25]. All these challenges have motivated us to conduct a survey study to find out the event detection techniques available to the researchers, the existing data collection and event detection and analysis tools available, and existing research challenges with respect to event detection and analysis. To support our study, a Twitter case is taken in order to elaborate all these scenarios.

The organization of this chapter is as follows: Sect. 2, briefs on the different types of OSN. Section 3, outlines the different category of events which occurs in OSN. Section 4, details the different event detection techniques available along with the inherent challenges of the same. The case study with respect to Twitter platform is given in Sect. 5. Analyzing the role of Twitter in different categories of events along with the event detection techniques used by the researchers for churning the insights from this platform are also described. Section 6, deals with the existing data collection tools and the event detection and analysis tools available for Twitter platform. Lastly, Sect. 7, concludes the survey.

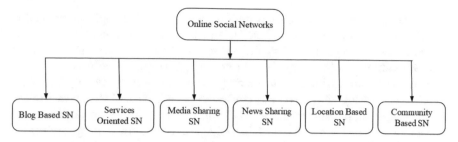

Fig. 3 OSN categories

2 Types of Online Social Networks

OSN can be categorized into several types based on different forms of UGC flows, through the corresponding network structure [26, 27].

The different categories of existing OSN's are outlined in Fig. 3 and their respective definitions & examples are discussed as follows:

Blog Based SN: A blog based SN is a service based on OSN. A blog is a type of a personal website in which bloggers (i.e. person who writes a post) can depict their own thoughts and opinions through a post [28–33]. Bloggers can influence one another through their postings and perform various activities in the blog world. Thus, variety of relationships gets created between the bloggers, resulting in the formation of a social network called blog based social network. Plethora of services are provided by these blog based SN to the bloggers. To mention a few, one is blog scraping through which one can scan through a huge number of blogs, copying content and generate his own blog. The other is trackback in which a blogger writes a new post related to someone else's post in their own blogs or refer back other blogs. These functions of scraping and trackback cause information diffusion over the blog world [34]. Few examples of blog based SN can be Blogger [30], WordPress [35] etc.

Services Oriented SN: Service Oriented SNs offer web-based services that allow individuals to construct a public or semi-public profile within a bounded system, articulate a list of other users with whom a connection is shared, view and traverse list of connections within the system. The nature and nomenclature of these connections can vary from one OSN site to other [36]. It includes a two-way communication and content sharing. These categories of SNs can be further divided into OSN like Facebook.com [4], microblogging site like Twitter [5] and professional SNs like LinkedIn [37], where each of them support millions of members on their respective platforms. Services-oriented SN also includes sites offering other services, such as Elftown.com offering service for fans and fantasy and science fiction [38] and Ravelry.com [39] which is for fans of knitting, both forms a part of the professional category, with small numbers of members but connected with a common interest.

Media-sharing SN: Social network services which are based on posting, sharing and commenting on contents like video, images, audio, etc. are considered to be in this category. When user creates multimedia content, it is uploaded/shared on the SN site to notify his/her network about the content. The notification can be shared explicitly by the user or by the website where UGC was created. Posts are generally composed of elements from the content (video, photos or images, audio) or sometimes appended with a link to the original content. A study by [40] has divided UGC into two categories: *direct share* where content is created on service oriented social networking site directly and *indirect share* where content is created in external websites/web services which are exclusively used as SNs for sharing services and included in service oriented SNs as a link to the external site. Few examples of SN sharing services are Youtube [3] for video sharing, Flickr [41] for photo sharing and Podcast [42] for audio sharing.

News-sharing SN: SNs have transformed the way news is consumed and shared by individuals. Traditionally, news was meticulously produced and disseminated amongst the public in a curated format [43]. Previously, audiences were also habituated in receiving their news only after a defined number of hours, in a defined format such as newspapers [44]. But nowadays, people exists within a constant buzz of attractive and ambient news, available anytime, anywhere on almost any device, sourced by both professionals and individual users [44]. This is an unprecedented shift from the traditional news media to social media. News is crowdsourced where any user can be a reporter during any kind of events. *News-sharing SN* is a platform on which this crowdsourced news is reflected from the users. The friends, family and acquaintances of an individual in their social network are populating and curating the news, which can be seen by any individual. In turn, individuals are consuming and re-sharing that content in their own SNs. Users can also favorite news, leaving it for the rest of the users in his/her network to see and accordingly like or vote it. Digg [45], Reddit [46] etc. are some of the most popular social news sites available for the users.

Location Based Social Networks (LBSN): SNs which insists on the users to share their location to enable location embedded information sharing amongst the people in their network structure, is a LBSN. It refers to a social structure where people are connected through same location mention as well as through their location-tagged media content, such as photos, video, and texts. Users of the same location normally share similar knowledge content, typically having common interest, behavior and attitude [47]. Location of a user indicates his current location at a given timestamp. These services operate through location-aware smartphones, where users explicitly record their presence at a location via an application. This action is termed as 'checkin', which is a message shared through a dedicated SN in near-real time. The geographical location of a user is normally represented as a venue in a LSBN, be it a meaningful place on streets, shop, park or building. The taxonomies of venues have evolved over the time from massive user participation and have also become widespread for urban areas throughout the world [48]. Few examples of LSBN include Facebook [4], Foursquare [49] etc.

Community-Based SN: Web communities which consists of members with strong identity ties based on race, nation, religion, interests, gender and so on are Community-based SNs. This variant of SN also sometimes serves as reshaping the existing offline communities. Sense of belongingness in these platforms sometimes gives rise to micro community's social network [27]. Passion-centric network also falls under this category of SN, as it gathers people who share common interests or hobbies, e.g. Dogster, CarDomain, etc. Google groups [50] can be an example for this community-based SN.

3   Event Categories

The relevance or significance of the term 'event' differs across domains, ranging from social computing to Topic Detection and Tracking (TDT). Accordingly, the definition of event also varies. An event, in case of OSN platform, can be defined as a particular incident/occasion that is entangled with some specific time or location [51]. Accordingly, Wang et al. [52] has defined event in his study as: "*An occurrence causing changes in the volume of text data that discusses the associated topic at a specific time*". So, it is evident from this definition that occurrence of an event can be popularly featured by the related topic involved, time of occurrence, and sometimes with the related entities or people and the location where it happened. So, an event when expressed through an OSN, will be eventually identified through all these mentioned parameters. Event is also defined in [53] as *a specific incident having a defined scope, with a firm beginning and ending*. Dong et al., defined event as a real world happening, which is realized through a sudden increase in the volume of textual data of expression in an OSN platform. Event was also defined through photos instead of text in a study by Chen and Roy [54]. According to the study, an event is realized through a set of photos if all the photos express the same specific incident with semantically consistent content, clicking time is within a certain time segment and the source of the photos denotes the same location. Events will be dealt with in this study covering two aspects: *event categories* and *event detection* as summarized in Fig. 4.

   The various types of OSN platforms is used in an extensive way by the users during different events like earthquake, forest fires, epidemics, elections etc. to express their mind, creating substantial amount of UGC. Researchers can get significant amount of information by churning the explosive UGC during these different kinds of events. This sizeable information could be regarding the people who are affected by the event, location and time of the event, extent of damage created and effects on surrounding environments. For example, during Queensland floods, data from OSN described the whole situation revealing the yacht sinking on Brisbane River, reopening of the port, incident of bull shark on a flooded street etc. [55].

   Post considering the different types of events that frequently happens, get reflected in OSN and generates massive data or information. As shown in Fig. 5, events can be categorized broadly into three groups, given as follows [56]:

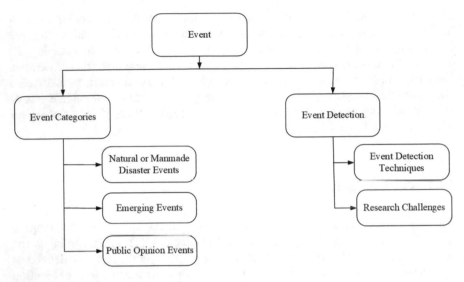

Fig. 4 Event and its aspects covered

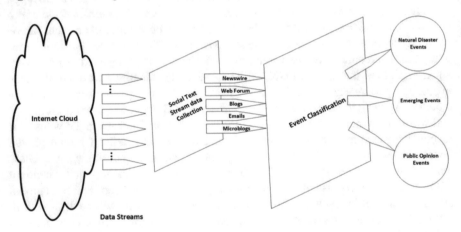

Fig. 5 Event detection and classification from data streams

Natural or Manmade Disaster Events (NMDE): The world has witnessed the occurrence of frequent and series of huge natural and man-made disasters in the recent years such as Hurricane Katrina in USA, The tsunami in Indonesia, Haiti and Asian earthquakes, earthquake and tsunami in Japan, Mumbai terrorist attack in India and World trade centre tragic event in New York. The disasters, whether natural or manmade come without any warning in general and take lives of hundreds and thousands of people [57]. These disasters result in increasing communication among the people for various reasons like to contact family and friends in the disaster zones, seek information regarding potential sources of food, shelter, transportation and others.

In recent disaster situations, OSN platforms such as Facebook, Twitter, Flickr, and Youtube have played an important role in breaking news about the disasters. OSN serve as platform for the users to share information, knowledge, and even seeking for help or support. There are plethora of research studies which shows the role of social networking services and their problem solving capability during natural as well as man-made disasters [58–61]. During disasters when all the conventional communications media generally stop functioning, the social media networking services remains active for the users. Thus, these OSN platforms functions as a live network of monitoring active sensor systems for detection of world disaster events in real life scenarios [62–66]. Additionally, these SN systems can provide the needful resources required to connect people recovered from disasters. There exist variety of communities or groups which get formed in most OSN platforms during and post disaster, who continue discussions on situation awareness, emergency needs and knowledge sharing among others. For example, during earthquake in Japan on Friday, March 2011, millions of users were able to connect links and resources on OSN platforms, taking a form of global voice through international community of bloggers in multiple languages. A recent study conducted by American Red Cross Society, reveals the immense popularity of different OSN platforms like Facebook, Youtube, Myspace, Flickr and Twitter, once natures strikes in various forms of natural or man-made disasters [67–69]. These OSN platforms can be analysed across three different time frames: pre, during and post disaster events. It is required to plan ahead during pre-disaster phase, to communicate, share and control important information during disasters and to coordinate recovery processes post disasters. In all these cases, the most important task is to extract important insights on the events which are meaningful from the user point of view.

Emerging Events (EE): Emerging events can be termed as 'trends' that appear in stream of OSN data and so real-time in nature. Trends are typically driven by breaking news and general topics that attract the attention of a large fraction of OSN users [70]. Thus, trend detection is of high importance and significance to news reporters and analysts, as churning this data can yield them with fast-evolving news stories. For example, at the announcement of Michael Jackson's death on June 25, 2009, Twitter was immediately flooded with enormous volume of connected discussions and comments. Trend detection is also important for online marketing professionals and opinion tracking companies, as trends or emerging events point to topics that capture the general public attention. Trend analysis mandatorily require real-time trend detection for a live stream of data from OSN platform where topics of discussion shift dynamically with time. The other requirement of the system is to be scalable, supporting this massive document streams over the defined period of time.

Public Opinion Events (POE): Focused events that are capable of inducing self-interested responses or comments from the users or people, either positive, negative or neutral way, through an OSN platform, are called Public Opinion Events. Public opinion was subject to constant potential influence and flux since the term 'CNN effect' was coined to describe the impact of a 24-h news cycle. Also, it has been noted that public opinion is a very slippery concept to define. There are numerous definitions which prevails for public opinion [71, 72]. The orthodox contemporary view of public opinion, as clearly expressed by early pollsters [73] is that it is the cumulative preferences of individual citizens. So from this, it is evident that public opinion is something which pollsters aim to discover. But, some critics have argued against this definition, claiming that public opinion research manufactures public opinion rather than simply discovering it [74]. Despite this critical views, the Gallupian paradigm of public opinion only dominate the journalistic, political and general imagination [75, 76].

The traditional method of charting public sentiments through survey based research was slow, constrained through questionnaire and expensive. Measurement of public opinions via OSN 'big data' removes all those limitations of past methodology. Measurement is continuous and instantaneous, providing a live monitor of public mood that can be tracked through a presidential debate, an election campaign, a foreign policy crisis, a full parliamentary term or a new product launch. The data from Facebook, Twitter is capable of capturing the shades and flavours of public opinion.

4 Event Detection

Event detection has drawn considerable attention of researchers to automatically understand, extract and summarize the happenings of an event in different fields like bio-surveillance, safety, health, economics etc. [77]. In the field of disease outbreaks, DARPA [78] has estimated that a two-day gain in the detection time can reduce the extent of fatalities by factor of six. Also, studies in [78, 79] has showed that a hotspot violent crime can be detected 1–3 weeks prior to the event by detecting the clusters of leading indication crimes like disorderly conduct, assault etc. In another study in the field of safety, detection of anomalous clusters of pipe breakage is done supporting the monitoring system of city's water distribution [78]. Environmental monitoring is done through remote sensors which are used for continuous observation of a certain place or domain, generating large volumes of data to analyse the same [80]. So it is evident that event detection in OSN contributes much to understanding or predicting the events.

The primary motivation of event detection analysis study got initiated by a project called Topic Detection and Tracking which was initiated as a joint venture of DARPA, CMU and Dragon systems [81]. According to one of the issues of this project, event is distinguished from the term topic by the property of time. For example, "Mumbai Terrorist Attack on 26 November, 2008" [82] is considered to be an event but "terrorist attacks" is a more general topic. This difference can be extended by incorporating the spatial or location aspect in the definition also [81].

4.1   Event Detection Techniques

According to [55], any input message by a user posted to an OSN can be considered as his observation on a real-world happening at a certain location and time. This observation is event element. Given a set of event elements say E, an event is defined as a subset E_i of E, provided all the event elements in the set Ei refers to the same real-world occurrence. Any event can be described by the tuple given as follows:

$$E_i = \{M_i, L_i, T_i, N_i) \text{ where } `i' => \text{same real-world occurrence}$$

In the above tuple, M_i represents data or message, L_i represents location in terms of longitude and latitude, T_i denotes time of occurrence of the event and N_i represents the embedded network structure associated with the event data.

Depending on these four features of real time streaming data of OSN platform, the existing event detection techniques can be categorized into four dimensions [55, 83] as follows:

(a) *Thematic Dimension (TH)*: This dimension includes methods/techniques which are used to detect events defined by analysing the semantic and contextual features of the OSN text stream [1]. OSN text stream data is a collection of informal text communication data which gets generated over time and each piece of text in the stream consists of some social attributes named author/user, reviewer, sender, and recipients. Each piece of text also carries part of the semantics, explaining the information about the real world events [84]. Basically, the textual content description of a specific message or a number of related messages, refer to the keywords of social messages after the specific stop words are excluded from them [55].

The Thematic Event Detection Techniques are as follows:

(i) *Term and Named Entity Vector*: TDT (Topic detection and Tracking) program is responsible for developing technologies that search, organize and structure multilingual textual news from a variety of broadcast news media. New Event Detection (NED), a subtask of TDT, is concerned with the developing of systems that is capable of detecting the first story on the topic of interest. For example, if a topic is regarding 'multi-storied building on fire', the first story on the topic could be the article which published the news. The other stories on the same topic could be the discussion regarding locality damage, missing people, salvaging efforts, financial impact and so on. A good NED system is the one which correctly identifies the first story article without delay [85, 86]. Detection of new stories is done by the comparison of the story in hand with the all past existing stories using similarity calculation metric like cosine, Hellinger distance, etc.

(ii) *Clustering Algorithm*: Clustering is a task of finding groups of similar documents in a given collection of documents, extensively used in the field of text mining. Different levels of granularity can be observed in case of text clustering where clusters can be formed out of documents, paragraphs, sentences or terms. For example, Cutting et al. [87] has used clustering technique to produce a table of contents of a large collection of documents. Various software tools used for implementation of common clustering algorithms are Lemur and BOW [88]. The text clustering algorithms are divided into agglomerative, partitioning and probabilistic clustering algorithms types [89].

(iii) *Incremental Clustering Algorithm*: Traditionally, the batch document clustering mechanism required all documents to initiate the process and cluster the document collection through multiple iterations over them. But, with the huge generation of informational textual documents now in World Wide Web in the form of Newswire, Blogs, Microblogs etc., the batch clustering is impossible with respect to the sheer volume of documents and also time consuming. This gave rise to the requirement of an incremental clustering algorithm which process the documents as soon as they arrive in real time. The incremental algorithm can cluster the text documents as they arrive into an information cluster hierarchy. A well-known incremental hierarchical clustering algorithm COBWEB [90] got used in previous researches in non-text domain. Study by [91] applied a variant of COBWEB to textual documents. This is less time consuming too as they make very few passes over the entire dataset and decide the cluster of an item as they see it.

(iv) *Gaussian Mixture Model*: This is a probabilistic model used to represent normally distributed subpopulations within the overall population. Normally, for mixture models, it is not required to know which subpopulation a data point belongs to, allowing the model to learn the subpopulations automatically. For example, suppose we are modeling the data of human heights, height is typically modelled as a normal distribution for both males and females. Given only the height data without any gender assignments for each data point, the distribution of all heights would follow the sum of two scaled (different variance) and shifted (different mean) normal distributions. This is Gaussian Mixture Model (GMM), where the model can have more than two components. A more practical example given by study in [92] states that suppose a primary requirement of medical diagnosis is to identify the range of values for an appropriate feature set which distinguishes between abnormal from normal cells. The measurable factors may be shape, color, texture, size, motility, chemical composition etc. The statistical tendency of abnormal cells cannot be easily characterized by any structured density. Hence, a mixture model consisting of a number of component densities, can be used in this case to construct a satisfactory model for distinguishing different classes of cells.

(v) *Binomial Distribution*: A study by [93] proposed hot bursty events detection in a text stream, where the text stream consists of a sequence of chronologically ordered documents. In this study, hot bursty event is a minimal set of bursty features that occur together in certain time windows with strong support of documents in the text stream. The objective of the study was to detect all the bursty features to detect the bursty event which is a partially supervised text classification procedure [94, 95]. For example, SARS (Special Severe

Acute Respiratory Syndrome) can be termed as a bursty event with a set of bursty features such as sars, outbreak, respire, pneumonia etc. This event was reported in four hot periods in major English newspaper, South China Morning Post, in Hongkong, in different temporal diversities from April 2003 to 11 January 2014. The complete procedure of event detection was categorized into three steps [93]: Bursty features identification, Bursty features grouping, Hot periods of bursty events determination. Binary distribution was used to model Bursty feature identification using hyper-geometric distribution. Binomial distribution is used to model the probability of occurrence of the feature f_j in the time window W_i, $P_g(n_{i,j})$ given as:

$$P_g(n_{i,j}) = \binom{N}{n_{i,j}} p_j^{n_{i,j}} (1 - p_j)^{N - n_{i,j}}$$

where N is the number of documents in a time window. The number of documents can be different in each window, but rescaling of the same is possible so that all N_i become all the same. It is done by adjusting the frequency of documents having the bursty features in all time windows.

(vi) *Discrete Fourier Transformation (DFT)*: A research study conducted in [96] on a problem of analysing word trajectories in both time and frequency domains, having the ultimate aim to identify important and less reported, periodic and aperiodic words. A set of words having same trends can then be further grouped together to report an event in an unsupervised manner. The document frequency of each word across time is treated like a time series, where each element is the *Document Frequency-Inverse Document Frequency* (DFIDF) score at a time point. In fact, in many predictive event detection tasks, vast set of potential features are present only for a fixed set of observations (which are obvious bursts). A small number of features in these are expected to be useful. The research studied a novel problem of analysing feature trajectories for event detection through identifying distributional correlations among all features by spectral analysis, where DFT comes into existence. For example, a high correlation between words like Easter and April can be found by plotting their normalized DFIDF data, depicted through heavy overlap of two word curves. This gives an indication that both of them belong to the same event 'Easter feast' which is an important aperiodic event over 1 year.

(vii) *Naïve Bayes Classifier*: A study by [97] demonstrated how to use Twitter to automatically obtain breaking news only from tweets posted by the Twitter users through a system *TwitterStand*. The system also provides a map interface for the users to read this news through clustering tweets according to geographic location. Now, tweets are basically noisy, as most of them carry a very least significant information as per as the contextual content is concerned. The limitation on the size of the tweet is also responsible for the same, as very little can be conveyed in 140 characters which is the upper limit of the size of a tweet. The study had to distinguish between news and noise by applying filter on inbound tweets to the system, with the exception on the tweets sourced by Seeders or the knowledgeable users regarding the event. The study aimed at finding a way to discard tweets which are not news, thus partially getting rid of noise data. Naïve Bayes Classifier [98] was used for the purpose which is already trained on a training corpus of tweets that have already been categorized as either news or junk.

Given a tweet 't', comprised of a set of words or terms $w_1, w_2, \ldots, w_k$, the probability that 't' is a junk, denoted by $P(J|w_1, w_2, \ldots, w_k)$, can be expressed using Bayes theorem as:

$$P(J|w_1, w_2, \ldots, w_k) = p(J) \cdot \frac{p(w_1, w_2, \ldots, w_k|J)}{p(w_1, w_2, \ldots, w_k)}$$

Similarly, given tweet 't', the probability that 't' is news, given as $P(N|w_1, w_2, \ldots, w_k)$, can be expressed through Bayes theorem as:

$$P(N|w_1, w_2, \ldots, w_k) = p(N) \cdot \frac{p(w_1, w_2, \ldots, w_k|N)}{p(w_1, w_2, \ldots, w_k)}$$

Assuming that in both cases, there is complete independence of words in 't', the above equalities can be reduced to as follows:

$$P(J|w_1, w_2, \ldots, w_k) = \frac{p(J)}{Z} \cdot \prod_{i=1}^{k} \frac{p(w_i|J)}{p(w_i)}$$

$$P(N|w_1, w_2, \ldots, w_k) = \frac{p(N)}{Z} \cdot \prod_{i=1}^{k} \frac{p(w_i|N)}{p(w_i)}$$

where Z is a normalizing factor that ensures

$$P(J|w_1, w_2, \ldots, w_k) + P(N|w_1, w_2, \ldots, w_k) = 1$$

Now, dividing the above equalities by each other, and taking logarithm on both sides,

$$D = \log \frac{P(J|w_1, w_2, \ldots, w_k)}{P(N|w_1, w_2, \ldots, w_k)} = \log\left(\frac{P(J)}{P(N)}\right) + \sum_{i=1}^{k} \log \frac{p(w_i|J)}{p(w_i|N)}$$

If $D < 0$, the resultant tweet is a news else it is identified as junk and removed from dataset.

Here,

$$P(J) = \frac{\textit{Total number of words in tweets marked as junk in corpus}}{\textit{Total number of words in the corpus}}$$

$$P(N) = \frac{Total\ number\ of\ words\ in\ tweets\ marked\ as\ news\ in\ corpus}{Total\ number\ of\ words\ in\ the\ corpus}$$

$$P(w_i|J) = \frac{Number\ of\ times\ word\ w_i\ appears\ in\ tweets\ marked\ as\ junk\ in\ corpus}{Total\ number\ of\ times\ in\ appears\ in\ the\ corpus}$$

$$P(w_i|N) = \frac{Number\ of\ times\ word\ w_i\ appears\ in\ tweets\ marked\ as\ news\ in\ corpus}{Total\ number\ of\ times\ in\ appears\ in\ the\ corpus}$$

(viii) *Support Vector Machine (SVMs)*: SVMs are based on the Structured Risk Minimization principle [99] from the computational learning theory. The theory of structural risk minimization is to find a hypothesis h for which a lowest true error can be guaranteed. The true error of h is the probability that h will make an error on an unseen and randomly selected test example. In machine learning context, SVM is normally associated with learning algorithms which handles data for classification and regression analysis. It supports linear classification as well non-linear classification of data.

(ix) *Latent Dirichlet Allocation (LDA)*: The topic models normally assume the presence of latent (hidden) topics behind words in human language. For that reason, if an author uses the word *automobile* in a document and a user who is browsing uses a word *vehicle* in a query, topic models assume both of them referring to the same concept/topic *car*. Based on this assumption, topic models provide methods which infer those hidden/latent topics from the visible words in documents. Now, though PLSI [100] contains a good probabilistic generative model and also a statistical inference method, the method does not fit well with hidden or unobserved words and also suffers from overfitting problem.

Blei et al. [101] introduced the Latent Dirichlet Allocation model to solve this problem which not only assigns high probability to members of the corpus, but also assigns high probability to other similar documents.

(x) *Gradient Boosted Decision Trees*: The most common approach to data-driven modeling is to develop a single strong predictive model. A different approach is to build a bucket, or an ensemble of models for some learning task. A set of strong models like neural networks is also considered which collectively can yield better prediction. Practically, the ensemble technique adheres to combining a large number of relatively weak simple models to generate a strong prediction. Random forests [102], Neural networks [103] and boosting algorithms are of this ensemble technique category. The main idea of boosting is to add new models to the ensemble sequentially. At each iteration, a new weak, base-learner model is trained with respect to the error of the whole ensemble learnt from the beginning.

The first prominent boosting technique was mainly algorithm-driven [104]. The two extremes of these algorithms were either they outperformed all other algorithms or were inapplicable due to overfitting problem [105]. To establish statistical framework connection, a gradient-descent based formulation of boosting methods was derived [106, 107]. This formulation of boosting methods and the corresponding models were called Gradient Boosting Machines (GBM). Using GBMs, the learning procedure consecutively fits new models or base learners as already mentioned to provide more accurate estimate of the response variable. The principle idea behind this algorithm is to construct the new base-learners to be maximally correlated with the negative gradient of the loss function, associated with the whole ensemble.

(xi) *Conditional Random Fields*: The requirement of assigning labels to a set of observation sequences is present in many fields like computer vision, bioinformatics, computational linguistics or speech recognition. For example, in a natural language processing task of labelling the words in a sentence with their respective part-of-speech (POS) tags. In this example, an annotated text can be generated by labelling each word with a tag indicating the corresponding POS. Sometimes, prediction of large number of variables are required that depend on each other as well as other observed variables. For example, the performance of a cricket team as a whole depends on the health and accomplishment of each member of the team. Again, the health of the members might get affected by frequent inter-state trips or world trips which might in turn affect the net outcome of the games played. Bad performance in games might have an impact on the morale of the team. So, here there are multiple variables that depends on each other intricately. These type of problems are normally modelled by *Conditional Random Fields* (CRFs) [108]. An ordinary classifier predicts a label for a single sample without regard to neighbouring input samples whereas the CRF takes the context of the features into account. The CRFs address a critical problem faced by the graphical models like *Hidden Markov Model* (HMM) which has gained a lot of popularity in recent years. CRF model the conditional distribution directly where dependencies that involve only variables from input data play no role, thus having a much simpler structure than the joint distribution model of HMM.

CRFs are probabilistic framework for labelling and segmenting structured data, such as sequences, trees and lattices. Particularly, this is useful in modeling time series data where temporal dependency can be utilized in many different ways. CRF defines a conditional probability distribution over label sequences given a particular observation sequence rather than a joint distribution over both label and observation sequences like HMM. The primary assumption of CRFs is the relaxation of independence assumption. The independence assumption says that variables are not dependent on each other neither they are affected by each other.

(xii) *Etree using n-gram based content analysis techniques*: Event modelling is a challenging problem in OSN for various reasons [109]. Firstly, messages posted in OSN platform tends to be of short length, unstructured, informal and differing writing style. This data sparseness, dynamic vocabulary and lack of contextual data make the challenge even stronger. Secondly, diversity of data with respect to content, time periods, and formal or informal relationships [110] also makes it difficult for event modelling. Last but not the least, the real time generation of data which demands highly efficient, incremental procedures and tools to be there to cater to this new information as and when generated, for on time information retrieval.

Etree was developed by the study [109] to address these challenges, generating an effective event modelling solution in OSN framework. Etree model identifies the major themes of the event, the major information blocks or clusters and their hierarchical structure within the themes and also the causal relationships between the information blocks. The key contribution of the model are:

- Proposed a n-gram based content analysis technique for identifying core information blocks, processed from a large number of user gen-erated content.
- An incremental or hierarchical modeling technique was identified to detect and construct event theme structures at different granular le-vels, adjustable enough to get tuned as events evolve.
- Technique for identifying potential causalities between information blocks.

(b) *Temporal Dimension* (T): The intuition of using this temporal dimension for event detection in OSN is that the textual stream of data about the same event are expected to fall into a specific time interval in the history. Each message/text is attached to a time stamp, which shows the posted time of the messages indicating the approximate time of an event. So, the traffic patterns (high, lows, and peaks) of stream belonging to a definite event topic can be used to extract events in a finer granularity [83].

The Temporal Event Detection Techniques are as follows:

(i) *Wavelet Transformation*: *Event Detection with Clustering of Wavelet based signals* (EDCoW) is a *model* [111] *used* for detecting events by analysing text streams in Twitter. EDCoW builds signals for individual words which captures only the bursts in the words appearance. The signals can be computed by wavelet analysis and also requires less space for storage. All the trivial words are filtered out seeing the corresponding signal auto-correlations. EDCoW then measures the cross correlation between signals. The events are detected by clustering signals together by modularity-based graph partitioning, which can be done through scalable eigenvalue algorithm. The big events are differentiated with trivial ones through quantifying event's significance which is calculated through two factors- the number of words and the cross correlation among the words relating to the event.

The wavelet analysis provides a brief idea of measurements about when and how the frequency of the signal changes over time [112]. The wavelet is a quickly vanishing oscillating function. As compared to the sine and cosine function of Fourier analysis, which are localized in frequency but extend infinitely in time, wavelets are relatively localized in both time and frequency.

Wavelet transformation is the core of the wavelet analysis. For wavelet transformation, signal is converted from the time domain to the time-scale domain where scale can be considered as the inverse of frequency. A signal is decomposed into a combination of wavelet coefficients and a set of linearly independent basis functions. These independent basis functions, also termed as wavelet family, are generated by scaling and translating a chosen mother wavelet $\psi(t)$. Scaling is stretching or shrinking $\psi(t)$, while translation means moving the mother to a different temporal position without changing its shape.

(ii) *TSCAN*: There are existing studies on effective *Topic Detection and Tracking* (TDT), to detect and track all related documents [113–115]. But, it is difficult for the users to understand a topic thoroughly through all its connected documents. So, there was a urgent need for an effective summarization methods which can extract the core parts of the detected topics as well as portray the relationship between the core parts graphically. Topic anatomy is the combination of both the techniques which can express essential information about a topic in a structured fashion. TSCAN (*Topic Summarization and Content ANatomy*) [116] is a topic anatomy system which organizes and summarizes a temporal topic expressed through a set of documents. It models the documents of a topic as a symmetric block association matrix and treats each eigenvectors of the matrix as a theme embedded in the topic. Events and its summaries from each theme are then extracted from the eigenvectors. Finally, a temporal similarity function is applied to generate the event dependencies, which eventually facilitates in generating the evolution graph of the topic as a whole.

(iii) *Temporal and Dynamic Query Expansion Model*: Solutions for microblog search task in the recent research is minimal. The existing search tasks only allow to retrieve individual microblog posts against a query. So, a novel search task called microblog event retrieval was proposed in [117], which has the capability to retrieve a ranked list of structured event representation from a huge archive of historical microblog posts. Structured representation is a list of timespans during which some instance of event has occurred and was discussed on generating microblog stream. Also, a small set of relevant messages which provides a high level summary of the event is also present in the timespan. This model intends to leverage real-time first-hand information to deliver a novel form of search result to the informative users, consisting of relevant and event-related information.

The framework accepts a query as input and returns a ranked list of structured event representations. To perform this, the model is divided into two steps— timespan retrieval and summarization. The timespan retrieval step identifies the timespans when the events actually happened and the summarization step retrieves a small set of microblog messages from each timespan which can play the role of event summary.

(iv) *Spectral Analysis*: In multivariate time series, user behaviours are very common with oscillatory content, leading them naturally to spectral analysis [118]. The computation of spectrum of user i quantifies the overall activeness of a user in OSN. In order to calculate the spectral estimate of users (auto or cross-spectrum and coherence), firstly the Fourier transformation is applied on A_i, where A is the m-dimensional multivariate process, where each row denotes a user and each column indicates the activeness of these users on a specific day. A simple estimate of the spectrum can be obtained by taking a square of the Fourier transform of the data sequence. But, this estimate faces difficulties of bias and leakage. A combination of spectral analysis with multitaper technique [118] obtains a smoother Fourier-based spectral density with reduced estimation bias.

(v) *Hidden Markov Model (HMM)*: In the Naïve Bayes model, only a single output variable has been considered. In this HMM model [119], a sequence of output variables are considered, given as $\vec{y} = (y_1, \ldots, y_n)$ for an observation sequence $\vec{x} = (x_1, \ldots, x_n)$. Dependencies between single sequence positions are not taken into account. Also, there is only one feature present at each sequence position, namely the identity of the respective observation:

$$p(\vec{y}, \vec{x}) = \prod_{i=1}^{n} p(y_i) \cdot p(x_i | y_i)$$

Each observation x_i depends only on the corresponding class or output variable y_i at that respective sequences position. Due to this independence assumption, transition probabilities from one step to another are not considered in this model. But, in practical, transition probability does exist, resulting in limited performance of such models. So, in practice, it is wise to assume that there are dependencies between the observations at consecutive sequence positions. Thus, the state transition probabilities are added to model and is given as:

$$p(\vec{y}, \vec{x}) = \prod_{i=0}^{n} p(y_i | y_{i-1}) \cdot p(x_i | y_i)$$

The initial probability distribution is assumed to be as $p(y_o | y_{-1}) = p(y_0)$. Thus, the model of HMM is given as:

$$P(\vec{x}) = \sum_{y \in \mathcal{Y}} \prod_{i=0}^{n} p(y_i | y_{i-1}) \cdot p(x_i | y_i)$$

where $\mathcal{Y}$ is the set of all possible label/output sequences $\vec{y}$. Though HMM has considered the dependencies between output variables $\vec{y}$, to avoid complexity the model has not considered the conditional independence between the input variables $\vec{x}$. Conditional Random Fields (CRFs) considers the problem and addresses the same.

(vi) *Multivariate Event Detection Methods*: Multivariate refers to the vector created by each data point and recorded at some specific point of time [120]. For example, patient records from an Emergency Department for a vector with data of admission, time, gender, age, prodrome, home location, work location and much more. The analysis of data is done by dividing the learning and the test set with respect to time. The multivariate event detection involves detection of a change has happened followed by identification of the subgroup that has changed the most. There are few algorithms available for multivariate event detection on categorical data, emerging patterns [121], STUCCO [122], WSARE 2.0 [123], and WSARE 3.0 [123].

(c) *Spatial (S)*: Recently, the additional source of information which the OSN services like Foursquare, Gowalla, etc. are offering about user behaviour is one's geographic location, which attracts millions of users over a short period of time [124]. In general terms, the service offers the geographic mobility of individuals. The online interactions of Facebook tend to be spatially clustered and this geographic locality of interest is exploited by Facebook to improve service responsiveness with distributed proxies [125], in a company's email network, and also to partition email traffic across storage locations [126]. Spatial difference in content requests arising from OSN have been used to reduce latency and bandwidth costs associated to content delivery [33]. Also the spatial patterns observed in Facebook social connections is exploited to predict the geographical location of the users, given their friends locations [127]. The location associated with the profile of each user in OSN, are city name, country name, suburb name, postcode etc. The location of a user is mapped into a point (la, lo), where la denotes its latitude and lo signifies the longitude.
The Spatial Event Detection Techniques are as follows:

(i) *Locality Sensitive Hashing (LSH)*: A problem of similarity search involves a collection of objects (e.g. documents, images), characterised by a collection of relevant features and are represented as points in a high-dimensional attribute space. The queries also form points in this space, where it is required to find the nearest or most similar object to the query. The features of the objects of interest are represented as points in n-dimensional Euclidean space and a distance metric is used to perform indexing or similarity searching for query analysis [128]. The study [128] proves that in many applications of nearest neighbour search, an approximate answer is good enough than the exact one. Also, the approximate similarity search can be performed much faster that the exact one. The study proposed a new indexing method LSH for approximate nearest neighbour with a truly sublinear dependence on the data size even for high dimensional data. The key idea is to hash the points using several hash functions for each function so that the probability of collision is much higher for objects closer to each other that for the ones which are far apart.

(ii) *Kalman and Particle Filtering*: Statistical representation of locations facilitates in giving a user-friendly, integrated interface for location information and is a domain to apply techniques like Bayesian filtering. In location estimation, the state is a person's or object's location and location sensors provide observations about the state [129]. The state could portray a simple 2D position or a complex vector including 3D position, pitch, roll, yaw and linear and rotational velocities. The aim of Bayes filters is to calculate a sequential estimate of these beliefs over the state space which is conditioned on all information contained in the sensor data.

Kalman Filters are one of the most common Bayesian filter variant. These filter approximate beliefs by their first and second moment, which is virtually same to a unimodal Gaussian representation. Kalman filters are optimal estimators, provided the initial uncertainty is Gaussian and the observation model and system dynamics are linear functions of the state. Now, as most the systems are strictly non-linear, extended Kalman filters are used instead of normal Kalman filter, which make the system linear using first order Taylor series expansions. Another advantage of Kalman filters is its computational efficiency. But the trade-off is this efficiency can be experienced only for unimodal distributions. Kalman filters are the best choice if the uncertainty with respect to the location of the state is not so high which also limits sensors uncertainty. Though, it has got so many drawbacks, researchers have successfully applied them to various tracking problem where the filters were able to perform efficiently, providing accurate estimates, including some highly nonlinear environment.

Particle Filters is a probabilistic approximation algorithm implementing a Bayes filter, and also a member of the family of sequential Monte Carlo methods [129]. The Sequential Importance Sampling (SIS) algorithm is a Monte Carlo method which forms the basis for particle filters.

(iii) *Univariate Spatial Scan Statistic Approach*: A scan statistic is used to detect clusters in a point process [130]. Naus et al. [131] and many other researchers have studied the concept in one-dimensional setting. For a point process on an interval $[a, b]$, a window $[t, t + w]$ of fixed size $w < b - a$ is moved along the interval. Over all possible values of t, the maximum number of points in the window is recorded and compared to its distribution under the null hypothesis of a purely Poisson process. The space-time statistic is defined through a huge number of overlapping cylinders [132]. For each cylinder z, a Log Likelihood Ratio $LLR(z)$ is calculated and the maximum LLR calculated over all the cylinders is defined as the test statistic. The type of application will determine the collection of cylinders. The circular base of the cylinder explains the geographical area where radius varies continuously from zero to some upper limit which should be defined according to the application so that the circle contains at most 50% of the population at risk. The height of the cylinder expresses the time duration or span that can range of a single day to several years. For each choice of base circle, all choices of height are considered, giving rise to different shapes cylinders.

(iv) *Multivariate Spatial Scan Statistic Method*: The multivariate scan statistic solves the problem of simultaneous search and evaluation of clusters in more than one data set at a time. The steps followed are as follows [133]:

The LLRs are calculated for each cylinder per data set and it is noted whether observed number of cases is bigger or smaller than the expected ones.

For each cylinder, the log likelihood ratios for the data sets with more expected number of cases is added up. For that particular cylinder, this sum is one of the two likelihoods. The second likelihood is the sum of all the log likelihood ratios for the data sets with fewer than expected cases.

The maximum of all the summed log likelihood ratios, calculated over all the cylinders, constitutes the most likely cluster. This maximum is the definition of the multivariate scan statistic.

(d) *Network Structure* (*NS*): This dimension involves the techniques which studies the embedded network structure or social network information of an OSN, information flow pattern between users and connection between their messages. During an event, the purpose and point of view from different communities can be different and these communities may or may not be somehow socially connected [83]. In that respect, even the same keyword or topic of event may get interpreted in different ways in different communities. This proves the significance of using the network structure analysis which can give a proper analysis of the events discussed. The network structure information basically involves the links between the current users and their followers, and the connections between their messages.

The Network Structure Event Detection Techniques are as follows:

(i) *Dynamic Time Warping*: Majority of the work in indexing of very large time series databases has focused under the Euclidean distance metric [134, 135]. But many of the studies have argued that Euclidean distance measure is a very brittle distance measure [136, 137]. The need of the day is a method which will support an elastic shifting of the time axis, to accommodate sequences that are similar but out of phase. This need was fulfilled by [138] introducing *Dynamic Time Warping* (DTW) [139] to the database community. Though they demonstrated the utility of the approach, they also acknowledge that DTW has got its resistance to indexing and performance also degrades on very large databases.

(ii) *Graph Cut*: A graph $G = (V, E)$ can be partitioned into two or more disjoint sets A, B, where $A \cup B = V, A \cap B = \emptyset$ by removing the edges connecting the two parts. The total weight of the edges that have been removed gives the degree of dissimilarity between these two graphs. In graph theory, it is called cut and is given as:

$$cut(A, B) = \sum_{u \in A,\, v \in B} w(u, v)$$

The optimal bi-partitioning of a graph is that which minimizes this cut value. There can be exponential number of partitions depending on the number of vertices in the graph, but plethora of algorithms are proposed for finding the minimum cut of a graph.

The minimum cut algorithm takes an undirected graph 'G' as input where parallel or multiple edges are allowed. The goal is to compute a cut with fewest number of crossing edges (a min-cut). The graph cut algorithms can find its applications in splitting large graphs, community detection, weakness on a network detection and also weak ties detection tasks in the network structure of graphs [140].

Table 2 Event detection: dimension and techniques

Dimension	Techniques
Thematic	Term and named entity vector
	Clustering
	Incremental clustering
	Gaussian mixture model
	Binomial distribution
	Discrete Fourier transformation
	Naïve Bayes classifier
	Support vector machines
	Latent Dirichlet allocation
	Gradient boosted decision trees
	Conditional random fields
	ETree using n-gram based content analysis
Temporal	Wavelet transformation
	TSCAN using eigenvectors of a temporal block association matrix
	Temporal and dynamic query expansion model
	Spectral analysis
	Hidden Markov model
	Multivariate event detection model
Spatial	Locality sensitive hashing
	Kalman and particle filtering
	Univariate spatial scan statistic
	Multivariate spatial scan statistic
Network structure	Dynamic time warping
	Graph cut algorithm

A summarization of all the detection techniques as mentioned above, with respect to their corresponding dimensions is given in Table 2:

4.2 Research Challenges

Though identification of new events and getting important insights related to the new events is very important, but there are challenges which these event detection techniques confront in general [1, 56, 77, 78, 141]:

Domain Dependence: An event detection procedure or technique which is suitable for one domain might not be the same for the other domains and it is extremely situational-dependent [78]. For example, the selection of parameter, variables and output metrics for predicting the electoral poll results will not be the same as the prediction of any natural disaster event for example earthquake.

Time Constraint: An extreme timeline constraint is a timeline in which the event detection method should be able to identify event correctly. Depending upon the domain criticality, the timeline can range from seconds to several minutes. For example, detection of any terrorist attack or measuring indicators of imminent catastrophic disasters is critical applications where constraint on a specific timeline is an important consideration.

High True and Low False Alarms: High precision means that the event detection method provides a high true positive (i.e. correct detection) rate while providing a low false positive (i.e. incorrect detection) rate [142]. Generation of alarm for a true event in mission critical domains like healthcare, banking sectors is very crucial. Similarly, false alarm generation due to wrong event detection could involve huge monetary loss and should be considered as a serious challenge.

Diversified Data Sources: OSN has effectively contributed to huge explosion of diversified data consisting of unstructured data, textual documents, images, audio, video, relational data, multivariate records and spatio-temporal data [141]. Thus, event detection problem is encountered with determining what data is relevant for the event detection under study and the approach which must be opted to evaluate the data from selected sources.

Voluminous Data: Huge volume of data requires high-powered computing algorithm and immense storage space to store, access, filter and process all data within stipulated timeframe. For example, millions of tweets get generated each day in the twitter platform. So, to process this huge data against some particular event, event detection algorithms should incorporate dynamism and suitable running environment so that it runs uninterruptedly even after sudden voluminous increase in the OSN data during some bursty events.

Authenticity and Missing Data: The event detection techniques should consider the inaccuracies and incompleteness of the raw sensor data [143]. For example, the position or location information in terms of longitude and latitude is very likely to be inaccurate or missing. The environmental activity information may have limited confidence level. So, event detection algorithms should consider this underlying incompleteness, inaccuracies and confidence levels while detecting the events in OSNs.

Handling Anomalous Behavior: Historical data from OSN consists of the mixture of both normal and anomalous event data mixed together [144]. The event detection method should first learn the predictable behavior pattern of the event as well as should be capable of detecting and extracting deviated patterns from the raw data. This can be applied for example cases such as predicting the number of customers entering a bank or predicting the number of freeway traffic accidents per day.

5   Case Study: Twitter Platform

Twitter, a microblogging site, was developed in 2006 and has become one of the fastest growing Online Social Network (OSN) site since its inception. It has rich source of short messages up to 140 characters known as *tweet*, which get published and exchanged between Twitter users, especially during the events [145]. A reply to a tweet can be done using markups like a retweet abbreviated as RT, '@' followed by a user identifier to address the user, and '#' followed by a word to tag any event

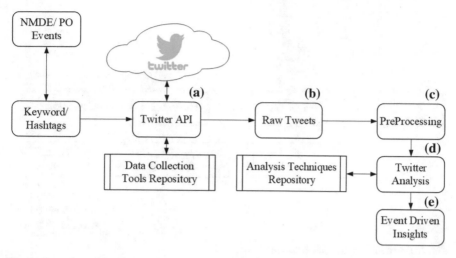

Fig. 6 Detection of disaster/public opinion events and analysis in Twitter

or trend. In Twitter, users are free to follow any person whom they are interested at and this process makes them aware of every message posted by the people they are following. A followee has no compulsion to reciprocate back to the follower by following them in turn. This makes the links in the Twitter social network directed [146]. Also, it makes Twitter easier to apply different concepts of graph theory for analysis purpose by the researchers.

According to Internet Live Statistics, around 500 million tweets get generated per day [147]. This huge data facilitates the different types of OSN Analysis (OSNA) done by the researchers. Twitter provides Application Programming Interface (API) to allow researchers and the data analysers to collect and access this huge volume of tweets in a variety of data format for their usage [25]. The tweets can be posted by the users through a mobile platform enabling a real-time dissemination of information to a large crowd through Twitter network. This makes Twitter ideal for broadcasting breaking news, emergency response and recovery during disasters [148]. A follower-following topology analysis of Twitter shows a non-power law follower distribution, a short effective diameter and low reciprocity is a deviation from the known social networks characteristics [149]. The one-to-many communication style which is supported by Twitter platform has made it easy for the users to simultaneously connect many people at a time to spread an upcoming issue or event. Also, the re-tweets almost instantly get into the next hop, which makes information diffusion much faster [149].

The standard procedure for collecting/extracting Twitter data using Twitter API in real-time are shown in Figs. 6, 7 and 8 for NMDE/POE, EE and Prediction of EE respectively. The description of each modules are as follows:

Module (A): Analysis of events involves collection of raw tweets as its first step. Twitter APIs are used to collect raw stream of tweets from Twitter platform. The

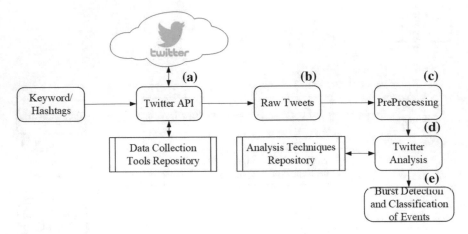

Fig. 7 Detection of emerging events in Twitter

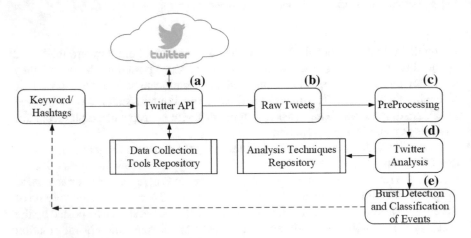

Fig. 8 Prediction of emerging events in Twitter

data collection tool repository consists of different tools available (free/paid) having inbuilt Twitter APIs. Generally, Few trending keywords/hashtags related to the respective events are given as an input by the user at the initial stage to filter and collect relevant raw tweets. In case of NMDE/POE, trending keywords/hashtags related to the event are used. For EE, random keywords/hashtags are used to collect hose collection of tweets. In Prediction of EE, the initial keywords/hashtags are selected similar manner as in the case of EE. But, the learning of new keywords from each iteration is done and is given as input to the next iteration for capturing all emerging events in a particular timeframe.

Module (B): This module receives the filtered streaming data in the form of raw tweets from Twitter platform, which comes directly from the Twitter API. Generally, tweets are received in JSON format. The data undergoes parsing, followed by encapsulation. After encapsulation of each message, it is en-queued in the memory to get considered for the next module in the procedure.

Module (C): This module does the filtering of spam tweets through classification procedure, discards messages written in language not required by the system for event analysis. This module also involves treatment of special characters and separation marks followed by standardization of data, which involves upper and lower case conversion and replacement of special characters. The filtered and standardized tweets are then grouped into sets of data, according to some criterion like timestamp of creation etc. If timestamp of creation is followed, each set of tweets correspond to a particular time window, which is chronologically sent to the next module in pipeline for processing. Any incoming tweet is included in the current time window for further processing.

Module (D): The incoming pre-processed tweets are processed and analysed in this module by extracting the informative keywords, which reflects the required type of event. A repository of event detection techniques is used to select appropriate algorithms/techniques for relevant analysis over the cleaned tweets. The relevant analysis and the appropriate algorithms/techniques will differ from one event detection problem to another.

Module (E): For events like NMDE/POE, important insights are driven post analysing the informative keywords. For events like detection of emerging events, bursty keywords are extracted by discarding the non-bursty keywords. The burst detection of keywords are followed by trend detection of the emerging event, normally following the timeline graph of the keywords. In case of prediction of emerging event, these complete steps of all modules will get repeated in an iterative manner, until a whole list of trending keywords is learnt by the system to declare an emerging event.

5.1 Twitter Platform for Detection of Disaster Events

Twitter has shown to be highly effective when it comes to disaster relief and rapid communication during a natural disaster. It has been reported by Twitter that over 2 million tweets had been posted during the Hurricane Sandy disaster in 2012. Recent studies related to the disasters in Yushu [150], Japan [151], Chile [145], and Haiti [152], showed the usefulness of Twitter data. Sakaki et al. in [63], used tweets to detect the earthquake occurrence via an event notification system developed by the researchers. This temporal cum spatial study involves a semantic analyzer [63] to find out relevant data and creating the training data by *Support Vector Machine* (SVM) classifier. Each user was considered to be a sensor and the tweet as sensory information. As event detection problem can be reduced to an object and location detection in ubiquitous computing environment [153], the study uses Kalman filter and particle filter which are widely used for location estimation. This work considered data volume, location, and event transpire over time and space. In another study by Asktorab et al. [154], a tool was proposed called *Tweedr* to extract real-time information significant for the first responders during disasters as well as perform analysis post disaster to produce reports on damage or casualty information. The geo-location and specific keywords of tweets are also used by Kumar et al. [155] to help the first responders through awareness during disasters. The social network measures like betweenness and global centrality was used by Cheong et al. [156] to identify important clusters and individuals during 2010–2011 Queensland floods in Australia. Machine learning classification methods were also used to monitor tweets during Hurricane Irene and found that number of Twitter messages is correlated with the peaks during hurricanes.

Twitter is also extensively used for studying Topic models which are tools to identify latent text patterns in the online content [157]. Kireyev et al. in [158], analyzed tweets collected against a crisis event using topic clustering and their new technique called dynamic corpus refinement was used to generate datasets that are more related to the disaster event under study. A partially supervised model was presented by Ramage et al. in [159] called *Labeled Latent Dirichlet Allocation* (LLDA) model that maps tweets into dimensions. These dimensions correspond to substance (topics about events, ideas, things and people), social characteristics (social topics), style (broader trends) and status (personal tweets). This model takes both users and tweets into account. Their results show that models trained on aggregated messages generate better performance in real-world scenarios. In [160], there is a comparative study between Twitter based topic models with the ones found in traditional media. They found Twitter acts in many ways similar to social media. But there are differences—(a) Twitter acts as an invaluable source for 'entity-oriented' topics that have low coverage in other sources of media. (b) Twitter has lesser involvement in international news but is very efficient in helping spreading important news.

Research Gaps:

Tracking of Multiple Event occurrences during disasters: Scenarios of multi events occurring during disasters are very common. For example, during earthquake, evacuation, relief measures, missing people's query—all can be considered as sub events which need to be tracked simultaneously. Multiple events occurrence like this are not considered to the extent in the existing literature. More advanced probabilistic models are needed to support multiple event occurrences which are more practical and realistic in nature.

Authenticity of event information: The existing algorithms for detecting disaster events do not ensure the authenticity and trustworthiness of the available information related to the detected events. New algorithms are needed not only to authenticate the detected events but also the information generated from those events.

Detection of Influential nodes: The content based analysis including the underlying OSN link structure is important for link prediction between the users in a network. It also facilitates in determining the influential nodes in the network. Most of the algorithms have not considered the underlying topology of the network for detecting user traffic patterns and finding the influential nodes in an OSN while analyzing a disaster event.

Real-time tweet analysis: Real time analysis of tweets is also an open area of research which has not been considered by most of the researchers for detecting a disaster event. The occurrence of a disaster event incorporates multiple other parallel events to happen. Individuals require reliable and instant information regarding the trends of the event happening and the diffusion of the effect of the event, which is a challenging task from the researcher's point of view. That is why real time analysis of tweets is significant in analyzing disaster events.

5.2   Twitter Platform for Detection of Emerging Events

The topic detection and real-time topic discovery technique was introduced by Cataldi et al. in [161]. The topics are described as a set of terms. Terms have a life cycle and a term or set of terms is considered emergent if its frequency increases in a specified time interval or time frame and was relatively a rare event in the past. They also weight content based on the PageRank of the authors and introduce a topic graph where users can identify semantically or meaningfully related emergent topics and keywords. A method for identifying controversial events was introduced in [162]. They formalize controversial events and approach the problem using regression methods. Their feature set includes Twitter-based features (e.g. linguistic, structural, buzziness, sentiment, and controversy) and External features (e.g. News Buzz and Web News Controversy). Luo et al. in [163], proposed indexing and compression techniques to speed up event detection without sacrificing detection accuracy. Another pilot study in [164], dealt with traditional First Story Detection (FSD) in the microblog streaming data setting. The study took a constant time to process each new document along with a constant space which is achieved by modified version of locality sensitive hashing (LSH) called streaming FSD system.

Automatic online detection of event can be categorized as a big data task which requires large scale and intensive real-time stream processing. This is exploited by a study in [51], who proposed an automatic distributed real-time event detection from large volume of tweets and also can scale to any volume of input stream without causing any degradation in the performance. Lexical key processing is used to distribute the computational cost of a single document over Storm distributed stream processing platform [165] instead of categorizing the document stream itself.

A study by [166] classified the event detection in three categories—*specific event detection* [151, 167], which pivots around some disaster or some special keywords, *specific person related event detection* [168], which were detected through specific person names or celebrity names and *general events* [151], which are detected through presence of hot keywords in microblogs. The study presented a framework for event detection from microblog messages containing three modules: microblog crawler and filtering, microblog event detection and event prediction. Weng et al. [111] detected event based only on temporal information of the events. The study used wavelet transformation to fit the temporal information of each word. Modularity-based graph partitioning algorithm was used to form events. A study defining event by incorporating a composition of multiple event elements over content, time, location and network structure was done by Zhou et al. [55]. Accordingly, the study proposed a new graphical model called location-time constrained topic to capture social microblogs over content, time and location. They also proposed a complementary measure to find the similarity between messages with content, time, location and the links in the network structure. Finally, they implemented a similarity join over microblog of Twitter and design hash-based index scheme to improve the efficiency of event detection.

Summarization of text in event detection is about properly displaying or visualizing the real-time information about events. The leading search engines simply display the relevant searches matching queries in reverse chronological order. Chakraborti et al. in [169], formalized this problem as a first study by giving a solution based on learning the underlying hidden state representation of the event via *Hidden Markov Model* (HMM). Normally, most search engines give query specific answers at one shot finding the most recent ones. This approach is acceptable for one-shot events like earthquake. But for events like ongoing game of American football, this approach is unsatisfactory because the recent tweets or posts will display repeated information and most users will be interested in summarization of the game results till date. A modified HMM known as SUMMHMM was proposed, which refers to event summarization (SUMM) by detecting stages or segmenting an event by HMM. The segmentation is based on the event time-line, depending on both the volume of the tweet-stream and the word distribution used in tweets. Each such segment represents one 'sub event', a semantically distinct portion of the full event. Then the key tweets were picked from each segment to describe that segment which is interesting enough and combine them together to finalize the summary of the event. Gong et al. in [170], proposed a method of computing relevance score to each sentence in the document and then picking sentences having the best scores. More complex methods are based on latent semantic analysis, HMM, and deep natural language analysis. Shamma et al. in [171], worked on Twitter timeline to explore the structure and content of media events using a video footage and Twitter activity during two broadly watched media events: the first 2008 USA Presidential debate and Inauguration of President Obama. They illustrated their approach with Statler which is a tool for identifying video content and in-position commentary from community annotation.

Research Gaps

Dynamic update of topics/events without any predetermined mention of event: Most of the studies in the Twitter platform consider the number of topics to be predetermined. For example, researchers opt for a conscious detection of events during a specific timeline where they are aware of the events happened. In that case, a query based or keyword based search are used to support the event detection. But, in real-time scenario, the events which are going to occur are not always known to the users. So, further studies are required to develop techniques where the number of topics will be updated dynamically along the timeline without any mention of the predetermined events.

Dearth of techniques coupling textual, spatial, temporal along with social/network structure: There does not exist any standard model or structure for event definition in microblog platform. This leads to different definitions of events made by the researchers in context to their research problem. Also, social or network structure properties of microblog platforms should be exploited more to increase the granularity of event detection. Most of the event detection in microblog platform happened with textual content. But, to have a more accurate detection, textual along with temporal, spatial and social/network structural properties should be considered by the applied techniques.

5.3 Twitter Platform for Public Opinion Events

User opinions constitute a major part of public opinion content and is normally created during events like electoral polls, product launch etc. These opinions have got a clear relevance for marketers, social scientists, customers, etc. and have got substantial information regarding the people who are influenced by the event and also opinions from general public [172]. The relevance of these opinions can be found out by sentiment detection, as a part of opinion mining. Sentiment detection [172] includes identifying and aggregating polar opinions—i.e. positive or negative statements about facts. According to Pang et al. in [173], since 2001, opinion mining and sentiment analysis was a field of interest among researcher. Go et al. in [174], used Twitter for the sentiment analysis. They constructed corpora by using emoticons to obtain positive and negative sentiments and then used several classifiers. The best result was obtained by the Naïve Bayes classifier with a mutual information measure for feature selection. But the method showed a bad performance when all the three classes of sentiments viz. positive, negative and neutral were considered. Present studies are mostly mathematical and the common classification algorithms for detection of opinion polarity using the collection of annotated text data are Naïve Bayes, SVM, K-Nearest neighbours, etc. [175]. A study by Aliza Sarlan [176], used both lexicon based method and machine learning techniques to find the polarity of the tweets. A Graph propagation algorithm and Lexicon Building algorithm were made by Bakliwal et al. [177] to find negation and elongation in tweets. Finally, a study by Bahrainian et al. [178] is worth mentioning for their use of slang dictionary for features selection purpose. The study used SentiStrength lexicon to the sentiment-bearing words and tagged negative words with 'NEGATE'. As a baseline method the study used SVM classifier combined with a novel hybrid method.

Research Gaps

Authenticity of Opinions: The presence of spam or fake reviews in public opinion can change the trend of any event related to public opinion. The detection of spam and fake reviews, mainly through the identification of duplicate tweets, the comparison of qualitative with summary reviews, the detection of outliers among tweets, and the reputation of the reviewer should be considered by researchers to ensure the trustworthiness and authenticity of the opinion mining results.

Algorithms using semantic analysis through lexicon of words: Most of the existing algorithms analysed the events considering the lexicon of words. But the accuracy of the algorithms in detecting appropriate events can be improved by incorporating semantic analysis through lexicon of words with known sentiment for sentiment classification.

5.4   Twitter Platform for Prediction of Emerging Events

The trend analysis of Twitter text stream identifies the real-time triggering events listed by most popular bursty terms or keywords in a data stream [179]. The events are then analyzed following a time line using keyword graphs [180], link-based topic models [181] and infinite state automations [182]. Leskovec et al. in [183], used memes and Twitter data as data streams for event analysis. Swan et al. in [184], dealt with constructing overview timelines of a set of new stories. The basis for this real-time emerging topic detection is the real-time bursty keyword detection [70]. Statistical methods and data distribution tests can be used to detect *bursty keywords* [182]. Twitter Monitor, a system that performs detection of topic trends i.e. emerging topics or breaking news in the twitter stream was introduced by Koudas et al. [70]. This was based on the detection of *bursty keywords* and performed in two steps. Firstly, the set of bursty keywords were found by computing the occurrence of individual keywords in tweets. Secondly, this system grouped keyword trends based on their co-occurrences. They introduced the algorithm of Queueburst to detect the burst. For grouping the set of related *bursty keywords*, they used an algorithm Group burst which evaluates the co-occurrences in recent tweets. But it requires an intensive preprocessing step for determining its optimal parameter settings for each keyword and also for global variables. These parameter settings must be computed with a historical dataset. Weng et al. in [147], presented a different approach to the problem by the system EDCoW (Event Detection with Clustering of Wavelet-based Signals). EDCoW builds individual word signals by applying wavelet analysis to word frequencies. They used it to analyze online discussions about the Singapore General Election of 2011. Bursty keyword Detection model known as *Window Variation Keyword Burst Detection* [179] was proposed which created a detection model depending on time-window analysis. They computed keyword frequencies, normalized by relevance and compared them in adjacent time windows. Another relevant study is that of Naaman et al. [185]. The dataset used by them consists of over 48 million messages posted on Twitter between September 2009 and March 2010 by 855,000 unique New York users. For each tweet in this dataset, they recorded its textual content, the associated timestamp and the user ID. In this work the authors make two contributions for interpreting emerging temporal trends. Firstly, they develop taxonomy of trends found in data, based on a large Twitter message dataset. Secondly, they identify important key features by which trends can be categorized.

Research Gaps

Requirement of algorithms for semantically coherent events: Present algorithms do not include procedures for handling multiple semantically coherent events along with the main event. New fast and efficient algorithms are required for detection of semantically coherent events on Twitter.

Techniques for detecting of unpredicted bursts and multi-profiles patterns: Some criterions are crucial in detecting trends and emotions. They are unpredicted bursts detections in Twitter as well as user multi-profiles patterns. Appropriate algorithms should be proposed to resolve these gaps.

Support applications for multi-lingual data: Existing algorithms does not support multi-lingual data as an input from the OSN platform. There is a requirement of support applications with multi-lingual language processing capability, at the same time maintaining privacy and anonymity.

Requirement of algorithms exploiting the Twitter structure: The topology or the structure of Twitter is important to identify influential nodes in a network. New algorithms are required considering the network structure influence while determining the trends of the data stream.

Advanced real-time search algorithms: Present algorithms consider collection of all tweets in a particular timeline to be analysed to predict an event. Advanced real time search algorithms need to be developed to get more relevant and accurate event related data against user queries. This will reduce the processing time of data to a large extent.

A summarization of techniques used across different types of research on Twitter by the researchers is given in Table 3. A tools perspective survey with respect to the data collection and event detection in Twitter is presented in the next section. This will provide a thorough understanding to the researchers regarding type of data that can be available from Twitter through tools, and also the event detection tools available to detect different types of events.

6 Existing Tools for Twitter Platform

6.1 Data Collection Tools

A survey on the common existing tools with reference to the Twitter platform which facilitates in collecting data from Twitter and detects and analyse events is given as follows. The tools mentioned as follows are active only for Twitter platform and discussed on the basis of type of license (opensource/subscribed), features and purpose.

Table 3 Techniques used by researchers on Twitter

Twitter domain	Event detection features	Event detection techniques used
Disaster events	Thematic, Temporal, Spatial, Network structure	SVM, HMM, Semantic analyzer, Kalman filter, Particle filter, Topic clustering, LDA, Author-topic model, Betweenness and Global centrality
Detection of emerging events	Thematic, Temporal, Spatial, Network structure	PageRank, Regression, Indexing and compression techniques, LSH, Wavelet transformation, Modularity-based graph partitioning, Location-time constrained topic model, Similarity join and hash-based indexed scheme, HMM, SUMMHMM, Latent semantic analysis
Public opinion events	Thematic	Naïve Bayes classifier, SVM, K-nearest neighbour, Lexicon-based and machine-learning based technique, Graph propagation algorithm, Lexicon building algorithm
Predicting emerging events	Thematic, Temporal, Spatial	Statistical methods and data distribution tests, Queueburst algorithm, Groupburst algorithm, Wavelet transformation and analysis, Window variation keyword burst detection

Foller.me [186]:

About: It is an open-source Twitter analytics application that offers rich insights about any public Twitter profile. Near real-time data about topics, mentions, hashtags, followers, location is available from this tool.

Features: The tool gathers the requested user profile along with latest tweets, analyse the tweet contents and generate the topic usage in the form of tag clouds so that one can easily rank the keywords in order of priority. It provides a small table along with the tweets, which portrays tweets, retweets, tags, replies, mentions, inks, media and more.

Purpose: It facilitates to spot for spammers and automated accounts. It can also give the join date, time-zone and followers ratio of the profile. The tool can help the users in making decision with respect to whether to follow somebody or not in Twitter.

Twitter Archiving Google Spreadsheet (TAGS) [187]:

About: It is an open-source free Google sheet template which allow its users to setup and run automated collection of search results from Twitter.

Features: Once set up is done for the tool, one can run multiple archives without repeating the authentication again. Favorited tweets of the users can be archived. With first time login, the tool extract 3,000 favorites and one can have a setup to keep collecting more. Google new sheet have better capacity and performance with the option to create new sheet versions of TAGS archives. It also creates an auto-updating archive of 18,000 tweets, based on hash tags or keywords.

Purpose: It can be used for filtering out spam and creating pagination.

Truthy [188]:

About: Truthy.indiana.edu is a sophisticated subscription-based new Twitter-based tool, created by a leading group of researchers from Indiana University, that combines data mining, social network analysis and crowdsourcing to uncover deceptive tactics and misinformation.

Features: The tool extract thousands of tweets per hour in search of political keywords, isolate patterns of interest and then insert those memes (ideas or patterns passed by imitation) into Twitter APIs to obtain more information about the meme's history.

Purpose: This tool has challenged the astroturfers, Twitter-bombers and smear campaigners through uncovering a number of abuses such as robot-driven traffic to politician websites and networks of bot accounts controlled by individuals to promote fake news.

Tweet Archivist [189]:

About: It is a subscription-based tool which facilitates in collection of tweets based on hashtags or keywords and data can be downloaded in the excel or pdf format.

Features: It helps in analysing the data in finding the influential users, words, URLs and hashtags. It provides tracking of hashtag campaign, capturing tweets during a conference, concert and sports event, monitor any brand on Twitter and perform academic research. Tracking of hashtag campaign involves understanding how a hashtag travelled through the tweetsphere. The tool can also show how many people has seen it.

Purpose: Users are very curious to know about conferences, workshops or training. Normal public will start tweeting about the conference event, detailing scenario before the conference, during the conference and post the conference. This tool helps the users to manage this lots of data from Twitter regarding conferences.

TweetStats [190]:

About: An opensource tool, provides statistical figure or graph on the Twitter Stats of an user.

Features: The tool is capable of reflecting on how long a specific Twitter handle is around, the time of the day when most tweeting happens on that handle, persons getting the most retweets, tweet cloud and follower-followee ratio. All these things can be accomplished against a Twitter username which has to be provided to the tool.

Purpose: Organizations who requires a social media monitoring for their business can opt for this tool.

Twiangulate [191]:

About: This is an open-sourced tool which makes use of Twitter data to find experts and users through discovering which accounts are followed by key people in scribe swarm, gossip empire, artful sphere or judicious gaggle.

Features: This tool understand clout through determining which influential and celebs follow an account using 'reach' search. Reveal accounts which are most followed by members of a group. Alerts are generated whenever any user account wins important new followers, mentions and retweets. The tool also can pinpoint key community members who are often unknown to outsiders by requesting a network map.

Purpose: The tool can be used for identification of hidden influencers in a certain community of OSN.

Twitonomy [192]:

About: This is an open-sourced tool which can be used to analyse any Twitter account, whether of any other user or of the competitors in the corporate world. One has to connect to his/her twitter account to work on this tool and then enter the Twitter address of self or some other user.

Features: The tool can provide the details of lists that the user is a part of, followers the user is engaged with, overall statistics of the user detailing how many tweets sent, the time of the day tweets are sent, how often content is retweeted, content that is retweeted the most, favourite tweets, download and print tweets, users replied to and mentioned, hashtags used etc.

Purpose: This tool can be used by organizations for social media monitoring purpose.

TweetNest [193]:

About: This is an open-sourced tool which can fetch browsable, searchable and easily customizable archive and backup for the tweets of any user account/profile page in Twitter.

Features: This tool provides a simple visualization of the account tweets as they include thumbnails of media like photos hosted on Twitpic.

Purpose: This tool is hosted on user's own web server, so any customization required can be performed with the whole lot of tweets.

iScience Maps [194]:

About: It is an open-sourced researching twitter content. This tool refers to a set of web application designed to help researchers mine and conduct OSNA of Twitter content.

Features: This is a browser-based tool, leveraging a simple interface that allows users to examine content from both a global and local level.

Purpose: Using global search, researchers can analyze worldwide trends over a period of time chosen by them. Results are in the form of random sample of geolocated tweets fetched via Twitter API and then assigned to the appropriate global region based on the iScience Maps taxonomy. Once search procedure is over, the users are also capable of building upon temporal nature of this data to create animations that show change over time. The local search on the other hand, allows the user to consider global search results and study occurrences for more specified geographic areas. Results from both local as well as global search features are downloadable in .xls or .csv format.

Chorus [195]:

About: It is an open-sourced, evolving, data harvesting and visual analytics suite designed to facilitate and enable social science research using Twitter data.

Features: The Chorus package consists of two distinct programs: i. Tweetcatcher which is a desktop and allows to sift Twitter for relevant data in two distinct ways: either by topical keywords appearing in Twitter conversation widely (i.e. semantically driven data) or by identifying a network of Twitter users and following their daily Twitter lives (i.e. user driven data), ii. Tweetvis is a Chorus-TV offering a visual analytic suite for facilitating both quantitative and qualitative methods to Twitter data. Visual analytics is an interdisciplinary computing methodology combining methods from data mining, information visualization, human-computer interaction and cognitive psychology. Two features offered through Tweetvis are the Timeline Explorer and Cluster Explorer. Timeline Explorer provides users an opportunity to analyse Twitter data across time and visualize the unfolding Twitter conversation according to various metrics (including tweet frequency, sentiment, semantic novelty and homogeneity, collocated words and so on). On the other hand, Cluster Explorer allows users to delve into the semantic and topical makeup of their dataset in a way that is significantly less reliant on the chronological ordering of topics. This explorer represents semantic similarity on a 2D map, which displays the semantic similarity of intervals, tweets, and terms as their proximity to each other in the cluster map.

Purpose: This tool provides access to interval level, tweet level and term level visualisations and provides the means for users to explore the different topics prevalent within their dataset and trace relationships between them via topical nodes.

DiscoverText [196]:

About: It is a subscription based tool which offers powerful text analytics, data science, human coding and machine learning features including instant access to Gnip Powertrack 2.0 for Twitter, historical Twitter, and the free twitter Search API.
Feature: It is a cloud based software which quickly evaluate large amount of text, survey and Twitter Data.
Purpose: This tool provides thorough text analytics performed on the tweets and provide insights on this huge data to the organizations.

Followthehashtag [197]:

About: This is a freemium application which means one can use it for free, but with some limits. It can obtain 1500 tweets or to last 7 days. Premium package involves tracking of data for 30–120 days up to 200,000 tweets.
Features: The feature Geo Coverage made the users to understand where the content is really relevant regardless the total number of tweets in a geo zone. Exporting of results is possible in excel format. PDF report formats can be obtained with custom title, company name etc. In Geo and Heatmap sections, one will get different ways to see geo and language related data. Satellite heat map shows tweets in a heat map over a satellite, Google maps heat map shows tweets in a heat map over a good maps view, Blob google maps chart shows tweets in a blob chart and country google maps chart shows tweets per country.
Purpose: It provides a dashboard for data analysts, social media managers and scientists for better understanding on sentiment analysis of the users, comparative study of insightful keywords through a deep search procedure of tweets which ensures coverage of maximum tweets within a timeframe.

All these above discussed tools access is sanctioned by Twitter specific Application programming Interface (API)'s to retrieve data from the content pool of the OSN. There are certain tools mentioned, which functions as a collection, analytics and visual tool. For example, Chorus is a tool which initiate the process with collection of data and also do the required analysis and visualization on the same. All the listed tools are used by the researchers and there are ample numbers of published work on the same. There are many other recent tools (i.e. Mozdeh, Netlytic, NVivo etc.) which offer either real-time cloud-based solution or analyse video, audio, webpages along with text, or active on more than one OSN platform (not exclusively on Twitter). All the above tools support the data collection module depicted in Module (A) of Figs. 6, 7 and 8.

6.2 Event Detection and Analysis Tools

Tools for detecting and analyzing events in Twitter are end-to-end tools which applies: (a) different natural language processing techniques on the raw data from the OSN platform for filtering; (b) perform different analysis for sentimental analysis, finding emerging patterns, running meaningful queries, trend analysis etc. on the filtered data and (c) finally incorporating techniques for meaningful visualization of the data. All the surveyed tools given in this section, are proposed for specific need to solve the research problems by researchers i.e. to find the pattern of some real time event or some activity in OSN. In doing so, each study tried to improve on the complexity of the algorithm employed to ensure consistent performance of the system in finding the pattern. All the tools support an end-to-end analysis of Twitter stream, covering module (A) to module (E), with reference to Fig. 6, Fig. 7 and Fig. 8 respectively. Apart from these tools, there are certain tools like Blogoscope, Memetracker, NodeXL, Twitris, Ushahidi etc. which are effective in detecting events across different data streams including newswires, blogs, web forums apart from microblogs of Twitter. Following tools are discussed on the basis of types of events analysis supported (NMDE—Natural and Manmade Disaster Events, EE—Emerging Events, POE—Public Opinion Events) by each of them on Twitter platform.

a. **Trendsmap** (Trendsmap 2015) [198]

About: Trendsmap shows the latest trend in the Twitter from anywhere in the world. It can be used in agile and content market, brand management, crisis management and trend monitoring.

Features: This tool helps in finding historical trends by analyzing historical data as far as back in mid 2009s or using their extensive archive data. It involves sentiment analysis involving the most engaged areas through map. Relevant data is filtered depending on location, time language etc. They also provide a word cloud display for specific countries.

Purpose: Finds trends from historical or archival data of the company to provide word cloud for specific countries for NMDE/EE/POE.

b. **Tweedr** Ashktorab et al. [154]

About: It is a Twitter mining tool that extracts actionable information for disaster response during NMDE.

Features: The Tweedr pipeline consists of three main parts: classification, clustering and extraction. In the classification phase, variety of classification procedures like LDA, SVM etc. are used to identify tweets reporting damage or casualties. In the clustering phase, filters are used to merge the tweets that are similar to one another and finally in the extraction phase, tokens and phrases are extracted to report specific information about different classes of infrastructure damage, damage types and casualties.

Purpose: The main purpose of this tool is to provide relevant information to the disaster relief workers during NMDE category of events.

c. **ReDites** Osborne et al. [199]

About: ReDites is a tool which aids in real-time event detection, tracking, monitoring and visualization, specially designed for information analysts of security sector. It dealt with large scale data and tailors it to the security domain. The tool was implemented on the dataset of terrorist attack happened on September 2013. *Features*: Event processing comprises of four steps: (i) new events are detected from the first tweet itself, (ii) it tracks the event, searching for new posts relating to the single tweet and keeping an update on the incremental study of tweets, (iii) the events are organized and categorized for the security domain, geolocation is performed and detect for the sentiment is detected that evolve around that event, (iv) the produced stream is visualized, summarized, categorized for the information security analysts.

Purpose: Real-time event detection tool which support large scale incremental microblog streaming data for NMDE, EE and POE category of events.

d.   ***TopicSketch*** Xie et al. [200]

About: Leverages Twitter for automated real-time bursty topic detection as soon as it occurs. TopicSketch was tested on a stream containing over 30 million tweets and demonstrated the effectiveness and efficiency of its method.

Features: The tool provides temporally ordered sub-events that are more descriptive in nature and can also detect events with bursts over shorter duration of time. It's contribution can be divided in three stages—at the first stage, it proposed a data sketch which does a calculation on the total number of tweets, the occurrence of each word and the occurrence of each word pair which provides an early indication to the popularity of a tweet. Depending on this, a topic model is developed to infer on bursty topics whose dynamism overtime is calculated through data sketch. In the second stage, tool proposed a hashing-based dimension reduction technique using hashing to achieve scalability, and maintain quality of the events with proved error bounds. The tool is scalable to an extent of handling 300 million tweets per day which is close to the data which gets generated on twitter platform in a day.

Purpose: Dynamic update of topics/events without any predetermined event mention from a very large scale data. Useful for data generated from different types of events like NMDE, EE and POE.

e.   ***TweetXplorer*** Morstatter et al. [201]

About: It is an analyst's tool to gain knowledge on social big data through effective visualization techniques. The data set used was Hurricane Sandy through which they exhibit the working of the system.

Features: This tool explored the Twitter data by following the data lifecycle phases which includes 'Plan and prepare', 'Collect and process', 'Analyze and Summarize', 'Represent and Communicate' and 'Implement and Manage'. *Tweettracker* support is for the first two phases. The main functionalities of the tool are in creating meaningful queries, discovering interesting time periods, representing important tweets and users depending on retweet facility, communicating salient locations, and discovering user patterns. D3 Visualization toolkit is used with the Force directed layout method for the visualization process.

Purpose: To convey information of any NMDE visually so that necessary actions can be taken during crisis.

f. **TEDAS** (*Event Detection and Analysis System*) Li et al. [202]

About: TEDAS tool detects and analyse events through tweets from the Twitter platform. The dataset used by the researchers for the tool is crime and disaster related events, for example car accidents, earthquake, etc.
Features: The three important functionalities are detecting new events, ranking the events w.r.t. their importance and generating temporal and spatial patterns of the event. It works on both offline and online computing mode. It detects new events maintaining a rule-based approach. The classifiers, the meta information extractor and text search engines are used for extracting informative tweets and extracting location and temporal details from the same. The visualization of the event is done on the basis of a given keyword and the timeline within which the event visualization is expected. The implementation of this tool was done based on Java, PHP with the backend support of MySQL, Lucene, Twitter API and Google Maps API.
Purpose: Event detection through Twitter during NMDE.

g. **Twevent** Li et al. [203]

About: An online segment based bursty event detection system for tweets with a predefined topic name. The event detection was done based on the 4.3 million tweets published by Singapore-based users in June 2010.
Features: The tool detect the bursty tweet segments as event segments and then perform a clustering using their content similarity and frequency distribution. The non-overlapping segments formed from each tweet are the semantically meaningful information units. Every bursty segment is identified within a fixed time window based on frequency patterns. After finding out the candidate events, Wikipedia is used to identify the most realistic events to report the final identified events.
Purpose: To detect bursty event through bursty tweet segments as event segments for POE/EE category of events.

h. **Twitcident** Abel et al. [204]

About: It is a web based system which provides a framework for automatically filtering relevant information from social media streams, searching for events and analyzing information regarding this real-world incidents or crisis. This tool automatically connects to the emergency broadcasting services and start tracking and filtering the new incident or event from the social media streams.
Features: The incident detection module senses the incidents from the broadcast emergency services. When this new thread of incident is reported by Twitcident core framework, the tool starts collecting and aggregating related messages from the web and Twitter. These messages are further processed by the semantic enrichment module which features named entity recognition, classification of messages,

linkages between messages to external web resources and the extraction of meta-data. Additionally, users are also provided with a search option through which they could further receive filtered messages according to their need. The tool also provides graphical visualization of the evolution of the incident overtime or the geographical impact on the area of an incident.

Purpose: Enhance the user involvement through finer level of visualization as required by the users during NMDE/EE/POE situations.

i. *TweetTracker* Kumar et al. [155]

About: It is an application designed to facilitate the Humanitarian Aid and Disaster Relief (HADR) organizations to track, analyze and monitor the disaster related tweets from the Twitter platform. The aim to design this tool is to provide first responders achieve proper whereabouts about the disaster situation to decide on the relief measures. It has used the tweets of Cholera crisis happened in Haiti to validate the functioning of the tool.

Features: The disaster relief operations require real-time monitoring of the tweets within a very short span of time from the time of the disaster. This tool helps in real-time monitoring by analyzing the tweets from temporal, geo-spatial and topical perspectives. Twitter streaming API is used to collect tweets and filtering of informative tweets is done based on specific keywords, hashtags and geo-location of the tweets. Keywords are used to find the trends through the keyword trending engine and are expressed through tag clouds. The map is used for showing the geo-located tweets. This tool also works on the multi-lingual tweets through Google Translate to enhance understanding of the tweets.

Purpose: It involves a real time detection of events with pre-defined event mentions from users during NMDE situations.

j. *Twitinfo* Marcus et al. [205]

About: Twitinfo is a real-time tool for visualizing and summarizing events on Twitter. It allows to browse huge amount of tweets about different events like disaster, politics, sport etc. and use a timeline-based display to show the peaks of high tweets activity. According to expert opinion, this tool is appropriate to monitor long running events and also to identify the eyewitnesses.

Features: The tool can extract the tweets matching the keywords present in the query and output a graphical view of the timeline of the subevent through peaks of tweets reaching a certain number. It also highlights important terms and messages concerning the subevent, provides zoom in facility to the users, does a sentiment analysis displaying the event related user's sentiment's, and also the geographic distribution of tweets. It also provides the links between tweets in a cluster if it is portraying any relevance.

Purpose: Real time analysis of tweets for event detection during NMDE/EE/POE situations is the primary aim of this tool.

k. *TwitterMonitor* Mathioudakis and Koudas [70]

About: TwitterMonitor is a visualization tool which shows the trend of an event in real-time, using Twitter platform, along a specific timeline.

Features: The tool starts with detecting the bursty keywords in the streaming data of tweets which is treated as a starting point of detecting a new event. Queue-burst and Groupburst algorithms were developed to find out bursty keywords in real-time. The most correlated keywords were found through context extraction algorithms like *Principle Component Analysis* (PCA), *Singular Value Decomposition* (SVD) etc. using Latent Semantic Analysis which finds out the most correlated keywords within the event. Grapevine's Entity Extractor's algorithm is used to find the most frequently maintained entities in the trends. A chart depicts the evolution of popularity of event overtime along with the geographical origins of tweets. The trending topics of different events extracted are also indexed according to the volume or a recency score or the combined score of the both.

Purpose: To index trending topics (EE) depending on the volume and newness of tweet messages and its timeline based visualization.

1. **SensePlace2** MacEachren et al. [206]

About: Senseplace2 is a geo-visual analytics application that collects tweets attributed by place-time attribute information from the *Twitterverse* and support crisis management during disaster events through visually-enabled sense making from the available information.

Features: The tool uses a crawler, which via Twitter API collects tweets of interest using keyword and hashtags. Tweets and auxiliary metadata is stored in JSON format. This data is parsed and stored in a PostgresSQL database. Different distributed applications that need to analyze tweets for named-entities such as location, entities, hashtags etc. then work on the database and create respective tables. Lastly, a Lucene text index is generated which supports a full text retrieval of tweets within geographical region and data range. The visualization supports the analysts to explore, characterize and compare the spatial, temporal and geography associated with the topics and the entities of the tweets.

Purpose: Supports full text retrieval of tweets within a particular geographical region and data range during NMDE.

7 Conclusion

OSN has emerged as a significant platform for dissemination of information and interaction between users. Based on the forms of interaction, OSN is categorized into different types- blog based, services oriented, media sharing, news sharing, location based and community based. Researchers use OSN platform to achieve relevant information by churning the huge social streaming data generated by UGC, available during various kinds of events—NMDE, EE & POE. There are four dimensions in which events can be categorized—thematic, temporal, spatial and network structure. A comprehensive review of event detection techniques for each dimension is done, which would be helpful in detection, analysing, summarizing and predicting of events. The popular OSN platform—Twitter is considered as a case study for strengthening the concept of event detection in OSN. The role of Twitter platform in detecting different types of events—NMDE, POE & EE and predicting emerging events is studied in depth. A detail review of event detection techniques with respect to different dimensions applicable to Twitter data is done. An exhaustive study carried out of existing data collection and event detection & analysis tools exclusively used for Twitter platform, will provide readers ample knowledge about the tools, type of data extracted and use of tools.

The research challenges based on the review of literature for event detection techniques (in general and for Twitter platform in specific) were also identified. The insights gained in this chapter would be useful for the readers/researchers to contribute further on the open challenges as suggested for event detection techniques through novel tools and new techniques.

References

1. Goswami A, Kumar A (2016) A survey of event detection techniques in online social networks. Soc Netw Anal Mining 6(1):107
2. Cisco VNI (2017) Cisco visual networking index: forecast and methodology, 2016—2021. https://www.cisco.com/c/en/us/solutions/collateral/service-provider/visual-net working-index-vni/complete-white-paper-c11-481360.pdf. Accessed 6 June 2017
3. Youtube (2016) www.youtube.com. Accessed Dec 2016
4. Facebook (2016) www.facebook.com. Accessed Dec 2016
5. Twitter (2016) www.twitter.com. Accessed Dec 2016
6. Deitrick W, Hu W (2013) Mutually enhancing community detection and sentiment analysis on Twitter networks. J Data Anal Inf Process 1:19–29
7. Chen Z, Kalashnikov DV, Mehrotra S (2009) Exploiting context analysis for combining multiple entity resolution systems. In: Proceedings of the 2009 ACM SIGMOD international conference on management of data, pp 207–218. ACM
8. Goswami A, Kumar A (2017) Challenges in the analysis of online social networks: a data collection tool perspective. Wirel Pers Commun 97(3):4015–4061
9. Wassaerman S, Faust K (1994) Social network analysis in the social and behavioural sciences. Social network analysis: methods and applications. Cambridge University Press, Cambridge
10. Mislove A, Marcon M, Gummadi KP, Druschel P, Bhattacharjee B (2007) Measurement and analysis of online social networks. In: Proceedings of the 7th ACM SIGCOMM conference on internet measurement, pp 29–42. ACM
11. Flake GW, Lawrence S, Giles CL, Coetzee FM (2002) Self-organization and identification of web communities. Computer 35(3):66–70
12. Flake GW, Tarjan RE, Tsioutsiouliklis K (2004) Graph clustering and minimum cut trees. Internet Math 1(4):385–408
13. Girvan M, Newman ME (2002) Community structure in social and biological networks. Proc Natl Acad Sci 99(12):7821–7826
14. Hopcroft J, Khan O, Kulis B, Selman B (2003) Natural communities in large linked networks. In: Proceedings of the ninth ACM SIGKDD international conference on Knowledge discovery and data mining, pp 541–546. ACM

15. Newman ME (2004) Detecting community structure in networks. Eur Phys J B Condens Matter Complex Syst 38(2):321–330
16. Kaplan AM, Haenlein M (2010) Users of the world, unite! The challenges and opportunities of social media. Bus Horiz 53(1):59–68
17. Moriarty GL (2010) Psychology 2.0: harnessing social networking, user-generated content, and crowdsourcing. J Psychol Issues Organ Culture 1(2):29–39
18. Liu B (2012) Sentiment analysis and opinion mining. Synth Lect Human Lang Technol 5(1):1–167
19. Shelley GA, Bernard HR, Killworth PD (1990) Information flow in social networks. J Quant Anthropol 2(3)
20. Pang B, Lee L (2008) Opinion mining and sentiment analysis. Found Trends Inf Retr 2(1–2):1–135
21. Asur S, Huberman B (2010) Predicting the future with social network. In: 2010 IEEE/WIC/ACM international conference on web intelligence and intelligent agent technology (WIIAT), vol 1
22. Bakshy E, Hofman JM, Mason WA, Watts DJ (2011) Identifying influencers on twitter. In: Fourth ACM international conference on web search and data mining (WSDM)
23. Wen-ying SC, Hunt YM, Beckjord EB, Moser RP, Hesse BW (2009) Social media use in the United States: implications for health communication. J Med Internet Res 11(4):e48
24. Zuber M (2014) A survey of data mining techniques for social network analysis. Int J Res Comput Eng Electron 3(6)
25. Shin H, Byun C, Lee H (2015) The influence of social media: Twitter usage pattern during the 2014 super bowl game. Life 10(3)
26. Site of SEO Company. SEO Positive. http://www.seo-positive.co.uk/blog/different-types-of-socialnetworks/. Accessed 31 Dec 2014
27. Fraser M, Dutta S (2010) Throwing sheep in the boardroom: how online social networking will transform your life, work and world. Wiley
28. Daum Communications Corp. Daum blog. http://blog.daum.net/
29. Daum Communications Corp. Tistory. http://www.tistory.com/
30. Google. Blogger. http://www.blogger.com/
31. MySpace Inc. Myspace.com. http://www.myspace.com/
32. NHN Corp. Naver blog. http://blog.naver.com/
33. SK Communications Corp. Cyworld. http://www.cyworld.com/
34. Lim SH, Kim SW, Kim S, Park S (2011) Construction of a blog network based on information diffusion. In: Proceedings of the 2011 ACM symposium on applied computing, pp 937–941. ACM
35. WordPress.com. https://wordpress.com/. Accessed Dec 2015
36. Boyd, Ellison (2008) Social network sites: definition, history & scholarship. J Comput Med Commun 13:210–230
37. LinkedIn. https://in.linkedin.com/. Accessed Dec 2015
38. Elftown. http://www.elftown.com/. Accessed 20 Dec 2017
39. Ravelry. http://www.ravelry.com/account/login. Accessed Dec 2015
40. Saha S, Paul G (2013) On effective sharing of user generated content. In: Proceedings of the 11th Asia Pacific conference on computer human interaction, pp 114–118. ACM
41. Flikr. https://www.flickr.com/. Accessed Dec 2015
42. Podcast Alley. www.podcastalley.com/. Accessed Dec 2015
43. Wong LLY, Burkell J (2017) Motivations for sharing news on social media. Soc Media Soc 17:1
44. Hermida A (2014) Tell everyone: why we share and why it matters. Doubleday Canada, Toronto
45. Digg. www.digg.com. Accessed Dec 2015
46. Reddit. https://www.reddit.com/. Accessed Dec 2017
47. Zheng Y (2011) Location-based social networks: users. Computing with spatial trajectories. Springer, New York, NY, pp 243–276

48. Chorley MJ, Whitaker RM, Allen SM (2015) Personality and location-based social networks. Comput Hum Behav 46:45–56
49. Foursquare. https://foursquare.com/. Accessed Dec 2015
50. Google Groups. https://groups.google.com/. Accessed Dec 2015
51. Allan J, Papka R, Lavrenko V (1998) On-line new event detection and tracking. In: Proceedings of the 21st annual international ACM SIGIR conference on research and development in information retrieval, pp 37–45. ACM
52. Dou W, Wang X, Skau D, Ribarsky W, Zhou MX (2012) Leadline: interactive visual analysis of text data through event identification and exploration. In: 2012 IEEE conference visual analytics science and technology (VAST), pp 93–102
53. Setzer A (2002) Temporal information in newswire articles: an annotation scheme and corpus study, Doctoral dissertation, University of Sheffield
54. Chen L, Roy A (2009) Event detection from flickr data through wavelet-based spatial analysis. In: Proceedings of the 18th ACM conference on information and knowledge management, pp 523–532. ACM
55. Zhou X, Chen L (2014) Event detection over twitter social media streams. VLDB J Int J Very Large Data Bases 23(3):381–400
56. Madani A, Boussaid O, Zegour DE (2014) What's happening: a survey of tweets event detection. In: Proceedings of 3rd INNOV
57. Baas S, Ramasamy S, DePryck JD, Battista F (2008) Disaster risk management systems analysis: a guide book, vol 3. Food and Agriculture Organization of the United Nations
58. Srikanth A. Social media can solve many problems during natural disasters
59. Freberg K (2011) Crisis information curators & digital relief coordinators via social media: Japan Tsunami catastrophe brief report 2011. Presented to the National Center for Food Protection and Defense. Minneapolis, MN
60. Pohl D, Bouchachia A, Hellwagner H (2012) Automatic sub-event detection in emergency management using social media. In: Proceedings of the 21st international conference on world wide web, pp 683–686. ACM
61. Bailey NJ, Bevington JS, Lewis HG, Swinerd GG, Atkinson PM, Crowther R, Holland D (2007) From Buncefield to Tunguska: hazard and disaster modelling at the University of Southampton
62. Santos ADPD, Wives LK, Alvares LO (2012) Location-based events detection on micro-blogs. arXiv:1210.4008
63. Sakaki T, Okazaki M, Matsuo Y (2010) Earthquake shakes Twitter users: real-time event detection by social sensors. In: Proceedings of the 19th international conference on world wide web, pp 851–860. ACM
64. Lee R, Sumiya K (2010) Measuring geographical regularities of crowd behaviors for Twitter-based geo-social event detection. In: Proceedings of the 2nd ACM SIGSPATIAL international workshop on location based social networks, pp 1–10. ACM
65. Iyengar A, Finin T, Joshi A (2011) Content-based prediction of temporal boundaries for events in Twitter. In: 2011 IEEE third international conference on privacy, security, risk and trust (PASSAT) and 2011 IEEE third inernational conference on social computing (SocialCom), pp 186–191. IEEE
66. Samadzadegan F, Rastiveisi H (2008) Automatic detection and classification of damaged buildings, using high resolution satellite imagery and vector data. In: Proceedings of the international archives of the photogrammetry, remote sensing and spatial information sciences, vol 37, pp 415–420
67. Velev D, Zlateva P (2012) Use of social media in natural disaster management. Int Proc Econ Dev Res 39:41–45
68. Palen L (2008) Online social media in crisis events. Educ Q 31(3):76–78
69. Chae J, Thom D, Bosch H, Jang Y, Maciejewski R, Ebert DS, Ertl T (2012) Spatiotemporal social media analytics for abnormal event detection and examination using seasonal-trend decomposition. In: 2012 IEEE conference on visual analytics science and technology (VAST), pp 143–152. IEEE

70. Mathioudakis M, Koudas N (2010) Twittermonitor: trend detection over the twitter stream. In: Proceedings of the ACM SIGMOD international conference on management of data, pp 1155–1158
71. Childs H (1939) By public opinion I mean. Public Opin Q 3(2):327–336. https://doi.org/10.1086/265298
72. Stromback J (2012) The media and their use of opinion polls: reflecting and shaping public opinion. In: Holtz-Bacha C, Strömbäck J (eds) Opinion polls and the media: reflecting and shaping public opinion. Palgrave Macmillan, Basingtstoke, pp 1–23
73. Gallup GH (1939) Public opinion in a democracy. Herbert L. Baker foundation, Princeton University Press, Princeton, NJ
74. Bourdieu P (1979) Public opinion does not exist. In: Mattelart A, Siegelaub S (eds) News and the empowerment of citizens. International General, New York, pp 124–130
75. Herbst S (1993) Numbered voices: how opinion polling has shaped American politics. University of Chicago Press, Chicago
76. Moon N (1999) Opinion polls: History, theory and practice. University of Manchester Press, Manchester
77. Tork H (2011) Event detection. Thesis, LIAAD-INESC TEC
78. Neill DB, Wong WK (2009) Tutorial on event detection. In: proceedings of KDD 2009
79. Neill DB, Gorr WL (2007) Detecting and preventing emerging epidemics of crime. Adv Dis Surveill 4:13
80. Dereszynski E, Dietterich T (2007) Probabilistic models for anomaly detection in remote sensor data streams. In: Proceedings of the 23rd conference on uncertainty in artificial intelligence (UAI-2007), pp 75–82
81. Papka R, Allan J (1998) On-line new event detection using single pass clustering. UMass Computer Science
82. CNN Library (2015) Mumbai terror attacks fast facts. http://edition.cnn.com/2013/09/18/world/asia/mumbaiterror-attacks/. Accessed Jan 2016
83. Zhao Q, Mitra P (2007) Event detection and visualization for social text streams. In: International conference of weblogs and social media. ICWSM
84. Garofalakis M, Gehrke J, Rastogi R (eds) (2016) Data stream management: processing high-speed data streams. Springer
85. Brants T, Chen F, Farahat A (2003) A system for new event detection. In: Proceedings of the 26th annual international ACM SIGIR conference on research and development in informaion retrieval, pp. 330–337. ACM
86. Kumaran G, Allan J (2004) Text classification and named entities for new event detection. In: Proceedings of the 27th annual international ACM SIGIR conference on research and development in information retrieval, pp 297–304. ACM
87. Cutting, DR, Karger DR, Pedersen JO (1993) Constant interaction-time scatter/gather browsing of very large document collections. In: Proceedings of the 16th annual international ACM SIGIR conference on Research and development in information retrieval, pp 126–134. ACM
88. McCallum AK (1996) Bow: a toolkit for statistical language modeling, text retrieval, classification and clustering. http://www.cs.cmu.edu/mccallum/bow
89. Aggarwal CC, Zhai C (2012) A survey of text clustering algorithms. Mining text data. Springer, Boston, MA, pp 77–128
90. Fisher DH (1987) Knowledge acquisition via incremental conceptual clustering. Mach Learn 2(2):139–172
91. Sahoo N, Callan J, Krishnan R, Duncan G, Padman R (2006) Incremental hierarchical clustering of text documents. In: Proceedings of the 15th ACM international conference on information and knowledge management, pp 357–366. ACM
92. Zhuang X, Huang Y, Palaniappan K, Zhao Y (1996) Gaussian mixture density modeling, decomposition, and applications. IEEE Trans Image Process 5(9):1293–1302
93. Fung GPC, Yu JX, Yu PS, Lu H (2005) Parameter free bursty events detection in text streams. In: Proceedings of the 31st international conference on very large data bases, pp 181–192. VLDB Endowment

94. Fung GPC, Yu JX, Lu H, Yu PS (2006) Text classification without negative examples revisit. IEEE Trans Knowl Data Eng 18(1):6–20
95. Li X, Liu B (2003) Learning to classify texts using positive and unlabeled data. In: Proceedings of IJCAI, vol 3, no 2003, pp 587–592
96. He Q, Chang K, Lim EP (2007) Analyzing feature trajectories for event detection. In: Proceedings of the 30th annual international ACM SIGIR conference on research and development in information retrieval, pp 207–214. ACM
97. Sankaranarayanan J, Samet H, Teitler BE, Lieberman MD, Sperling J (2009) Twitterstand: news in tweets. In: Proceedings of the 17th acm sigspatial international conference on advances in geographic information systems, pp 42–51. ACM
98. Mitchell TM (1997) Machine learning, vol 45, no 37. McGraw Hill, Burr Ridge, IL, pp 870–877
99. Vapnik V (2013) The nature of statistical learning theory. Springer science & business media
100. Cha Y, Cho J (2012) Social-network analysis using topic models. In: Proceedings of the 35th international ACM SIGIR conference on Research and development in information retrieval, pp 565–574. ACM
101. Blei DM, Ng AY, Jordan MI (2003) Latent Dirichlet allocation. J Mach Learn Res 3:993–1022
102. Breiman L (2001) Random forests. Mach Learn 45(1):5–32
103. Hansen LK, Salamon P (1990) Neural network ensembles. IEEE Trans Pattern Anal Mach Intell 12(10):993–1001
104. Schapire RE (2003) The boosting approach to machine learning: an overview. Nonlinear estimation and classification. Springer, New York, pp 149–171
105. Sewell M (2011) Ensemble learning. Technical report, Department of Computer Science, University College London. http://www.csucl.ac.uk/fileadmin/UCL-CS/research/Research_Notes/RN_11_02.pdf
106. Friedman JH (2001) Greedy function approximation: a gradient boosting machine. Ann Stat 1189–1232
107. Freund Y, Schapire RE (1995) A desicion-theoretic generalization of on-line learning and an application to boosting. In: European conference on computational learning theory. Springer, Berlin, Heidelberg, pp 23–37
108. Lafferty J, McCallum A, Pereira FC (2001) Conditional random fields: probabilistic models for segmenting and labeling sequence data
109. Gu H, Xie X, Lv Q, Ruan Y, Shang L (2011) Etree: effective and efficient event modeling for real-time online social media networks. In: 2011 IEEE/WIC/ACM international conference on web intelligence and intelligent agent technology (WI-IAT), vol 1, pp 300–307. IEEE
110. Nallapati R, Feng A, Peng F, Allan J (2004) Event threading within news topics. In: Proceedings of the thirteenth ACM international conference on information and knowledge management, pp 446–453. ACM
111. Weng J, Lee BS (2011) Event detection in twitter. In: Proceedings of ICWSM, vol 11, pp 401–408
112. Kaiser G (2010) A friendly guide to wavelets. Springer Science & Business Media
113. Allan J, Carbonell JG, Doddington G, Yamron J, Yang Y (1998) Topic detection and tracking pilot study final report
114. Chen CC, Chen MC, Chen MS (2005) LIPED: HMM-based life profiles for adaptive event detection. In: Proceedings of the eleventh ACM SIGKDD international conference on knowledge discovery in data mining, pp 556–561. ACM
115. Yang Y, Pierce T, Carbonell J (1998) A study of retrospective and on-line event detection. In: Proceedings of the 21st annual international ACM SIGIR conference on research and development in information retrieval, pp 28–36. ACM
116. Chen CC, Chen MC (2008) TSCAN: a novel method for topic summarization and content anatomy. In: Proceedings of the 31st annual international ACM SIGIR conference on research and development in information retrieval, pp 579–586. ACM
117. Metzler D, Cai C, Hovy E (2012) Structured event retrieval over microblog archives. In: Proceedings of the 2012 conference of the North American chapter of the association for

computational linguistics: human language technologies, pp 646–655. Association for Computational Linguistics

118. Tang X, Yang C, Gong X (2011) A spectral analysis approach for social media community detection. Soc Inform 127–134

119. Klinger R, Tomanek K (2007) Classical probabilistic models and conditional random fields. TU, Algorithm Engineering

120. Wong WK, Neill DB (2009) Tutorial on event detection KDD 2009. Age 9:30

121. Dong G, Li J (1999) Efficient mining of emerging patterns: discovering trends and differences. In: Proceedings of the fifth ACM SIGKDD international conference on knowledge discovery and data mining, pp 43–52. ACM

122. Bay SD, Pazzani MJ (1999) Detecting change in categorical data: mining contrast sets. In: Proceedings of the fifth ACM SIGKDD international conference on knowledge discovery and data mining, pp 302–306. ACM

123. Wong WK, Moore A, Cooper G, Wagner M (2005) What's strange about recent events (WSARE): an algorithm for the early detection of disease outbreaks. J Mach Learn Res 6:1961–1998

124. Allamanis M, Scellato S, Mascolo C (2012) Evolution of a location-based online social network: analysis and models. In: Proceedings of the 2012 ACM conference on internet measurement conference, pp 145–158. ACM

125. Wittie MP, Pejovic V, Deek L, Almeroth KC, Zhao BY (2010) Exploiting locality of interest in online social networks. In: Proceedings of the 6th international conference, p 25. ACM

126. Karagiannis T, Gkantsidis C, Narayanan D, Rowstron A (2010) Hermes: clustering users in large-scale e-mail services. In: Proceedings of the 1st ACM symposium on cloud computing, pp 89–100. ACM

127. Backstrom L, Sun E, Marlow C (2010) Find me if you can: improving geographical prediction with social and spatial proximity. In: Proceedings of the 19th international conference on world wide web, pp 61–70. ACM

128. Gionis A, Indyk P, Motwani R (1999) Similarity search in high dimensions via hashing. In: Proceedings of VLDB, vol 99, no 6, pp 518–529

129. Fox V, Hightower J, Liao L, Schulz D, Borriello G (2003) Bayesian filtering for location estimation. IEEE Pervasive Comput 2(3):24–33

130. Kulldorff M (1997) A spatial scan statistic. Commun Stat Theor Methods 26(6):1481–1496

131. Neff ND, Naus JI (1980) The distribution of the size of the maximum cluster of points on a line. Amer Mathematical Society

132. Kulldorff M, Athas WF, Feurer EJ, Miller BA, Key CR (1998) Evaluating cluster alarms: a space-time scan statistic and brain cancer in Los Alamos, New Mexico. Am J Public Health 88(9):1377–1380

133. Kulldorff M, Mostashari F, Duczmal L, Katherine Yih W, Kleinman K, Platt R (2007) Multivariate scan statistics for disease surveillance. Stat Med 26(8):1824–1833

134. Agrawal R, Lin KI, Sawhney HS, Shim K (1995) Fast similarity search in the presence of noise, scaling, and translation in times-series databases. In: Proceedings of the 21st international conference on very large databases, pp 490–501

135. Chan FP, Fu AC, Yu C (2003) Haar wavelets for efficient similarity search of time-series: with and without time warping. IEEE Trans Knowl Data Eng 15(3):686–705

136. Aach J, Church GM (2001) Aligning gene expression time series with time warping algorithms. Bioinformatics 17(6):495–508

137. Bar-Joseph Z, Gerber G, Gifford DK, Jaakkola TS, Simon I (2002) A new approach to analyzing gene expression time series data. In: Proceedings of the sixth annual international conference on computational biology, pp 39–48. ACM

138. Berndt DJ, Clifford J (1994) Using dynamic time warping to find patterns in time series. In: KDD workshop, vol 10, no 16, pp 359–370

139. Keogh EJ (2002) Exact indexing of dynamic time warping. In: Proceedings of VLDB, pp 406–417

140. An P, Keck P, Kim T. Min-cut algorithms. http://www.comp.nus.edu.sg/~rahul/allfiles/cs623 4–16-mincuts.pdf

141. Kerman MC et al (2009) Event detection challenges, methods, and applications in natural and artificial systems. In: Proceedings of 14th international command and control research and technology symposium: C2 and agility

142. Dash D, Margineantu D, Wong WK (2007) Machine learning algorithms for event detection. Spec Issue Mach Learn J. Accessed 14 Mar 2008 (Springer)

143. Balazinska M (2007) Event detection in mobile sensor networks. In: National science foundation (NSF) workshop on data management for mobile sensor networks (MobiSensors) 2007

144. Ihler A, Hutchins J, Smyth P (2006) Adaptive event detection with time-varying poisson processes. In: The twelfth international conference on knowledge discovery and data mining (Association for Computing Machinery)

145. Mendoza M, Poblete B, Castillo C (2010) Twitter under crisis: can we trust what we RT? In: Proceedings of the first workshop on social media analytics, pp 71–79. ACM

146. Huberman BA, Romero DM, Wu F (2008) Social networks that matter: Twitter under the microscope. arXiv:0812.1045

147. Internet live stats. http://www.internetlivestats.com/. Accessed Feb 2018

148. Vieweg S (2010) Microblogged contributions to the emergency arena: discovery, interpretation and implications. In: Proceedings of computer supported collaborative work, pp 515–516

149. Kwak H, Lee C, Park H, Moon S (2010) What is Twitter, a social network or a news media? In: Proceedings of the 19th international conference on world wide web, pp 591–600. ACM

150. Qu Y, Huang C, Zhang P, Zhang J (2011) Microblogging after a major disaster in China: a case study of the 2010 Yushu earthquake. In: Proceedings of the ACM 2011 conference on computer supported cooperative work, pp 25–34. ACM

151. Long R, Wang H, Chen Y, Jin O, Yu Y (2011) Towards effective event detection, tracking and summarization on microblog data. In: International conference on web-age information management. Springer, Berlin, Heidelberg, pp 652–663

152. Gao H, Barbier G, Goolsby R (2011) Harnessing the crowdsourcing power of social media for disaster relief. IEEE Intell Syst 26(3):10–14

153. Hightower J, Borriello G (2001) Location systems for ubiquitous computing. Computer 34(8):57–66

154. Ashktorab Z, Brown C, Nandi M, Culotta A (2014) Tweedr: mining twitter to inform disaster response. In: Proceedings of ISCRAM

155. Kumar S, Barbier G, Abbasi MA, Liu H (2011) TweetTracker: an analysis tool for humanitarian and disaster relief. In: International conference on web and social media (ICWSM), 5 Jul 2011

156. Cheong F, Cheong C (2011) Social media data mining: a social network analysis of tweets during the 2010–2011 Australian floods. In: Proceedings of PACIS, vol 11, pp 46–46

157. Hong L, Davison BD (2010) Empirical study of topic modeling in twitter. In: Proceedings of the first workshop on social media analytics, pp 80–88. ACM

158. Kireyev K, Palen L, Anderson K (2009) Applications of topics models to analysis of disaster-related twitter data. In: NIPS workshop on applications for topic models: text and beyond, vol 1, Canada, Whistler

159. Ramage D, Dumais ST, Liebling DJ (2010) Characterizing microblogs with topic models. In: Proceedings of ICWSM, vol 10, no 1, p 16

160. Zhao WX, Jiang J, Weng J, He J, Lim EP, Yan H, Li X (2011) Comparing twitter and traditional media using topic models. In: European conference on information retrieval, pp 338–349. Springer, Berlin, Heidelberg

161. Cataldi M, Di Caro L, Schifanella C (2010) Emerging topic detection on twitter based on temporal and social terms evaluation. In: Proceedings of the tenth international workshop on multimedia data mining, p 4. ACM

162. Popescu AM, Pennacchiotti M (2010) Detecting controversial events from twitter. In: Proceedings of the 19th ACM international conference on information and knowledge management, pp 1873–1876. ACM

163. Luo G, Tang C, Yu PS (2007) Resource-adaptive real-time new event detection. In: Proceedings of the 2007 ACM SIGMOD international conference on management of data, pp 497–508. ACM
164. Petrović S, Osborne M, Lavrenko V (2010) Streaming first story detection with application to twitter. In: Human language technologies: The 2010 annual conference of the North American chapter of the association for computational linguistics, pp 181–189. Association for Computational Linguistics
165. McCreadie R, Macdonald C, Ounis I, Osborne M, Petrovic S (2013) Scalable distributed event detection for twitter. In: 2013 IEEE international conference on big data, pp 543–549 (2013)
166. Zhao J, Wang X, Ma Z (2014) Towards events detection from microblog messages. Int J Hybrid Inf Technol 7(1):201–210
167. Huang J, Iwaihara M (2011) Realtime social sensing of support rate for microblogging. Database systems for adanced applications. Springer, Berlin, Heidelberg, pp 357–368
168. Popescu AM, Pennacchiotti M, Paranjpe D (2011) Extracting events and event descriptions from twitter. In: Proceedings of the 20th international conference companion on world wide web, pp 105–106. ACM
169. Chakrabarti D, Punera K (2011) Event summarization using tweets. In: Proceedings of ICWSM, vol 11, pp 66–73
170. Gong Y, Liu X (2001) Generic text summarization using relevance measure and latent semantic analysis. In: Proceedings of the 24th annual international ACM SIGIR conference on research and development in information retrieval, pp 19–25. ACM
171. Shamma D, Kennedy L, Churchill E (2010) Tweetgeist: can the twitter timeline reveal the structure of broadcast events. In: Proceedings of CSCW Horizons, pp 589–593
172. Gindl S, Weichselbraun A, Scharl A (2010) Cross-domain contextualisation of sentiment lexicons.
173. Pang B, Lee L, Vaithyanathan S (2002) Thumbs up?: sentiment classification using machine learning techniques. In: Proceedings of the ACL-02 conference on empirical methods in natural language processing, vol 10, pp 79–86
174. Go A, Huang L, Bhayani R (2009) Twitter sentiment analysis. Entropy 17 (2009)
175. Rao N, Srinivas S, Prashanth M (2015) Real time opinion mining of twitter data. Int J Comput Sci Inf Technol 6(3):2923–2927,
176. Sarlan A, Nadam C, Basri S (2014) Twitter sentiment analysis. In: International conference on information technology and multimedia (ICIMU), Putrajaya, Malaysia, 18–20 Nov 2014
177. Akshat B, Arora P, Kapre SMN, Singh M, Varma V (2012) Mining sentiments from tweets. In: Proceedings of the 3rd workshop on computational approaches to subjectivity and sentiment analysis. Association for Computational Linguistics, pp 11–18, Jeju, Republic of Korea
178. Bahrainian S-A, Denge A (2013) Sentiment analysis and summarization of Twitter data. In: IEEE 16th international conference on computational science and engineering
179. Jheser G, Poblete B (2013) On-line relevant anomaly detection in the twitter stream: an efficient bursty keyword detection model. In: Proceedings of the ACM SIGKDD workshop on outlier detection and description, pp 31–39. ACM
180. Sayyadi H, Hurst M, Maykov A (2009) Event detection and tracking in social streams. In: Proceedings of ICWSM
181. Lin, CX, Zhao B, Mei Q, Han J (2010) PET: a statistical model for popular events tracking in social communities. In: Proceedings of the 16th ACM SIGKDD, pp 929–938. ACM
182. Kleinberg J (2003) Bursty and hierarchical structure in streams. Data Min Knowl Disc 7(4):373–397
183. Leskovec J, Backstrom L, Kleinberg J (2009) Meme-tracking and the dynamics of the news cycle. In: Proceedings of the 15th ACM SIGKDD, pp 497–506. ACM
184. Swan R, Allan J (2000) Automatic generation of overview timelines. In: Proceedings of the 23rd annual international ACM SIGIR conference on research and development in information retrieval
185. Naaman M, Becker H, Gravano L (2011) Hip and trendy: characterizing emerging trends on Twitter. J Am Soc Inf Sci Technol 62, no. 5 (2011)

186. Foller.me. https://foller.me/. Accessed 20 Dec 2017
187. #TAGS. https://tags.hawksey.info/. Accessed Dec 2017
188. Truthy. http://cnets.indiana.edu/blog/tag/truthy/. Accessed 20 Dec 2017
189. Tweet Archivist. http://www.tweetarchivist.com/. Accessed Dec 2017
190. TweetStats, http://www.tweetstats.com/. Accessed December, 2017
191. Twiangulate. http://twiangulate.com/search/. Accessed Dec 2017
192. Twitonomy. http://www.twitonomy.com/. Accessed Sept 2014
193. Tweetnest. https://github.com/graulund/tweetnest. Accessed Sept 2014
194. iScience Maps. http://datadrivenjournalism.net/resources/iscience_maps. Accessed Dec 2017
195. Chorus. http://chorusanalytics.co.uk/. Accessed Dec 2017
196. Discovertext. https://discovertext.com/. Accessed Dec 2017
197. Followthehashtag. http://www.followthehashtag.com/. Accessed Dec 2017
198. Trendsmap. http://www.trendsmap.com. Accessed Dec 2017
199. Osborne M, Moran S, McCreadie R, Von Lunen A, Sykora MD, Cano E, Jackson T (2014) Real-time detection, tracking, and monitoring of automatically discovered events in social media. In: Proceedings of 52nd annual meeting of the association for computational linguistics: system demonstrations, pp 37–42. ACL 2014
200. Xie W, Zhu F, Jiang J, Lim EP, Wang K (2013) Topicsketch: realtime bursty topic detection from twitter. In: 2013 IEEE 13th international conference on data mining, pp 837–846
201. Morstatter F, Kumar S, Liu H, Maciejewski R (2013) Understanding twitter data with tweet-xplorer. In: Proceedings of the 19th ACM SIGKDD international conference on knowledge discovery and data mining, pp 1482–1485
202. Li R, Lei KH, Khadiwala R, Chang KCC (2012) TEDAS: a twitterbased event detection and analysis system. In: 2012 IEEE 28th international conference on data engineering (ICDE), pp 1273–1276
203. Li C, Sun A, Datta A (2012) Twevent: segment-based event detection from tweets. In: Proceedings of the 21st ACM international conference on information and knowledge management, pp 155–164
204. Abel F, Hauff C, Houben GJ, Stronkman R, Tao K (2012) Twitcident: fighting fire with information from social web streams. In: Proceedings of the 21st international ACM conference companion on world wide web, pp 305–308. https://doi.org/10.1145/2187980.2188035
205. Marcus A, Bernstein MS, Badar O, Karger DR, Madden S, Miller RC (2011) Twitinfo: aggregating and visualizing microblogs for event exploration. In: Proceedings of the ACM SIGCHI conference on Human factors in computing systems, pp 227–236
206. MacEachren AM, Jaiswal A, Robinson AC, Pezanowski S, Savelyev A, Mitra P, Blanford J (2011) Senseplace2: geotwitter analytics support for situational awareness. In: 2011 IEEE conference visual analytics science and technology (VAST), pp 181–190

Printed in the United States
By Bookmasters